Land Surface Remote Sensing in Urban and Coastal Areas

Remote Sensing Observations of Continental Surfaces Set

coordinated by
André Mariotti

Land Surface Remote Sensing in Urban and Coastal Areas

Nicolas Baghdadi
Mehrez Zribi

ISTE PRESS

ELSEVIER

First published 2016 in Great Britain and the United States by ISTE Press Ltd and Elsevier Ltd

ISTE Press Ltd
27-37 St George's Road
London SW19 4EU
UK

www.iste.co.uk

Elsevier Ltd
The Boulevard, Langford Lane
Kidlington, Oxford, OX5 1GB
UK

www.elsevier.com

British Library Cataloguing-in-Publication Data
A CIP record for this book is available from the British Library
Library of Congress Cataloging in Publication Data
A catalog record for this book is available from the Library of Congress
ISBN 978-1-78548-160-4

Contents

Foreword

I have been entrusted by ISTE Science Publishing with the responsibility for a multidisciplinary editorial line: Earth System – Environment, and within this framework it gives me great pleasure today to present a set of books dedicated to the topic of remote sensing, compiled and edited by Nicolas Baghdadi and Mehrez Zribi.

Both the content and the organization of this collection have largely been inspired by reflections, analyses and prospective works conducted by almost 200 authors and researchers with a high level of international expertise in this discipline.

This community, which is recognized for its scientific merit, has sought to expand its research activities under the direction the two editing authors founded on a solid effort in the area of acquisition and wider dissemination of knowledge within this field.

This represents a community characterized by the firm commitment to adopting a holistic or even an ecosystem approach within the context of an interdisciplinary science of the Earth system. In this scientific context where the complexity of natural systems is compounded with the complexity of societies, the authors have given careful consideration to depicting a finalizable and "public" type of discipline, open to decision makers, managers and all those beyond the scientific community who are interested in the future of our planet.

Two main tools are necessary in order to satisfy the requirements in terms of understanding and characterizing our environment and its evolution: process modeling and observation.

Remote sensing observations in conjunction with measurements and modeling constitute a discipline that makes it possible to understand the functional properties of the observed system and their dependence on its structural properties. This is one of the key disciplines that allow the analysis and provide access to the understanding of the functioning of our environment: in general, this is dedicated to aspects such as the analysis of climate change, the effects of anthropogenic and demographic pressure, natural disasters, the increasing decline in resources (water, etc.), the degradation of biodiversity across all environments, desertification, the need to nourish the planet (for example, mapping of crops and yield prediction), etc.

Analyzing and understanding these different types of problems is rendered possible:

– by analyzing the detected, structural and functional objects (soils, hydrosystems, vegetation, etc.);

– by understanding the main basic processes, which incorporate these main elements: water flow covering all scales and compartments, erosion, meteorology, crop development, soil pollution, etc;

– by developing indicators in order to evaluate the short-, medium- and long-term evolution of all environmental compartments and variables.

The importance of these scientific questions has led to a general mobilization of the international organizations by means of various international conventions and agreements to protect the environment and meet the specific requirements in terms of observation. Various international networks have been developed over the past few years with the purpose of conducting continuous measurements. However, these punctual measurements could not provide sufficient spatiotemporal monitoring, in particular in difficult-to-access regions. Within this context, spatial observation could be implemented to its full potential, both by means of considerable progress in terms of instrumentation and by means of the development of effective data processing and analysis methods, data whose provision becomes increasingly free of rights.

Under the initiative of numerous space agencies (in particular European, North-American, Japanese, etc.), important space missions were launched for the purpose of conducting Earth observations, among which the following may be mentioned:

– Sentinel, within the framework of the Copernius program (formerly referred to as GMES for Global Monitoring for Environment and Security) implemented in numerous areas such as land and marine environment monitoring, emergency management (for example, natural disasters) and climate change monitoring (radar and optical imaging);

– Landsat;

– ALOS, launched by the Japan Aerospace Exploration Agency, in particular for deforestation monitoring;

– SMOS and SMAP, in particular for the global mapping of soil moisture, etc.

Although remote sensing represents a field in which specialist knowledge is required in order to conduct a better analysis and interpretation of data, this programmatic development is undoubtedly associated with a significant progress with respect to the implementation of space-based Earth observations at the level of an increasing number of laboratories across both developed and emerging nations. This development is likewise associated with new disciplines and thematic backgrounds, among which numerous areas of the humanities and social sciences which enrich and extend the primarily physical foundations of remote sensing may be mentioned.

It would be superfluous to list all remote sensing applications along with the disciplines and scientific questions which adopted this concept, as this would also inevitably result in regrettable omissions: nonetheless, the extensive implementation of spatial observation grants the latter a strong interdisciplinary status.

The launch of new large-scale space missions, the higher degree of convenience, including financial convenience, as well as the access to data will facilitate an intensification and generalization of the use of spatial observation data and products: new scientific subjects, new users (managers, decision makers, etc).

The high demand for educational material containing updated information on the various remote sensing concepts and methods and the main applications thereof, in particular at the level of continental surfaces, are derived therefrom.

It is within this framework that this collection of books is proposed, which aims to provide researchers, students in masters, engineer and PhD programs, as well as decision makers, engineers specialized in management services on a territorial, departmental, regional or national scale and players in the decision-making authorities with a tool which incorporates both the foundations of the physical principles underlying various spatial applications and the implementation methods and exemplification at the level of various applications based on spatial observation.

In these six volumes, Nicolas Baghdadi and Mehrez Zribi have mobilized almost 200 internationally recognized researchers to propose a comprehensive "toolkit", describing the latest scientific methods and actions in terms of the implementation of spatial observation.

The first two volumes describe the physical principles underlying various techniques which cover the frequency spectrum ranging from visible to microwaves. The third volume illustrates the agricultural and forestry applications of spatial observation. The fourth volume presents the applications of spatial observation in the field of continental hydrology. The fifth volume is dedicated to the observation of urban and coastal areas, whereas the final volume presents the implementation of spatial observation within the context of risk assessment and understanding.

Thanks are due to Nicolas Baghdadi and Mehrez Zribi for taking the time to draft, harmonize and partially edit these volumes and committing to this effort in terms of putting this modern and high-quality knowledge across and making it accessible to a diverse and vast scientific audience.

I wish to thank them both for their altruism, perseverance and devotion in service of the success of this endeavor.

André MARIOTTI
June 2016

Acronyms

2D	Two dimensions
3D	Three dimensions
4AOP	Automatized Atmospheric Absorption Atlas Operational
6S	Second Simulation of Satellite Signal in the Solar Spectrum
AATSR	Advanced Along-Track Scanning Radiometer
ACORN	Atmospheric Correction Now
ADC	Analog-to-digital converter
ADCP	Acoustic Doppler Current Profiler
ADEME	French Environment and Energy Management Agency
ADEOS	Advanced Earth Observing Satellite
AERONET	Aerosol Robotic Network
AET	Actual Evapotranspiration
AFRITRON	African Tropical Rainforest Observation Network
AGB	Above-Ground Biomass
AGNES	Agglomerative Nesting
AHS	Airborne Hyperspectral Scanner
AHT	Astronomical High Tide
AIEM	Advanced Integral Equation Model
AirSAR	Airborne Synthetic Aperture Radar
ALB	Airborne LiDAR Bathymeter
ALEXI	Atmosphere-Land Exchange Inverse
ALS	Airborne Laser Scanning

AltBOC	Alternate Binary Offset Carrier
AMARTIS	Advanced Modeling of the Atmospheric Radiative Transfer for Inhomogeneous Surfaces
AMMA	African Monsoon Multidisciplinary Analysis
AMSR	Advanced Microwave Scanning Radiometer
ANA	*Agência Nacionalde Aguas* (Brazilian National Water Agency)
AOL	Airborne Oceanographic LiDAR
APD	Avalanche Photodiode
API	Antecedent Precipitation Index
APOM	Aerosol Plume Optical Model
ARVI	Atmospherically Resistant Vegetation Index
ASAR	Advanced Synthetic Aperture Radar
ASCAT	Advanced Scatterometer
ASDF	Averaged Square Difference Function
ASI	*Agenzia Spaziale Italiana* (Italian Space Agency)
ASI	Agriculture Stress Index
ASTER	Advanced Spaceborne Thermal Emission and Reflection Radiometer
ATCOR	Atmospheric and Topographic Correction
ATLAS	Advanced Topographic Laser Altimeter System
ATM	Airborne Topographic Mapper
ATREM	Atmospheric Removal (atmospheric correction method)
ATSR	Along Track Scanning Radiometers
AUC	Area Under Curve
AVHRR	Advanced Very High Resolution Radiometer
AVIRIS	Airborne Visible/Infrared Imaging Spectrometer
AWiFS	Advanced Wide Field Sensor
BEIDOU/ COMPASS	Chinese Navigation Satellite System
BELA	BepiColombo Laser Altimeter
BGB	Below-Ground Biomass
BIL	Band Interleaved by Line
BIOMOD	Biomolecular Design (Computer platform implemented in R)
BIP	Band Interleaved by Pixel

BLUE	Best Linear Unbiased Estimator
BOC	Binary Offset Carrier
BPSK	Binary Phase Shift Keying
BPT	Binary Partition Tree
BRDF	Bidirectional Reflectance Distribution Function
BRGM	*Bureau de Recherches Géologiques et Minières* (French Geological Survey)
BSA	Back Scatter Alignment
BSQ	Band Sequential
BUI	Build-up Index
BWI	Bassist Wetness Index
C/A	Coarse Acquisition
CaCO3	Calcium Carbonate
CAIC	Consistent Akaike's Information Criterion
CALIOP	Cloud-Aerosol LiDAR with Orthogonal Polarization
CALIPSO	Cloud-Aerosol LiDAR Pathfinder Satellite Observation
CanEx-SM	Canadian Experiment for Soil Moisture
CART	Classification and Regression Trees
CASI	Compact Airborne Spectrographic Imager
CBERS	China–Brazil Earth Resources Satellite
CBOC	Composite Binary Offset Carrier
CCD	Charged Coupled Devices
CCDAS	Carbon Cycle Data Assimilation System
CCI	Climate Change Initiative
CCRS	Canada Center for Remote Sensing
CDF	Cumulative Density Function
CDMA	Code Division Multiple Access
CDOM	Colored Dissolved Organic Matter
CEC	Cation-Exchange Capacity
CEM	Constrained Energy Minimization
CEOS	Committee on Earth Observation Satellites
CERES	Crop Environment Resource Synthesis

CESBIO	*Centre d'Études Spatiales de la Biosphère* (Center for the Study of the Biosphere from Space)
CFC	Chlorofluorocarbons
CFD	Constant Fraction Discriminator
CFFDRS	Canadian Forest Fire Danger Rating System
cGNSS	Conventional Global Navigation Satellite Systems
CHAMP	Challenging Minisatellite Payload (German geosciences satellite)
CHL	Chlorophyll Content
CHM	Canopy Height Model
CHRIS	Compact High Resolution Imaging Spectrometer
CIRAD	Agricultural research for development (France)
CLASlite	Carnegie Landsat Analyse System Lite
CLM	Community Land Model
CLMGW	Community Land Model with a Ground Water Parameterization
CLPX	Cold Land Processes Field Experiment
CLS	Collecte Localisation Satellite
CLSM	Catchment Land Surface Model
CM	Code Moderate (for GNSS)
CMC	Canadian Meteorological Centre
CMEM	Community Microwave Emission Model
CNES	*Centre National d'Études Spatiales* (French Space Agency)
CNRS	*Centre National de Recherche Scientifique* (National French Center for Scientific Research)
COD	Controlled Origin Designation
COG	Center of Gravity
CONUS	Contiguous United States
COST	Cosine Estimation of Atmospheric Transmittance
CR	Continuum Removed
CRI	Carotenoid Reflectance Index
Cryosat	Satellite Radar Altimeter
CSA	Canadian Space Agency
CSR	Centre for Space Research
CTA	Classification Tree Analysis

CTFC	Center for Tropical Forest Science (Cameroon)
CTFS	Center for Tropical Forest Science
CTMF	Cluster Tuned Matched Filter
CW	Continuous Waves
CWFIS	Canadian Wildland Fire Information System
CyGNSS	NASA's Cyclone Global Navigation Satellite System
CZCS	Coastal Zone Color Scanner
CZMIL	Coastal Zone Mapping and Imaging LiDAR
DAM	Dry Aerial Mass
DART	Discrete Anisotropic Radiative Transfer
DATAR	Interministerial Delegation for Territorial Planning and Regional Attractiveness
DBH	Diameter at Breast Height
DC	Drought Code
DCA	Dual Channel Approach
DCM	Digital Canopy Model
DDM	Delay Doppler Map
DEGRAD	Forest Degradation program
DEIMOS	Deep Imaging Multi-Object Spectrograph
DEM	Digital Elevation Model
DEOS	Delft Institute of Earth Observations and Space Systems
DERD	Double bounce Eigenvalue Relative Difference
DETER	*Detecção de Desmatamento em Tempo Real* (Near real-time deforestation detection system)
DGPS	Differential Global Positioning System
DHSVM	Distributed Hydrology Soil-Vegetation Model
DIACT	Inter-ministerial Agency for Spatial Planning and Competitiveness
DIANA	Divise Analysis
DIC	Digital Image Correlation
DIMAP	Digital Image MAP
DInSAR	Differential SAR Interferometry
DisALEXI	Disaggregated Atmospheric Land Exchange Inverse
DLIS	Desert Locust Information Service

DLR	*Deutsches Zentrum für Luft und Raumfahrt* (German Space Agency)
DM	Dry Matter
DMC	Disaster Monitoring Constellation or Duff Moisture Code, depending on the application
DMRT	Dense Media Radiative Transfer
DMSP	U.S. Air Force Defense Meteorological Satellite Program
DMU	De Monfort University
DOAS	Differential Optical Absorption Spectroscopy
DOS	Dark Object Substraction
DOY	Day-of-Year
DPSS	Diode-Pumped Solid-State
DS	Diffuse Scatterer
DSM	Digital Soil Mapping
DSM	Digital Surface Model
DTC	Dry Troposphere Correction
DTED	Digital Terrain Elevation Data
DTM	Digital Terrain Model
DVI	Difference Vegetation Index
DW	Dry Weight
DWBA	Distorted Wave Born Approximation
EAARL	Experimental Advanced Airborne Research LiDAR
ECDC	European Centre for Disease Prevention and Control
ECMWFMMT	European Centre for Medium-Range Weather Forecasts Mobile Mapping Technology
ECV	Essential Climate Variables
EDF	*Électricité de France* (French electric company)
EEA	European Environment Agency
EFA	Effective Field Approximation
EFFIS	European Forest Fire Information System
EGNOS	European Geostationary Navigation Overlay Service
EID-Méditerranée	Interdepartmental Agreement for Mosquito Control on the Mediterranean Coast
EKF	Extended Kalman Filter

ELBARA	ETH L-Band Radiometer
ELUE	Effective Light Use Efficiency
EM	Electromagnetic
ENEA	*Energia Nucleare ed Energie Alternative* (Italian National Agency for New Technologies, Energy and Sustainable Economic Development)
EnKF	Ensemble Kalman Filter
ENMAP	Environmental Monitoring and Analysis Program
ENSO	El Niño Southern Oscillation
ENVEO	Environmental Earth Observation Information
ENVISAT	Environmental Satellite
EOS	Earth Observing System
EPICA	European Project for Ice Coring in Antarctica
EPS	European Polar System
EQeau	Model developed by INRS-Ete with the objective to extract the soil water content from SAR images
ERM	Exact Repeat Missions
ERS	European Remote-sensing Satellite
ESA	European Space Agency
ESCAT	ERS Scatterometer
ESSA	Environmental Science Services Administration Satellite
ET	Evapotranspiration
ETM	Enhanced Thematic Mapper
EUFAR	European Facility for Airborne Research
EUMETSAT	European Organization for the Exploitation of Meteorological Satellites
EVASPA	Evapotranspiration Assessment from Space
EVI	Enhanced Vegetation Index
EWT	Equivalent Water Thickness
EZW	Embedded Zerotrees of Wavelet transforms
FAA	Federal Aviation Administration (USA)
FAI	Floating Algae Index
FAO	Food and Agriculture Organization of the United Nations
FAPAR	Fraction of Absorbed Photosynthetically Active Radiation

FBD	Fine-Beam Double Polarization
FBP	Fire Behavior Prediction
Fcover	Fraction of Vegetation Cover
FDTD	Finite Difference Time Domain
FFMC	Fine Fuel Moisture Code
FFT	Fast Fourier Transform
FI	Fine Particles Index
FIPAR	Fraction of Intercepted Photosynthetically Active Radiation
FLAASH	Fast Line-of-sight Atmospheric Analysis of Spectral Hypercubes (Atmospheric correction)
FLOE	Fish LiDAR Oceanographic Experimental
FMA	Fore-Mid-Aft
FMCW	Frequency Modulated Continuous Waves
FOTO	Fourier-based Textural Ordination
FOV	Field of View
FSA	Forward Scatter Alignment
FST	Spectro-Transfert functions
FT	Fourier Transform
FW	Full Waveform or Fresh Weight, depending on the application
FWHM	Full Width at Height Maximum
FWI	Fire Weather Index
FWS	Fraction of Water Surface
GAGAN	GPS Aided GEO Augmented Navigation system
GAI	Green Area Index
GALILEO	European Global navigation satellite system
GCOS	Global Climate Observing System
GCP	Ground Control Points
GDAL	Geospatial Data Abstraction Library
GDR	Geophysical Data Record
GEO	Group on Earth Observations
GEOCAPI	Geostationary Ocean Color Advanced Permanent Imager
GEOGLAM	Group on Earth Observations Global Agricultural Monitoring
GEOSS	Global Earth Observing System of Systems

GEOSUD	Geoinformation for Sustainable Development (France)
GEOTIFF	Geostationary Earth Orbit Tagged Image File Format
GEROS-ISS	GNSS Reflectometry, Radio Occultation and Scatterometry Onboard International Space Station
GF	Green Fraction
GFO	Geosat Follow-On (radar altimeter satellite)
GFZ	*GeoForschungs Zentrum* (German research centre for Geosciences)
GHG	Green House Gase
GIAM	Global Irrigated Area Map
GIEWS	Global Information and Early Warning System on Food and Agriculture
GIM	Global Ionospheric Model
GIMMS	Global Inventory Modeling and Mapping Studies
GIS	Geographic Information System
GLAI	Green Leaf Area Index
GLAS	Geoscience Laser Altimeter System
GLCM	Grey Level Co-occurrence Matrix
GLDAS	Global Land Data Assimilation System
GlobCover	Global Land-Cover Map
GLONASS	Global Navigation Satellite System (Russia)
GLP	Global Land Project
GLRT	Generalized Likelihood Ratio Test
GLS	Global Land Survey (Landsat)
GM	Geodetic Mission
GMES	Global Monitoring for Environment and Security
GMFS	Global Monitoring of Food Security
GMM	Gaussian Mixture Model
GNSS	Global Navigation Satellite System
GNSS-R	Global Navigation Satellite System – Reflectometry
Go	Giga-Octet (1,000,000,000 octets)
GOCE	Gravity field and Steady-state Ocean Circulation Explorer (ESA)
GOES	Geostationary Operational Environmental Satellite
GOFC Fire IT	Global Observation of Forest Cover Fire Implementation Team

GOM	Geometrical Optics Model
GORS	GNSS Occultation Reflectometry Scatterometry
GPCC	Global Precipitation Climatology Centre
GPP	Gross Primary Production
GPR	Gaussian Process Regression
GPS	Global Positioning System
GRACE	Gravity Recovery and Climate Experiment (satellite)
GRDC	Global Runoff Data Center
GRGS	Space Geodesy Research Group
GtC	Gigatonnes of Carbon (10^9 tons)
GUS	Ground Uplink Stations
GVI	Difference Vegetation Index
GWIS	Global Wildfire Information System
Ha	Hectare (= 10,000 m^2)
HCFC	Hydro-chlorofluorocarbons
HEC-RAS	Hydrologic Engineering Center River Analysis System (hydraulic model)
HgCdTe	Mercury Cadmium Telluride
HiRI	High Resolution Optical Imager
HITRAN	High-Resolution Transmission Database
HOG	Histogram of Oriented Gradient
HPC	High Performance Computing
HRG	High Resolution Geometry
HSC	Height-Scaled Crown
HSCOI	Height-Scaled Crown Openness Index
HSR	High Spatial Resolution
HYMAP	Hyperspectral Mapper (airborne hyperspectral sensor)
HYPXIM	Hyperspectral-X Imagery
HYSPEX	Hyperspectral Imaging System
HyspIRI	NASA's Hyperspectral Infrared Imager
IASI	Infrared Atmospheric Sounding Interferometer
ICA	Independent Component Analysis
ICARE	Inversion Code for urban Areas Reflectance Extraction

ICESat	Ice, Cloud and Land Elevation Satellite
ICF	Interferometric Complex Field
ICP	Iterative Closest Point technique
IDAN	Intensity Driven Adaptative Neighborhood
IEM	Integral Equation Model
IFN	French National Forest Inventory
IFOV	Instantaneous Field of View
IGN	French National Geographic Institute and Forest Information
iGNSS	Interferometric GNSS
IHS	Intensity, Hue, Saturation
IMU	Inertial Measurement Unit
INERIS	French National Institute for Environmental Protection and Industrial Risks
InGaAs	Indium Gallium Arsenide
INPE	*Instituto Nacional de Pesquisas Espaciais* (Brazilian Institute of Space Research)
INRA	French National Institute for Agricultural Research
INSAR	Interferometric Synthetic Aperture Radar
InSb	Indium antimonide
INSEE	French National Institute for Statistics and Economic Studies
IOD	Indian Ocean Dipole
IPCC	Intergovernmental Panel on Climate Change
IPT	Interference Pattern Technique
IRD	French Research Institute for Development
IRNSS	Indian Regional Navigational Satellite System
IRSTEA	French National Research Institute of Science and Technology for Environment and Agriculture
ISBA	Interactions Soil-Biosphere-Atmosphere (model)
ISDC	Integrated System Data Center
ISI	Initial Spread Index
ISODATA	Iterative Self-Organizing Data Analysis Technique
ISRO	Indian Space Research Organisation
ITC	Individual Tree Crown
ITCZ	Intertropical Convergence Zone

ITG	Institute of Theoretical Geodesy
IWPB	Institute of Water Problem of Bishkek, Kyrgyzstan
Jason	Radar altimeter
JAXA	Japan Aerospace Exploration Agency
JECAM	Joint Experiment for Crop Assessment and Monitoring
JPEG	Joint Photographic Experts Group (image format)
JPL	Jet Propulsion Laboratory
JRC	European Commission's Joint Research Centre
KBR	K-Band Microwave Ranging
KLT	Karhunen–Loeve transform
LaDAR	Laser Detection And Ranging
LADS	Laser Airborne Depth Sounder
LAGEOS	Laser Geodynamics Satellite
LAI	Leaf Area Index
LANDSAT	LAND + Satellite
LAUVA	Airborne Ultraviolet Aerosol LiDAR
LAX	Maximum LAI
LBAS	Local Based Augmentation System
LCCS	Land Cover Classification System
LCLU	Land Cover / Land Use
LDAS	Land Data Assimilation System
LEGOS	Laboratory for Studies in Geophysics and Spatial Oceanography (France)
LEnKS	Local EnKF Smoother
LEO	Low Earth Orbit
LEP	Leading Edge Position
LEWIS	L-band for Estimating Water In Soils
LFMC	Live Fuel Moisture Content (%)
LHCP	Left Hand Circular Polarization
LiDAR	Light Detection and Ranging
LISAH	Laboratory for Soil, Agrosystems and Water Systems (France)
LISFLOOD-FP	Two-dimensional Hydrodynamic Model
LMA	Leaf Mass per Area

LMM	Linear Mixed Model
LOADDT	Spatial planning and territorial development
LOLA	Lunar Orbiter Laser Altimeter
LOV	Villefranche Oceanography Laboratory (France)
LPCA	Laboratory for Physico-Chemistry of the Atmosphere
LPRM	Land Parameter Retrieval Model
LRM	Low Resolution Mode
LSCE	Climate and Environment Sciences Laboratory (France)
LSM	Land Surface Model
LSSM	Least Squares 3D Surface Matching
LST	Land Surface Temperature
LULCC	Land Use and Land Cover Change
LUT	Look-Up Table
LWIR	Long-Wave Infrared
LZW	*Lempel-Ziv-Welch* (compression algorithm)
MACCS	Multisensor Atmospheric Correction and Cloud Screening processor
MARS	Monitoring of Agriculture with Remote Sensing
MATISSE	Advenced Earth Modeling for Imaging and Scene Simulation
MaxEnt	Maximum Entropy Method
MBOC	Multiplexed Binary Offset Carrier
MCT	Mercury Cadmium Telluride
MEB	Microwave Emission of the Biosphere
MERIS	Medium Resolution Imaging Spectrometer
MESA	Monitoring of Environment and Security in Africa
METEOSAT	METEO + Satellite
METOP	Meteorological Operational Satellite Programme/Advanced Scatterometer
METRIC	Mapping Evapotranspiration at High Resolution with Internalized Calibration
MGVI	MERIS Global Vegetation Index
MHz	Mega-Hertz (= 1,000,000 Hz)
MIMR	Multichannel Microwave Imaging Radiometer

MIPERS	Multistatic Interferometric Polarimetric Electromagnetic model for Remote Sensing
MIR	Middle Infrared
MISDc	*Modello Idrologico Semi Distribuito in Continuo* (Continuous rainfall-runoff model)
MISR	Multi-angle Imaging Spectro Radiometer
MISTIGRI	Microsatellite for Thermal Infrared Ground Surface Imaging (CNES, France)
MLR	Multiple Linear Regression
MLS	Mobile LiDAR Scanner
MMD	Minimum–Maximum Difference
MMS	Mobile Mapping Systems
MMU	Minimum Mapping Units
MMV	Mobile Mapping Vehicle
MNDWI	Modified Normalized Difference Water index
MNF	Maximum Noise Fraction
Mo	Mega-octet (1,000,000 octets)
MODCOU	Hydrogeological model
MODIS	Moderate Resolution Imaging Spectroradiometer
MODTRAN	Moderate Resolution Atmospheric Transmission
MOLA	Mars Orbiter Laser Altimeter
MPE	Maximum Permissible Exposure
MSAS	Multi-functional Satellite-based Augmentation System
MSG	Meteosat Second Generation
MSI	Moisture Stress Index
MSI	Multispectral Instrument (Sentinel-2)
MTSAT	Multi-functional Transport Satellites
MVSA	Minimum Volume Simplex Analysis
MWIR	Mid-Wavelength Infrared
NAOMI	New AstroSat Optical Modular Instrument
NASA	National Aeronautics and Space Administration (USA)
NBR	Normalized Burn Ratio
NCC	Normalized Cross-correlation
NCEP	National Centers for Environmental Prediction

Nd:YAG	Neodymium-doped Yttrium Aluminium Garnet
NDSI	Normalized Difference Snow Index
NDVI	Normalized Difference Vegetation Index
NDVITM	Normalized Difference Vegetation Index Threshold Method
NDWI	Normalized Difference Water Index
NEBN	Noise Equivalent Beta Naught
NEDT	Noise Equivalent Delta Temperature
NEE	Net Ecosystem Exchange
NEF	Noise Equivalent Flux
NEM	Normalized Emissivity Method
NEP	Net Ecosystem Productivity
NEP	Noise Equivalent Power
NIR	Near Infrared
NLES	Navigation Land Earth Station
NLRI	Near Laser Ranging Investigation
NMC	National Meteorological Center (USA)
NMF	Non-negative Matrix Factorization
NOAA	National Oceanic and Atmospheric Administration
NOHD	Nominal Ocular Hazard Distance
NORUT	Norut Northern Research Institute (Norway)
NPV	Non-photosynthetic Vegetation
NPW	Numerical Weather Prediction models
NSC	NarynSyrdarya Cascade
NSCAT	NASA Scatterometer
NSIDC	National Snow and Ice Data Center (USA)
NWP	Numerical Weather Prediction
OA	Overall Accuracy
OBIA	Object Based Image Analysis
OLCI	Ocean and Land Colour Instrument
OLI	Operational Land Imager
OLS	Operational Linescan System
OM	Organic Matter
OMI	Ozone Monitoring Instrument

ONERA	French Aerospace Research Agency
ONF	French National Forest Office
OS	Open Service
OSCAT	OceanSat-2 Scatterometer
OTB	OrfeoToolBox
PA	Producer's Accuracy
PACE	Pre-Aerosol, Clouds and Ocean Ecosystems
PAH	Polycyclic Aromatic Hydrocarbons
PAI	Plant Area Index
PARIS-IoD	Passive Reflectometry and Interferometry System In Orbit Demonstrator
PCA	Principal Components Analysis
PERSIANN	Precipitation Estimation from Remotely Sensed Information using Artificial Neural Networks
PET	Potential Evapotranspiration
PFT	Phytoplankton Functional Types
PGN	Permanent GNSS Network
PHR	Panchromatic High Resolution
PIF	Pseudo-Invariant Features
piGNSS-R	Partial Interferometric GNSS-R
PLF	Polarization Loss Factor
PLOF	Local Land Tenure Plan (France)
PLSR	Partial Least Squares Regression
PMT	Photomultiplier Tubesv
POA	Polarization Orientation Angle
PODAAC	Physical Oceanography Distributive Active Data Center
POLDER	Polarization and Directionality of the Earth's Reflectances
PolInSAR	Polarimetry-interferometry Synthetic Aperture Radar
PolSAR	Polarimetry SAR
POLYMER	Polynomial Based Algorithm Applied to Meris
POM	Physical Optics Model
PPCDAM	Plan for Preventing and Controlling Deforestation in Amazônia Legal
PRESS	Prediction Sum of Squares

PRF	Pulse Repeat Frequency
PRISMA	*Precursore Iperspettrale della Missione Applicativa* (Italian hyperspecral mission)
PRN	Pseudo Random Noise
PROBA	Project for On-board Autonomy
PRODES	*Programa de Cálculo do Desflorestamento da Amazônia* (Brazilian Amazon Deforestation Monitoring Program)
PROSPECT	Leaf Optical Properties Spectra (radiative transfer model)
PRS	Public Regulated Service
PS	Persistent Scatterers
PSI	Persistent Scatterers Interferometry
PSR	Penalized-Spline Regression
PV	Photosynthetic Vegetation
PVI	Perpendicular Vegetation Index
PWC	Plant Water Content
QBO	Quasi-Biennal Oscillation
QCA	Quasi-Crystalline Approximation
Qgis	GIS software (open source)
QPSK	Quadrature Phase Shift Keying
QUIKSCAT	Quick Scatterometer (NASA)
QZSS	Quasi-Zenith Satellite System
RADAR	Radiodetection and Ranging
RAF	French Altimetric System
RAINFOR	Amazon Forest Inventory Network
RAMSES	ONERA Airborne Multi-frequency SAR Imaging System (France)
RAN	Royal Australian Navy
RANSAC	Random Sample Consensus
RCA	Radio Corporation of America
RCM	RADARSAT Constellation Mission
RCS	Radiometric Control Sets
REDD	Reduction of Emissions from Deforestation and Forest Degradation

REDDAF	Reducing Emissions from Deforestation and Degradation in Africa (European Project)
RENAG	French National GNSS Permanent Networks
RESIF	French Seismological and Geological Network
RF	Random Forests (classifier)
RFI	Radio Frequency Interference
RG	Relative Greenness
RGB	Red Green Blue
rGNSS-R	Reconstructed GNSS-R
RHCP	Right Hand Circular Polarization
RMSD	Root Mean Square Difference
RMSE	Root Mean Square Error
RPAS	Remotely Piloted Aircraft System
RPC	Rational Polynomial Coefficients
RPD	Ratio of Performance to Deviation
RPIQ	Ratio of Performance to Inter-Quartile range
RTK	Real Time Kinematic
RTM	Radiative Transfer Model
RUE	Rain Use Efficiency
RVI	Radar Vegetation Index
RVoG	Random Volume over Ground
RXD	Reed-Xiaoli Detector
SAFY	Simple Algorithm For Yield estimate
SAM	Spectral Angle Mapper
SAMIR	Satellite Monitoring of Irrigation (model)
SAR	Synthetic Aperture Radar
SAR in	SAR interferometric
Saral/Altika	Radar Altimeter (French–Indian altimetry mission)
SASS	SEASAT Advanced Scatterometer System
SAVI	Soil-Adjusted Vegetation Index
SBAS	Satellite Local-based Augmentation System
SBAS	Small Baselines
SCA	Snow Cover Area
SCARAB	Scanning Radiometer for Radiation Balance

SCF	Snow Cover Fraction
SDC	Snow Depletion Curve
SDS	Science Data System (USA)
SEAS	Survey of the Environment Assisted by Satellite
SeaWiFS	Sea-viewing Wide Field-of-view Sensor (satellite)
SEBAL	Surface Energy Balance Algorithm for Land
SEBS	Surface Energy Balance System
SEC	Standard Error of Calibration
SEKF	Self-extended Kalman Filter
SEP	Standards Error of Prediction
SER	Section Efficace Radar
SERD	Single bounce Eigenvalue Relative Difference
SEVIRI	Spinning Enhanced Visible and Infrared Imager
SFCW	Stepped Frequency Continuous Waves
SfM	Surface-from-Motion
SFT	Strong Fluctuation Theory
SGBM	Semi-Global Block Matching algorithm
SHALOM	Spaceborne Hyperspectral Applicative Land and Ocean Mission
SHI	State Hydrological Institute (St. Petersburg, Russia)
SHOALS	Scanning Hydrographic Operational Airborne LiDAR Survey
SHOM	French Navy's Hydrographic and Oceanographic Service
SID	Spectral Information Divergence
SIERRA	Spectral Reflectance Image Extraction from Radiance with Relief and Atmospheric Correction
SIFT	Scale Invariant Feature Transform
SIGMA	Simulation Innovation for Global Monitoring of Agriculture
SLA	*Scanner LiDAR aérien* (aerial LiDAR scanner)
SLC	Single Look Complex
SLR	Single-Lens Reflex
SM	Soil Moisture
SMA	Spectral Mixture Analysis
SMAC	Simplified Method for Atmospheric Correction

SMAP	Soil Moisture Active and Passive mission (Radiometer)
SMAPVEX	SMAP Validation Experiment
SMEX02	Soil Moisture Experiment 2002
SMF	Spectral Matched Filter
SMLR	Stepwise Multiple Linear Regression
SMMR	Scanning Multichannel Microwave Radiometer
SMOS	Soil Moisture and Ocean Salinity mission (satellite)
SNAS	Chinese Satellite Navigation Augmentation System
SNR	Signal to Noise Ratio
SNSB	Swedish National Space Board
SNV	Standard Normal Variate
SPAD	Single-Photon Avalanche Diode
SPM	Small Perturbation Model
SPM	Suspended Particulate Matter
SPOT	Satellites for Earth Observation
SRM	Snowmelt-Runoff Model
SRTM	Shuttle Radar Topography Mission
SSA	Surface Specific Area
SSC	Soil Surface Characteristics
S-SEBI	Simplified Surface Energy Balance Index
SSM	Soil Surface Moisture
SSMI	Special Sensor Microwave Imager (satellite)
SST	Sea Surface Temperatures
STF	Spectrotransfer Functions
STICA	Socio-technical Information and Communication Arrangements
STICS	Crop model
SUCROS	Simple and Universal Crop Growth Simulator
SUHI	Surface Urban Heat Islands
Suomi-NPP	Suomi National Polar-Orbiting Partnership
SURFEX	Surface model platform (Météo France)
SVAT	Soil–Vegetation–Atmospheric Transfer
SVM	Support Vector Machine
SVMR	Support Vector Machine Regression

SVR	Support Vector Regression
SWAT	Soil and Water Assessment Tool
SWE	Snow Water Equivalent
SWI	Soil Wetness Index
SWIR	Short-wave Infrared
SWOT	Surface Water Ocean Topography (satellite)
SYSIPHE	Airborne hyperspectral imaging system
T/P	Topex/Poséidon (Franco-American altimeter)
TDR	Time Domaine Reflectometry
TEC	Total Electron Content
TES	Emissivity Separation algorithm
THEIA	French Land Data Centre
THIRSTY	Thermal Infrared Spatial System (satellite project by CNES and NASA)
TIFF	Tag Image File Format
TIN	Triangular Irregular Network
TiO_2	Titanium dioxide
TIR	Thermal Infrared
TISI	Temperature Independant Spectral Indices
TLS	Terrestrial Laser Scanning
TM	Thematic Mapper
TMBOC	Time Multiplexed Binary Offset Carrier
TNT2	Topography based Nitrogen Transfer and Transformation
To	Tera-octet (1,000,000,000,000 octets)
TOA	Top of Atmosphere
TOC	Top of Canopy
TomoSAR	Tomography SAR
Topex/ Poseidon	Radar altimeter
TOPLATS	Topographic Land Atmosphere Transfer Scheme
TRIP	Total Runoff Integrating Pathways
TRMM	Tropical Rainfall Measuring Mission (satellite)
TSAVI	Transformed Soil Adjusted Vegetation Index

TSEB	Two-Source Energy Balance (model)
TTL	Transistor-Transistor Logic
TWAP	Transboundary Water Assessment Program
TWS	Terrestrial Water Storage
UA	User's Accuracy
UAA	Utilized Agricultural Area
UAV	Unmanned Aerial Vehicles
UHI	Urban Heat Islands
ULICE	Ultraviolet LiDAR for Canopy Experiment
UNEP	United Nations Environment Program
USDA	United States Department of Agriculture
USGS	United States Geological Survey
USO	Ultra-Stable Oscillator
UTC	Coordinated Universal Time
UTM	Universal Transverse Mercator
UV	Ultra-violet
VARI	Visible Atmospherically Resistant Index
VCA	Vertex Component Analysis
VD	Virtual dimensionality
VHF	Very high frequency
VHI	Vegetation Health Index
VHSR	Very High Spatial Resolution
VIC	Variable Infiltration Capacity
VIIRS	Visible Infrared Imager Radiometer Suite
VIS	Visible
VISAT	Video, Inertial, and Satellite GPS
Vis-NIR	Visible and Near Infrared
VITO	Flemish Institute for Technological Research (Belgium)
VLA	Very Large Array
VLBI	Very Large Baseline Interferometry
VOS	Volatile Organic Compound
VPD	Vapor Pressure Deficit
VSDI	Visible and Shortwave Infrared Drought Index

VWC	Vegetation Water Content
WALID	Water LiDAR Simulation Model
WASS	Wide Area Augmentation System
WDI	Water Deficit Index
WGHM	Water GAP Global Hydrology Model
WGS	World Geodetic System
WMA	Winter Metric Anomaly
WMO	World Meteorological Organization
WMS	Wide-area Master station
WRS	Wide-area Reference Stations
WSI	Water stress index
WTC	Wet Troposphere Correction
XML	Extensible Markup Language
ZSSD	Zero-mean Sum of Squared Difference

Introduction

For a long time, the dynamics of urban and coastal areas have been the focus of administrators and decision makers in charge of public policy. In the case of urban areas, administrators are particularly concerned with evaluating their density and the extent of their growth. More recently, we have seen new questions regarding sustainable development: air quality, green spaces, etc. For coastal areas, a better understanding of anthropogenic pressure and the impact of climate change is necessary in order to design? elaborate? a plan for territorial organization, especially for residential areas and tourist infrastructure. The effect of climate change on coastline evolution is a key area of study.

The aim of this volume is to outline the leading thematic applications of remote sensing in urban and coastal areas. This volume, featuring internationally renowned scientists, will bring readers up to date with the latest developments and will describe the potential challenges in research and development for the years to come. It has been written for research teams in remote sensing, and geophysics postgraduate students (engineering schools and Masters) and PhD students.

The first part highlights various applications in urban environments with the first chapter focusing on the use of remote sensing in urban planning (urban expansion, light pollution, air quality, etc.). The second chapter analyzes urban scenes using a mobile mapping vehicle and object identification. The next chapter introduces the use of remote sensing in urban and periurban environments within the framework of territorial

---

Introduction written by Nicolas BAGHDADI and Mehrez ZRIBI.

decision making, with examples given from France and territories in developing countries.

The second part of this volume begins with a chapter on the use of remote sensing for dealing with the ocean color. The following chapter considers the physical parameters that are attainable using LiDAR in coastal and continental waters, in particular bathymetry and altimetry, as well as the characterization of the water column's properties. The next chapter focuses on coastal applications and the increasingly frequent use of airborne topographic LiDAR for studying coastal dynamics, especially identifying coastlines and estimating sediment balances.

The third part is made up of two chapters and analyzes the dynamics of mangroves, one of the most fragile types of forest on the planet.

We would like to thank everyone who contributed to the preparation of this volume, especially the authors of each of the chapters, as well as the members of the scientific community for reviewing chapters. This project was carried out with the support of the IRSTEA (French Institute for Research in Science and Technology for the Environment and Agriculture) the CNRS (French National Center for Scientific Research) and the CNES (French National Center for Space Studies).

We would also like to thank our families for their support which was vital for the completion of such a project. Finally, we would like to thank Mr André Mariotti (Professor Emeritus, Pierre and Marie Curie University) and Mr Pierrick Givone (Scientific Director, IRSTEA) for their support.

1

Optical Remote Sensing in Urban Environments

1.1. Introduction

1.1.1. *The urban system*

Cities today face a variety of issues: attractiveness and economic development, living conditions and urban redevelopment, the quality of life of citizens and the environmental conditions of the urban system as a whole.

These challenges reflect the situations in urban territories where the economic development and population growth required for stabilizing the urban system come into conflict with the promotion of esthetic urban improvements appropriate for social cohesion and ensuring the safety of users while guaranteeing a reduction of the environmental impact caused by urban spread [WGI 14, WEB 15, MAC 07, WIL 11]. The process of urbanization, which has significantly transformed our countryside over the last hundred years, was the most important factor of the 20th Century and saw the urban population surpass the rural population in the majority of Western countries.

The process of urbanization can be defined as "particular forms and structures of space being occupied by the population [RHE 14]" – in other

---

Chapter written by Xavier BRIOTTET, Nesrine CHEHATA, Rosa OLTRA-CARRIO, Arnaud LE BRIS and Christiane WEBER.

words the transformation of space (natural, agricultural, forest, etc.) through human activity. A recurrent problem is the objective demarcation of the urban system which is partly artificial and difficult to transpose from one case to another (conurbation, administrative boundaries, etc.). The determinants can vary in time and space (from one country to another[1]). In fact, identifying the place of study can sometimes get in the way of properly understanding the dynamics of urban growth and expansion, because like any other system, it is not exempt from external factors which influence its internal dynamics. For instance, environmentally evaluating the quality of the water supply can be carried out at various locations in the urban system: where the water is extracted from, where it leaves the urban system and in the city center. The quality of the supply can only be determined based on sections that have been artificially developed.

What makes the urban space so complex is that it only exists insofar as it is "built, worked and used through social ties" [CAS 72]. It can be thought of as a construction where the organization and the structure impact upon urban society but also as one element of a much wider ecosystem on which it depends and/or which it influences. This system incorporates political, economic, social and, increasingly, environmental dimensions and we therefore see it as a complex and heterogeneous ecosystem whose social, technical and natural components must be studied in order to understand the way in which it functions [MAT 06]. Taking these dimensions into account is necessary for urban studies, whether this is in terms of evaluating environments, modeling dynamics or outlining evolution scenarios.

Over the past few years, the focus has been on gaining a better understanding of how the urban ecosystem works. This has taken place under different investigations:

– a study focusing on the social, economic and spatial aspects of the urban system in order to consider the environmental aspect;

– another study which focuses more on the interactions between the urban ecosystem as a whole and the natural systems on which it depends. This study examines both the fluctuations (energy resources, food, raw

1 The space considered here rarely corresponds to the administrative body as defined by the National Institute of Statistics and Economic Studies (INSEE) but more generally to the morphological urban zone taking into account the expansion of built-up zones built in addition to the employment area.

materials, etc.) and the state of these systems (the state of biodiversity, for example).

Both studies introduce different but compatible problems; for example, the quality of living conditions will be analyzed in the first study by studying the behavior and aspirations of citizens related to the presence or absence of amenities; the second study will focus on disparities in areas featuring particular environmental phenomena (pollution, heat islands, presence of vegetation, etc.). Based on the problems that have been explored, it is necessary to establish a comprehensive framework for spatial and temporal analysis. This framework is essential for detecting, characterizing and identifying a given phenomenon [SAI 06]. It makes it possible to consider the state and dynamic of the urban ecosystem, as well as how its different elements interact with each other. The trio space*time*function [PEU 95] summarizes the complexity of these approaches:

– Space outlines the states of the urban ecosystem. These can be determined according to the general characteristics present, such as buildings, population and employment. This space is characterized through spatial, social and even environmental attributes at a number of levels that are often intertwined.

– Depending on the temporal scale that is chosen, urban dynamics can give us an indication of, for example, changes in land use (LU). These are made available through evolution processes that can sometimes be highlighted by the causes of change. Thus, urban wasteland constitutes a stage in the modification of soil usage (green zone to construction zone) and of a function (a strip of garden to an office block). Increasing demand for urban housing can lead to construction dynamics and increased housing stock which in turn can lead to a rise in urban growth.

– How the urban ecosystem interacts with the natural systems on which it is reliant depends on a number of factors: for instance, the characterization and the use of surface water in the city, urban climate zones [STE 06] or the part played by vegetation in a city's energy patterns.

Thus, new studies have been undertaken and knowledge of a territory as well as how to monitor it is increasingly important. As a result, there is an increasingly pressing need to collect, assimilate and analyze high-quality geographical data which can be put to a variety of uses.

In 2010, Sebari and Morin [SEB 10] demonstrated how satellite imaging could be used for urban studies (*Measuring the evolution of the urban environment, Mapping the LU, Urban planning, Revised cartography* or *Updated automatic cartography*) following on from the challenges outlined by Donnay *et al.* [DON 00]. Taking these environmental challenges into account has altered the way in which information is gathered by introducing the need to identify the natural components of the urban matrix[2] and how they interact with the population and activity in a way that is both prospective and retrospective.

Thanks to the acuteness of the information obtained, satellite imaging makes it possible to take into account policy choices which are environmental (protecting habitats, improving green spaces and waste ground), patrimonial or urbanistic (property management and protecting sensitive zones) or pertaining to landscape (reorganizing riverbanks, minimizing the break-up of habitats and improving focus areas). Satellite imaging can therefore be thought of as a medium for (re)presentation, a tool for consultation or even for cooperation. The images provided are used at different stages of the process: detection and observation, how results affect decision making, evaluating action taken and monitoring this action for future reference.

1.1.2. *The urban environment*

The urban environment is a heterogeneous mineral environment, made up of buildings that vary in terms of compactness and morphology[3]. Transport networks, waterways and green spaces intertwine in the built matrix and this leads to ruptures in the urban landscape. Seen from above, these landscapes

2 The term *urban matrix* refers to the set of components which make up a system (buildings and infrastructure, vegetation, water surfaces, bare earth or waste ground). It is possible to distinguish between the gray matrix (buildings and infrastructure), the natural matrix (vegetation and water) and the brown matrix (soil and waste ground).

3 *Urban morphology* is the study of urban shapes. Urban morphology aims to study the urban make-up beyond a simple analysis of the architecture of buildings and to identify underlying patterns and structures. Urban morphology studies a city's shapes and characteristics (its roads, its plots of land, how areas are divided, density and functions) and the main factors behind it: topography, history, cultural influence, economics, town planning as well as the technological or even the energy context. It is based on different levels of the urban area: buildings, blocks, the urban fabric, the city and its surroundings.

can normally be determined by groups of buildings which appear similar at first sight, but which vary according to their distance from the center(s) by general morphology, the layout of neighborhoods, the space between buildings, how buildings are grouped around a more significant building such as a church or a theater or a square, or even along an avenue, the presence of other ruptures (roads, railway lines and canals) and the presence of vegetation within a built-up area and/or in parks and at the roadside. Depending on their size, their shape or the material used for roofing, buildings can disturb the impression of uniformity. Various studies have successfully defined categories for LU following criteria based on the density of the area, how far it is from the city center and the minimal mapping unit. Within the framework of the CORINE Land Cover (CLC) program, efforts were made to standardize the different categories of LU on a Europe-wide level. From this point onwards, the typology put forward has been expanded and adapted, notably for urban spaces where the minimal zones for representation with CLC (25 ha) are not consistent with the urban scale ($1/5000^e$ to $1/20000^e$).

On this scale, the characteristics of urban environments are as follows:

– areas that are mineral or not and distances between locations (and therefore the density);

– the volume and spatial arrangement of areas;

– the environment's components: built matrix and networks, natural matrix (depending on various strata and water surfaces);

– interior and continuous break lines: coastline, lakes, rivers, road and rail networks, walls or empty lots;

– gradients: from the center toward the periphery. Given that city limits are difficult to identify, the interactions of the city and its surrounding area makes readability complicated;

– materials used to cover mineral surfaces.

The Urban Atlas[4] program put forward by the European Environment Agency (EEA) proposes covering all towns in Europe with a population greater than 50,000 where the focus will be on functional urban areas

4 http://land.copernicus.eu/local/urban-atlas.

(functional division where the objective is statistical) defined by ESPON (2006)[5]. The characteristics of the different types of LU (Table 1.1) pick up the large lines above by using more specifically urban density and the distance between a continuous or discontinuous building, and one that is isolated. Artificially created surfaces can be obtained using the Fast Track Laser European Environment Agency (EEA). This product uses several databases as well as satellite data with a resolution of 2.5 m (initial or a fused with the panchromatic channel) ranging from the visible to near-infrared. This is one of the services that is linked to the GMES/Copernicus program[6].

Type of surface	Object	Shape and organization of the objects (among themselves and in relation to their environment)
Artificial areas	Urban fabric	Continuous urban fabric Discontinuous urban fabric (dense, average and less dense) Isolated structure
	Commercial, industrial, public, military and private units	Commercial, industrial, public, military, transport and private units Roads, rail networks and related surfaces
	Ferry ports	
	Airports	
	Mines, building sites, waste disposal sites	Mineral extraction and waste disposal sites Building sites Wasteland areas
Non-agricultural green spaces	Green spaces Agricultural or semi-natural zones, wetlands	
Forest		
Water surfaces		

Table 1.1. *Characteristics of different types of land cover and land use (Urban Atlas, 2012)*

5 http://www.espon.eu/main/.
6 http:www.esa.int/Our_Activities/Observing_the_Earth/Copernicus/Overview.

In 2001 Puissant and Weber [PUI 01] were able to establish a way of naming urban objects that can be used for describing the urban matrix based on SPOT images (Table 1.2). In certain cases, the general organization of the built matrix also makes it possible to determine some of the functions of the environment: the presence of air or sea traffic, important road and rail networks, peripheral areas, etc.

Type of surface	Object	Shape and organization of the objects (among themselves and in relation to their environment)
Urban vegetation	Trees	Single trees Grove (clump of trees) Forest Line of trees
	Grass or shrub	Field (park or garden depending on the surrounding area or the type of sports field according to its shape)
Built-up areas	Building	Continuous - dense urban layout (most often facing the road) Discontinuous – large buildings Residential (depending on the size of a building and its surroundings) Isolated
Network	Road	Motorway or four-lane (and more) roads Dual carriageway
	Rail network	Railway lines Rail depot
Industrial zones, airports/sea ports, service	Buildings and infrastructure	Building layout, carparks, proximity to networks
Interstitial mineral areas	Parking Place Building area	
Water pond	Rivers, canals, ponds, lakes etc.	Shallow water networks

Table 1.2. *Typology of useful urban objects (taken from [PUI 01])*

Satellite images can thus be thought of as one aspect of a wider information network that is used for studying the urban ecosystem. Changes caused by the digital boom have radically changed the situation since the end of the 20th Century. The potential for benefiting from localization, description and valorization in 2D or 3D via paying or free networks has led to the spread of digital media and their use as a spatial reference. The contribution of various sources of information makes it possible to reuse satellite images alongside existing databases for a given area. Since then, satellite imaging has been one of a number of elements used (national or international databases, shared collective data, online information, of variable internal and external quality). An additional advantage of satellite data is the stability of the data provided and the capacity to obtain different data modulated by the number of sensors, the spectral range or the spatial resolution. Through the changing ways in which the Earth is observed, satellite imaging meets a variety of demands (observation, measurement, modeling, representation and application) while continuing to reframe the debate over how we study the environment. The current range is expanding in order to meet the latest challenges concerning the anthropization of areas by making technical improvements which will enable better results.

1.1.3. *The main characteristics of the urban environment: geometric, spectral and temporal*

The urban environment is characterized by the heterogeneity of its shapes, materials and density. Furthermore, a city is made up of large buildings and their shapes create shadows which increase its complexity on the radiative understanding.

1.1.3.1. *Geometric characteristics of urban objects*

The geometric diversity of urban objects is caused by several factors such as density, the age of the city or its latitude. Since the 1980s [CLA 80, WEL 82], determining the optimal spatial resolution in order to identify urban elements has been a key issue for the new generation of sensors. While in North America a resolution of between 20 and 30 m is considered acceptable, a resolution of between 20 and 10 m, or even metric resolution has become the necessary threshold for the use of satellite imaging in town planning. Small [SMA 03] used IKONOS images with a spatial resolution of 1 m in order to determine the average size of urban objects across 14 cities worldwide and was able to show that it is somewhere between 10 and 20 m. Resolution of 5 m or below is considered necessary for the accurate representation of urban objects for the following applications: buildings, roads and their aging process [HER 05], the spread of urban vegetation [WEN 07, JEN 12] and land planning [PUI 03, PUI 04, WAN 07a].

Following a study of useful urban objects in Strasbourg (France), Puissant and Weber [PUI 01] were able to draw up a table of optimal resolutions using simulated SPOT images (XS and P) with resolutions ranging from 5 to 1.36 m (Table 1.3). The results indicate that a fine resolution of 1.36 m is necessary in order to identify "dense urban" buildings, while a resolution between 2.5 and 5 m is sufficient for identifying "large group" buildings; a resolution of 2.5 m is sufficient for identifying vegetation and roads.

Reference objects	10 m	5 m	2.5 m	1.36 m
Trees	X	X	X	X
Fields	X	X	X	X
Interstitial green spaces		X	X	X
Dense urban	X	X	X	X
Large groups	X	X	X	X
Roads	X	X	X	X
Railway lines			X	X
Interstitial mineral areas	X	X	X	X
Water	X	X	X	X
Shadow	X	X	X	X
Rate of well classified pixels	*8 classes 85 %*	*9 classes 89 %*	*10 classes 86 %*	*10 classes 80 %*

Table 1.3. *Urban typology and spatial resolution for the city of Strasbourg [PUI 01]*

1.1.3.2. *Spectral characteristics*

A city can be said to be made up of three types of areas: mineral areas (buildings, infrastructure and bare soil), vegetation and water. The spectral signature of these areas depends on their chemical composition, the acquisition time (day/night) and the surrounding environment. Spectral signature is influenced by several parameters which play a more significant role depending on the given spectral range: reflectance, emissivity and surface temperature.

These optical parameters (detailed in section 1.1.4) are characteristics of the chemical composition (spectral signature), the surface state (spectral and/or directional signature) and the time of measurement (directional signature) for a given material.

1.1.3.3. *Temporal characteristics*

The urban system offers a range of dynamics depending on the components being considered. Vegetation follows the seasonal vegetative cycle (soil, sunlight and weather conditions) while being strongly influenced by factors such as the shapes of buildings and the specific ecological conditions of a given environment.

The development of buildings covers various dynamics:

– the growth of built-up areas (residential, industrial and the different types of network) outside of the urban matrix through urban sprawl: the standardization of the production process has considerably shortened construction time. However, observing the creation of a built-up area is only possible by observing the change in LU. Of course, this is also connected to the size and localization of the modified surface. This is why development seems fairly slow, because on the one hand only groups of buildings can generally be observed while on the other hand, such changes are taking place in pre-existing construction units;

– the filling-in of built-up areas within the urban matrix: the construction of a new building where an old one previously stood or the use of waste ground tends to be difficult to observe without close monitoring;

– urban materials have their own intrinsic dynamics depending on the composition of the used material (aging or dirty tiles) as well as extrinsic dynamics relating to the use of new materials, including, for example, materials with a higher capacity for retaining heat.

We can see therefore that the system's heterogeneity is an essential element in the temporal dimension of both the urban matrix and the dynamics related to its components.

1.1.4. *Optical properties of urban materials*

1.1.4.1. *Spectral signatures*

The range of different urban materials of various chemical compositions leads to a wide variety of spectral signature in the optical domain. Several authors have carried out studies in order to obtain spectra of urban materials in the reflective field [BEN 01, HER 04, SOB 12a] as well as in the optical field [BRI 06]. By way of an illustration, Figure 1.1 and Figure 1.2 shows the spectral signature of urban materials in the range (0.4–2.5 μm) (and [3.0–14.0 μm]) respectively. In the range (0.4–2.5 μm), tile roofs and bricks have similar signatures. It is only possible to distinguish between these two types of surfaces using the spectral signature through the use of a spectral imaging device.

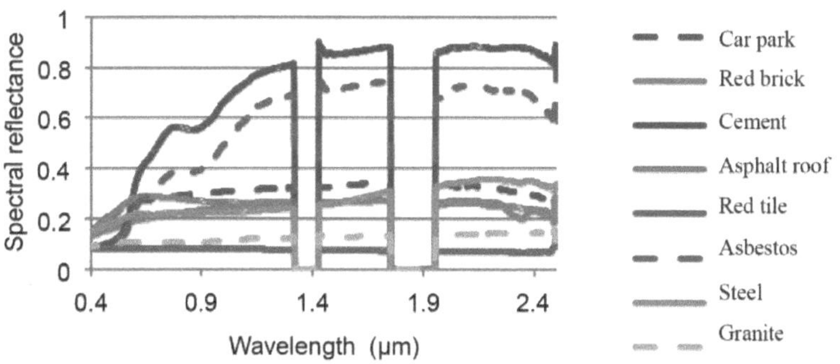

Figure 1.1. *Average spectral reflectance of urban materials measured in Madrid in the range [0.4–2.5 μm] [SOB 12a]. The absence of measurements around 1,4 and 1,9 nm corresponds to high water absorption bands in the atmosphere. For a color version of this figure, see www.iste.co.uk/baghdadi/5.zip*

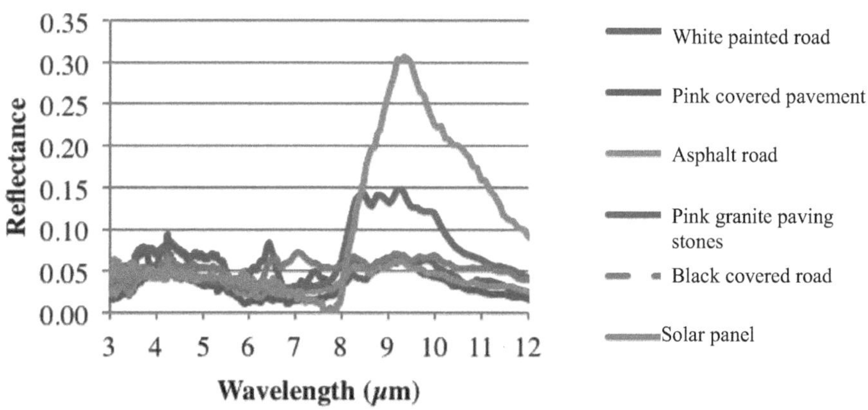

Figure 1.2. *Average spectral reflectance of urban materials in the range [3.0–14.0 μm] [BRI 06]. For a color version of this figure, see www.iste.co.uk/baghdadi/5.zip*

Figure 1.2 shows that surfaces such as roads and pavements covered with asphalt and minerals have a similar spectral behavior due to the dominant presence of hydrocarbon on the surface. There is an increase in reflectance for granite paving stones caused by the presence of the carbonate ion CO_3^{2-}. The presence of silicate in solar panels results in an increase in reflectance around 10 µm.

The high diversity in spectral signatures is used in hyperspectral imaging in order to distinguish between materials [JEN 12, BEN 01] and to estimate their state [HER 05]. As it is shown, similar spectral signatures can represent different kinds of objects (an asphalt roof and an asphalt road for example), but it is still possible to distinguish between them. This can be done using hyperspectral images combined with data obtained either from an airborne 3D LiDAR, or from a digital surface model (DSM) (detailed in section 1.2.1.2).

1.1.4.2. *Intraclass variability*

However, for the same material, there can be distinct variations in its signature and this leads to a change in the classification or unmixing techniques: this is called intraclass variability. There are several reasons for this variability:

– physical (difference in texture or roughness);

– use (for instance, the signature of asphalt will not have the same signature on a pavement as it will on a road due to different deteriorations);

– definition of classes (for instance, granite can include different colors and different components).

By way of an illustration, Briottet *et al.* [BRI 06] were able to quantify this variability for the city of Toulouse, summarized in Table 1.4 for a sample size of 20 cm. A maximum standard deviation is calculated corresponding to the maximum of standard deviations calculated for each of the different types of intraclass variability. The analyzed classes include different types of asphalt, cement, granite, brick and tile. The strong relative standard deviation (%), found for physical intraclass variability in the range (3.0–12.0 µm), is caused by the diversity of materials included in each class, but also due to the weak level of reflectance (given that the standard deviation is of the same magnitude as the reflectance, the resultant relative standard deviation, defined as the link between the standard deviation and the reflectance, is large).

Type of variability	Maximum standard deviation in the range (0.4-2.5 µm)	Maximum standard deviation in the range (3.0-12.0 µm) (%)
Physical	13	88
Use	30	21
Class definition	50	40

Table 1.4. *Maximum standard deviation of the intraclass variability of the reflectance across all classes measured in Toulouse in the optical domain*

Table 1.5 shows that the used classification used has a significant impact on variability. The influence of the texture of the material is highly variable depending on the given spectral domain.

Level 1	Level 2: Type of soil	Level 3: Type of material	Level 4: Surface material		Characteristic wavelengths (nm)
Artificial surfaces	Building/roof	Mineral material	Asbestos	Asbestos	580, 800, 2,323, (2,320; 944)
			Asphalt roof	Asphalt roof	1,700, 2,350, 11,300
			Clay roof (red)	Clay roof (red)	520, 670, 870, (iron ion absorption)
			Concrete roof	Concrete roof	A 2,300–2,370 (calcite and dolomite), A 2,170 (argile), 1,413 and 2,323 (portlandite, muscovite)
			Gravel	Gravel	Silicate 2,200, carbonate 2,200, iron oxide 520, 670, 870
			Plate glass window	Plate glass window	R 372, 400 A 2280
			Slate	Slate	R 791, A 940, 2,200, 2,260, 2,340
		Metallic material	Aluminum	Aluminum	A 840
			Copper	Copper	R 540–650, A 2,240, 2340
			Zinc	Zinc	A (strong) 550
					A 1,020

			Polyvinyl chloride	Polyvinyl chloride	Strong absorption of 1,200, 1,700, 2,170, 2,300
			Polyvinyl (PV)	Monocrystalline silicon(Si)	732, 1,100
				Gallium Arsenide (GaAs), Indium Gallium Phosphide (GaInP)	381, 887
				PV	732–1,067 (Optimal transmission window)
		Vegetal material	Green roof	Green roof	R 550, 750, A 450, 680, 980, 1,200, 1,400, 1,900
			Thatched roof	Thatched roof	Cellulose vibration A 2,100, 2,300
			Wooden roof	Wooden roof	Cellulose vibration A 2,100, 2,300
		Waterproof surface	Asphalt	Asphalt	Weak absorption 1,850, 2,315, 2,350
					1,700, 2,350, 11,300
			Cement brick	Cement brick	R 600–700, Iron oxides A1,410, A1,940–1,970, mini reflectance of gray cement 2,310–2,350
		Artificial water surface	Swimming pool	Swimming pool	A 1,340–1,480 and 1,770–1,970
			Garden pond	Garden pond	

Table 1.5. *Extract of the characteristics of urban materials [ZIN 15] (A: absorption; R: reflectance)*

1.1.4.3. *Directional variability*

With urban materials, there is also a wide variety of directional behaviors [BRI 16] which is due in large part to their roughness. Thus, a smooth surface such as a solar panel or polished granite will show marked specular reflection, while a rough material (such as tiles or rough casting) will have a maximum of reflectance in the backscattering direction.

Examples of directional behavior are given in Figure 1.3 where the directional reflectance of different artificial materials is outlined in polar representation. Thus, with concrete, asphalt and pink granite, the peak of reflection is in a forward direction. The measured granite has a smooth surface which explains the higher amplitude of the reflection peak. However, asphalt does not display a marked directional signature (Lambertian reflectance).

1.1.4.4. *Temporal variability*

The materials found in a city undergo natural changes (evolution of vegetation along a year), but they are also chemically damaged (pollution) and through use (changes to the surface). This modifies their spectral signature and can thus be used as an indication on their status. Herold *et al.* [HER 05] were able to estimate the aging and deterioration of roads using ground spectrometric measurements. These modifications are attributed to physical changes (through use) and chemical changes to the road surface.

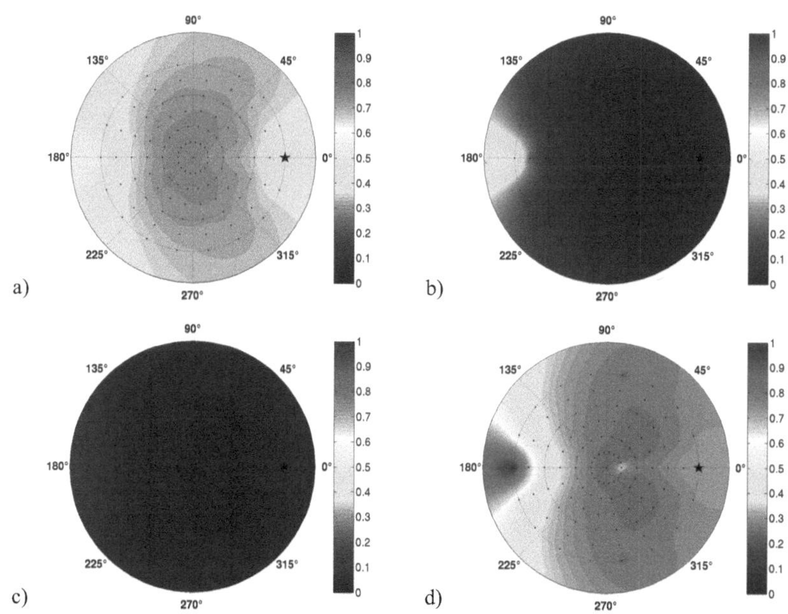

Figure 1.3. *Bidirectional reflectance of different urban materials at 550 nm for an incident zenith angle of 60° (size of the sample measured at nadir of 20 cm): a) concrete, b) red asphalt, c) asphalt and d) pink granite. In a polar representation, the length of the radius represents the zenith angle, the angle between this radius and the x axis is the azimuthal angle and the color represents the value of the bidirectional reflectance. The star indicates the position of the light source. For a color version of this figure, see www.iste.co.uk/baghdadi/5.zip*

1.1.5. *Spectral characteristics*

1.1.5.1. *Introduction*

In remote sensing, the interpretation of the signal acquired over an urban environment is very complex because of its specific structure which causes occultations, shadows, light reflected by facades and the directional effects of the materials present. These processes make it difficult to interpret the signal received and to automatically determine the materials present directly from the signal measured by a sensor. Figure 1.4 illustrates the radiative components present in remote sensing at high spatial resolution[7] optical imagery, in particular.

– shadows are classified separately (in black in Figure 1.4(a));

– roofs of the same material but with different slopes are classified separately (in blue, green, yellow and cyan in Figure 1.4(b)).

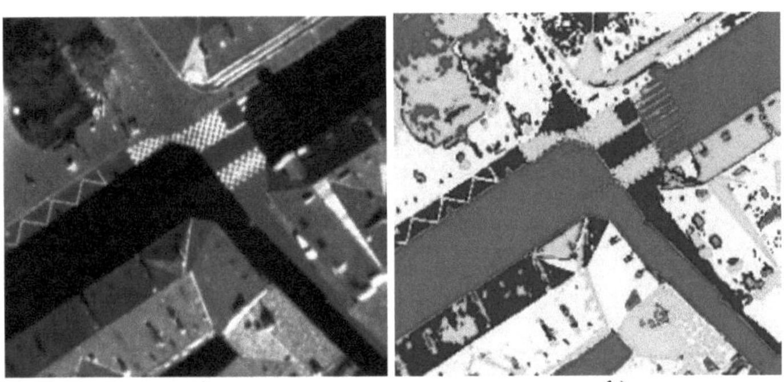

a) b)

Figure 1.4. *a) Colored composition of an image acquired for Toulouse by the multispectral airborne camera PELICAN (spatial resolution of 20 cm) and b) non-supervised classification for the same area (ONERA credit). For a color version of this figure, see www.iste.co.uk/baghdadi/5.zip*

A degradation of the spatial resolution (Figure 1.5) leads to a loss of geometric details and several objects can be contained in a single pixel. Furthermore, lower spectral resolution makes it more difficult to discriminate between different materials.

7 For decametric spatial resolutions, we will refer to high spatial resolution (HR) and for resolutions between metric and submetric we will refer to very high spatial resolution (VHR).

Figure 1.5. *Images of Toulouse at three spatial resolutions of 30, 8 and 2 m generated from data obtained from the airborne multispectral camera PELICAN during the Musarde experiment (ONERA/IGN credit, 2010). For a color version of this figure, see www.iste.co.uk/baghdadi/5.zip*

For applications such as the soil mapping [LU 06], the growth of cities or changes in LU [MAD 01, JAT 08], hyperspectral instruments with a spatial resolution of roughly 30 m are used. For this, the baseline information is the spectral reflectance. The advantage of these instruments is that their swath makes it possible to cover vast urban areas. However, the main disadvantage is that the images are made up of a large number of pixels and include a mix of several different materials. Various studies have been carried out in order to evaluate the contribution of hyperspectral imaging compared to multispectral imaging. Platt and Goetz [PLA 04] show that the classification of urban surfaces obtained from images taken by the spectro-imaging device AVIRIS is better than classification with Landsat ETM+ data. A number of authors [LU 06, WU 03] explain the limitations of multi-spectral sensors for characterizing impervious surfaces and discuss the potential of hyperspectral imaging. Cavalli *et al.* [CAV 08] obtain a similar conclusion in their classification works.

Identifying smaller objects (< 5 m) has become possible using panchromatic and multispectral devices such as SPOT 6/7 and Pléiades satellite sensors. However, such a spatial resolution comes at the expense of spectral quality (the number of bands is limited to 3 or 4 in the visible and near infrared spectrum with a typical bandwidth ranging from 60 to 100 nm). These methods rely on the use of features such as the spectral indices, the geometric shapes [DUR 08, ING 07] and the texture [ZHA 09]. These methods will be outlined in section 1.2.1.

The advent of new spatial missions which combine a hyperspectral camera with a spatial resolution of 8 m for the HYPXIM mission [CAR 14], (10 m for the SHALOM mission [BEN 14]) and a panchromatic camera with a spatial resolution of 1.8 m (4 m for SHALOM) will allow for a better understanding of urban environments. Thus, the joint use of images acquired through these two hyperspectral and panchromatic imaging devices makes it possible to generate a fused image with a high spectral resolution and a high spatial resolution. There are different techniques for this type of fusion. For instance, a comparison of eight methods of panchromatic and hyperspectral fusion was proposed in Loncan *et al.* [LON 15]. The best results were obtained with the methods developed by Laben and Brower [LAB 00], Yokoya *et al.* [YOK 12], Wei *et al.* [WEI 15] and Simoes *et al.* [SIM 15].

The use of airborne imaging rounds off the range of spatial technology through its capacity to acquire images simultaneously with a high spatial and spectral resolution. By way of an illustration, the AVIRIS[8], HYSPEX[9] or HYMAP[10] cameras, at their nominal altitude, have spatial resolutions between 2 and 4 m, while covering the range 0.4–2.5 μm. We are currently witnessing the arrival of new devices which are sensitive in the thermal infrared spectrum, such as the TELOPS[11] systems, or across the whole optical spectrum, such as the Sysiphe system [ROU 15], with compatible spatial resolutions for observing urban environments.

1.1.5.2. *Spectral preprocessing*

The data acquired by these devices are georeferenced images expressed in radiance units. To extract the perturbing effects of the atmosphere, whether spectral or temporal, preliminary preprocessing specific to urban environments is necessary in order to estimate the following parameters: reflectance, emissivity and/or temperature for applications.

The presence of 3D urban structures creates a lot of shadows which deteriorate the effectiveness of processing for applications such as the identification of urban materials [LAC 08, LEB 96], detecting objects [SHI 11], 3D reconstruction [CHE 11] or monitoring traffic [FLE 06]. To improve these results, two steps are necessary: shadows identification and

8 Aviris.jpl.nasa.gov.
9 http://www.hyspex.no/.
10 http://hyvista.com/?page_id=440.
11 http://telops.com/fr.

atmospheric correction, in order to estimate the optical properties for surfaces at all points of the image, in the sun and in shadow. We are able to justify this step because the different total incident radiation (for example, there is no direct solar radiation in shadow) means that different types of preprocessing will be applied.

Shadows can be characterized in the reflective domain by the following radiative properties:

– the levels of radiance in shadow are weaker than areas illuminated by the sun;

– the radiative effects of the environment caused by multiple reflections can no longer be disregarded in shadow [LAC 08];

– the spectral radiance measured in shadow decreases as the wavelength increases (decrease in the atmospheric diffusion);

– the spectral radiance measured in shadow depends on the spectral reflectance of the material.

Methods of identifying shadows rely on either one or several of these characteristics. For a detailed comparison of existing methods, we would refer the reader to studies carried out by Adeline *et al.* [ADE 13]. The limitations of current methods are due to the difficulty of separating water from shadow and identifying shadows in the vegetation when it is less dense.

Once the shadows have been detected, different methods of atmospheric correction have been developed in order to reduce the impact of perturbing radiative effects in the city–atmosphere system. In an urban environment, radiative exchange is more complicated than with a flat surface (see Chapter of [BAG 16]). Figure 1.6 illustrates the different irradiance E components incident to the surface at point B in shadow and at point C in the sunlight and radiance L components at sensor level. In the shadow, the direct irradiance E_D is absent, the downwelling atmospheric irradiance $E_{atm,d}$ is no longer integrated over the $2.\pi$ sr but instead over the sky solid angle Ω_{sky} delimited by the 3D shape of the buildings, the coupling irradiance E_C is also reduced by the presence of 3D structures. Finally, a new form of irradiance appears, caused by multiple reflections, E_{refl},, due to the 3D environment of the considered point.

The total incident irradiance, E_T is written as:

$$\begin{aligned}&\text{in the sunlight: } E_T = E_D + E_{atm,d} + E_c + E_{refl}\\&\text{in shadow : } E_T = E_{atm,d} + E_c + E_{refl}\end{aligned} \quad [1.1]$$

Given that the total radiance at the level of the sensor L_T is the sum of the incident radiance s, and supposing Lambertian reflectance, reflectance ρ at ground level is written as:

$$\begin{aligned}&\text{in the sunlight: } \rho = \frac{\pi.(L_T - L_{atm,m} - L_{envi})}{(E_D + E_{atm,d} + E_c + E_{refl})}\\&\text{in shadow : } \rho = \frac{\pi.(L_T - L_{atm,m} - L_{envi})}{(E_{atm,d} + E_c + E_{refl})}\end{aligned} \quad [1.2]$$

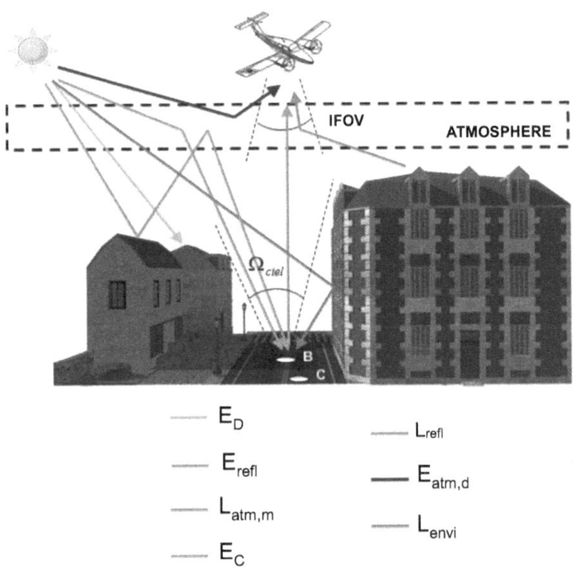

Figure 1.6. *Illustrations of the radiative components in E irradiance and L radiance for a surface in the shadow of a building (B) and for a surface in sunlight (C) [ADE 13] For a color version of this figure, see www.iste.co.uk/baghdadi/5.zip*

Different methods are then used in order to solve equation [1.2] depending on whether the point is in sunlight or shadow. The first type of method considers the ground as flat and no correction is possible in shadow. The chapter by Briottet in this set of books [BRI 16] provides a description of these methods. Nevertheless, Chen *et al.* [CHE 13] put forward an alternative by adding empirical geometric correction factors: weighting of the total descending atmospheric irradiance, calculated over the $2.\pi$ sr by an average of the solid angle of the sky, suppression of the terms L_{envi}, E_C and E_{refl}. This method provides better results but only for the visible domain (VIS). Other more empirical methods aim to equalize the grayscale between the areas in shadow and the areas in sunlight. The disadvantage of such methods is that it is not possible to correct the spectral effects caused by the presence of different radiative terms.

More recent methods (ATCOR-4 [RIC 03], ICARE [LAC 08]) have been developed using a DSM in order to calculate the different terms of equation [1.2] using a radiative transfer code like Modtran. Thus, Lachérade *et al.* [LAC 08] were able to obtain a maximum deviation of 0.04 in reflectance units [BAG 16] for the estimated reflectance in shadow, compared to experimental data. The main sources of error are due to a registration error between the image and the DSM, and the shadow caused by non-opaque trees which allows part of the direct solar irradiance to pass through, which is not taken into account in equation [1.2].

Methods of atmospheric correction in the thermal domain are less sensitive to the environment. Thus, the methods described in the chapter by Briottet [BRI 16] are used. Additional information relating to the urban environment is provided in section 1.3.3.

1.2. Main applications of optical remote sensing in urban environments

Since their arrival, HR and VHR sensors have been used for a wide range of applications:

– mapping land cover/land use (LCLU), revisiting and updating maps: this is most needed in developing countries, although it is still used elsewhere;

– urban mapping and planning: optical imagery must contribute with additional information that is not provided by databases and must be in accordance, geometrically and semantically with the rules of urban planning;

– land cover (LC) mapping and its evolution dynamic. There has been a significant revival in the use of these applications through modeling and the development of evolution scenarios.

More recently, the study of ecosystems has concerned itself with the challenges posed by sustainable development and biodiversity:

– some of these elements tend to be considered within the framework of the quality of ecosystems, habitats and general resources. This quality is determined by the surfaces and the natural or semi-natural components (vegetation strata, urban waste ground, river banks and waterways, etc.) but also using calculations of biomass [AVI 11, AVI 12, HER 11], storing or capturing carbon, the detection and deterioration of environments or their dispersion [MUR 13], etc. The image is then combined with ancillary data and/or used as an input variable in various models or methods of analysis;

– optical imaging has also recently been used to meet the demands of energy transition such as the thermal evaluation of buildings, identifying rooftops that are optimal for the installation of solar panels or the detection of new construction materials.

The various kinds of available optical technology, such as VHR, hyperspectral or infrared cameras, provide a wide range of possibilities for monitoring the urban socio-ecosystem. Three examples of such applications are outlined in the following sections, with reference to different imaging techniques:

– urban planning by using high spatial resolution imaging devices;

– biodiversity, especially green and blue belts and urban vegetation using spectral imaging devices in the reflective domain;

– monitoring heat islands using thermal infrared imaging.

1.2.1. *The use of very high spatial resolution multispectral imaging (VHR) for urban mapping and planning*

LC databases have been compiled at different levels (worldwide, European, national, regional and local) in order to meet societal, regulatory and scientific demands. Describing LC makes it possible to produce environmental indicators for the management of ecosystems and territories, especially for urban areas. Thus, an up-to-date LC database makes it possible to study territorial evolutions and to monitor certain natural or anthropic phenomena in order to implement related public policies related. These data are then used to evaluate the impact of these policies. A good example of the way in which this is used relates to territorial urbanization [DUP 12, SU 11,

PHA 11], with the monitoring of artificial sprawl and urban densification. Another example concerns the study of the spread of vegetation and how it has evolved in urban environments [ZHO 08, DEL 08, LAN 09, PHA 12]. Data regarding LC is also used for planning purposes, particularly within the framework of impact studies, where decisions are taken regarding important building sites, for example. Finally, LC data can be used as an input for climatological or hydrological simulations.

1.2.1.1. *Land cover databases*

The specifications of a LC database, especially for an urban environment, will relate to:

– its geometric accuracy, i.e. the minimal size of objects to be represented for each degree of classification legend. The minimal mapping units (MMUs) are defined by class;

– its semantic accuracy, i.e. the level of detail in its classification legend.

The classification legend of LU is most often organized hierarchically, ranging from fairly general classes to finer subclasses (for example, a general class called "artificial territories" might include a subgroup "communication networks", where third-level subgroups on roads and paths would be found).

In terms of LC, it is possible to distinguish between land cover (the description of elements without predicting their function) and LU (the role played by certain areas in terms of economic or human activity). For example, what would be termed a built-up area in terms of LC could be residential, commercial or industrial in terms of LU.

Data for LC in urban environments have therefore been compiled at different scales in order to meet different needs. The data are produced using photo-interpretation or semi-automatic techniques which require a phase of correction and validation by photo-interpretation. Existing LC databases describing urban environments will be presented in the following section, ranging from the general to the more specific. These databases are different in terms of their classification legend, their scale of use and the extent of the covered areas.

A summary is presented in Table 1.6.

1.2.1.1.1. European scale: Urban Atlas

The Urban Atlas database was created by the EEA as part of the European Copernicus program. It concerns exclusively the largest urban areas in Europe. The aim of Urban Atlas is to develop a shared terminology for every city in Europe in order to be able to easily compare them in a rigorous manner or to calculate relevant indicators (Figure 1.7).

Figure 1.7. *Extract from the land cover database Urban Atlas [source: EEA] (above: corresponding orthoimage [source: IGN])*

Two versions of the Urban Atlas database exist. The 2006 version dealt only with urban areas with a population greater than 100,000, while the 2012 version was extended to cover urban areas with a population greater than 50,000.

The minimal mapping unit for Urban Atlas was set at 0.25 ha, with a minimal size of 10 m for longer objects. Urban Atlas was produced using satellite images with a resolution of 2.5 m and topographical maps with a scale of 1:50000.

The classification legend of Urban Atlas is hierarchical, featuring three levels and a total of 17 different classes.

1.2.1.1.2. National scale: IGN's large-scale land cover database

In France, IGN[12]'s large-scale LC database (*Base de Données d'OCcupation du Sol à Grande Echelle* or BD OCS GE) is still being compiled[13]. With a metric level of accuracy and compatible with the other layers of the national geographic database (*Référentiel à Grande Echelle* – RGE), its purpose is to provide a new component of the large-scale reference geodatabase RGE.

The OCS GE is composed of two levels, one dealing with LC and the other with LU and each has its own hierarchical classification legend. For LC, they include four levels with a total of 14 different classes. The minimal mapping unit varies between 0.5 (particularly for built-up areas) and 2.5 ha.

The BD OCS GE (Figure 1.8) is produced by integrating existing data while also calculating new information from available data, before correcting and completing the process through photo-interpretation.

1.2.1.1.3. Regional scale: land cover in Ile-de-France

LC data Mode d'Occupation des Sols (MOS), provided by the Institute for urban planning and organization (*Institut d'Aménagement et d'Urbanisme* – IAU)[14] in Ile-de-France is one of the oldest databases relating to LC and also one of the most accurate, semantically and geometrically. In use since 1982, it was compiled through the photo-interpretation of aerial images and there have been eight different versions (1982, 1987, 1990, 1994, 1999, 2003, 2008 and 2012).

12 IGN: French National mapping agency.

13 IGN's OCS GE project began in 2012. Compiling the database is carried out in partnership with local organizations.

14 Paris based organization that consults on urban development.

Its legend is hierarchical divided into four levels, with the most detailed level containing 81 different classes. However, this classification legend mostly addresses LU (Figure 1.9).

Figure 1.8. *Extract from IGN's large-scale land cover database. From top to bottom – land cover, orthoimage of land cover and land use (source: IGN), For a color version of this figure, see www.iste.co.uk/baghdadi/5.zip*

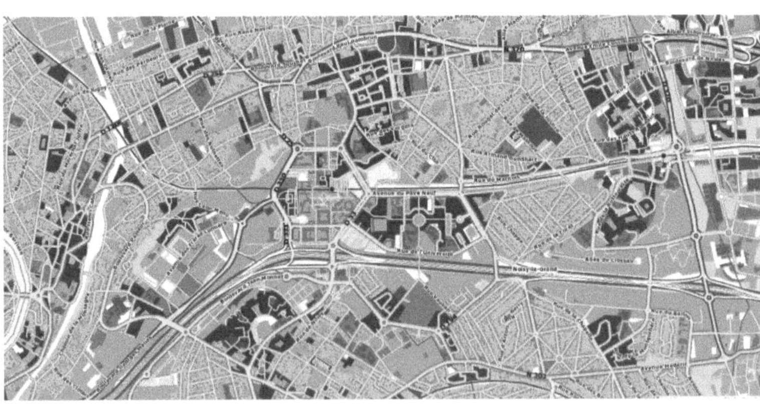

Figure 1.9. *Land use data for Ile-de-France (for the same area as Figure 1.7) (source: IAU-IDF) For a color version of this figure, see www.iste.co.uk/baghdadi/5.zip*

For urban planning purposes, the classification legend of the LC database must be as detailed as possible and includes the different types of urban objects (Table 1.6). These objects are then mapped in order to produce an inventory at a given date, but also to study their evolution. Within this framework, very high spatial resolution multispectral imaging (VHR) is able to produce submetric images, at least in panchromatic mode. At this spatial resolution, it is possible to clearly delineate useful urban objects and to improve how their classification in dense urban areas. It is therefore possible to extract urban objects using VHR multispectral optical data, especially when using supervised methods of classification.

Database	Version	Scale	Minimal mapping unit	Number of levels of classification	Number of classes	Data and production techniques
Urban Atlas	2006, 2012	Europe	0.25 ha	3	17	–Photo-interpretation and supervised classification –Satellite image 2.5 m –Topographic map 1:50000
OCS GE (land cover)	Still being compiled, varies depending on the region	France	0.25-2.5 ha	4	14	–VHR aerial (or satellite) orthoimages – Integration of existing databases – Photo-interpretation
Land use (MOS)	1982, 1987, 1990, 1994, 1999, 2003, 2008 and 2012	Ile-de-France	Undefined (<500 m^2)	4	81	– Aerial orthoimages – Topographic map – Photo-interpretation

Table 1.6. *Main characteristics of different urban land cover databases*

1.2.1.2. *Classification techniques using VHR multispectral images*

Given a typology of useful urban objects (such as the one in Table 1.1), supervised classification appears to be a technique that is well suited for mapping LC. This consists of training a classifier model using learning data.

These data are collected for each separate class through photo-interpretation, collected *in situ* or taken from existing databases, and it is characterized by spectral, textural or geometric information. A classifier model is calculated from these data and then makes it possible to predict the class of each observation (pixel or region) taken from the image. The model is therefore applied to the whole image in order to produce a LC map. Changes can then be detected by comparing this most recent version to a previous database in order to complete it, update it or to monitor a phenomenon over time.

The operational extraction of LC takes place over several stages described in Figure 1.10:

– calculating the image's spectral and textural features;

– extracting training areas either manually or using an existing database (especially when used for change detection);

– training: estimating the classifier model using training data and calculated attributes;

– classification: either by pixel or by region;

– detecting changes by making a comparison between the classification and the database requiring an update.

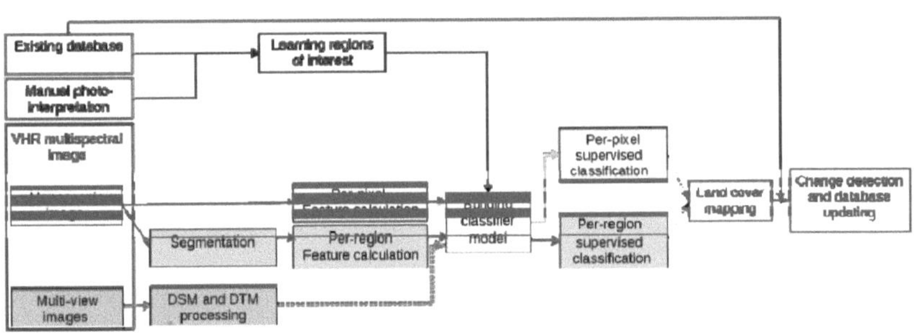

Figure 1.10. *Framework for classifying land cover and detecting changes in urban environments. In yellow: the pixel approach; in blue: the object approach; in orange: specific approach for multi-view images (source Bordeaux INP) For a color version of this figure, see www.iste.co.uk/baghdadi/5.zip*

Variations to this framework exist depending on whether or not the object concept is taken into account and whether the imagery is monoscopic or multi-view. Indeed for VHR imaging, it can be useful to segment the image in homogeneous regions before moving directly to the classification of these regions, which are also called superpixels: this is referred to as Object Based Image Analysis (OBIA) [BLA 10]. Where multi-view data is available, a DSM (i.e. an image which connects each pixel to its altitude) can be calculated and used as an input for classification, greatly improving the accuracy classification especially for the discrimination of ground and above-ground objects.

The most widely used classifiers in remote sensing are neural networks [PAI 15], the maximum likelihood, support vector machines (SVMs) [PAL 05] or random forests [BRE 01].

For large-scale processing, the process can be applied by tiles.

Figure 1.11 shows examples of data produced from Pléiades VHR images.

Figure 1.11. *Examples of data produced from two Pléiades images (in Tarbes, France): a) resampled and pansharpened orthoimage at 50 cm, b) an example of segmentation, c) Digital Surface Model (increasing altitudes from black to white) and d) an example of classification (shadows in black, buildings in red, road networks in gray, bare earth in yellow, high vegetation in bright green and low vegetation in pale green) (source: IGN). For a color version of this figure, see www.iste.co.uk/baghdadi/5.zip*

1.2.1.3. *Feature calculation and selection*

Whether an object-based or pixel-based approach is adopted, the classifier does not necessarily use the initial bands of the multispectral image, but transformed channels calculated from the initial bands (most of the time red – green – blue – near-infrared) of the image. These derived channels make it easier to distinguish certain classes, as is the case with the spectral index NDVI (Normalized Difference Vegetation Index) which is used to detect vegetation. There are several different kinds of these derived channels, or neo-channels, also known as features in case of classification:

– spectral features: these are the spectral indices such as the vegetation indices [HEN 12], brightness indices, water indices or bands transformed in another colorimetric space (Intensity/Hue/Saturation or L*a*b[15] spaces, for example) or all other indices obtained by a simple combination of initial bands;

– textural features: these are calculated at the pixel level (by integrating the information within a given neighborhood) or at the regional level (if we have access to a segmentation) and characterize the image's textural information (structure, roughness, contrast, etc.). We can cite, for example, the textural indices of Haralick [HAR 73] or Gabor [GRI 02], the morphological attribute profiles [DAL 10] or approaches based on descriptors such as the Histogram of Oriented Gradient or HoG [DAL 05];

– geometric features: features which are specific to the object-based approaches, outlined in section 1.2.1.4.

The use of well-suited derived channels often makes it possible to improve classification and to better distinguish between different classes. Selecting these features is highly important and will affect the accuracy of LC classification.

With this in mind, automatic feature selection methods can be used in order to identify optimal sets of features. Sometimes, these will be dependent on the classifier [GUY 02, BRE 01].

---

15 The L*a*b* color space includes all perceivable colors. Three figures characterize the colors: *lightness* L which is derived from the radiance and two parameters *a* and *b* which express the difference between the color and a gray surface of the same brightness.

1.2.1.4. *The contribution of object-based classification*

The pixel-based approach is easy to implement but has a tendency to lead to noisy classification. This then requires postprocessing in order to regularize the result in a given neighborhood to obtain a LC map. In order to solve this problem, different regularization techniques have been proposed. They include local, filter-type approaches such as majority vote or bilateral filtering [TOM 98], or more general methods based on probabilistic graphic models solved owing to optimization methods such as graph cuts [SCH 12].

This phenomenon is much more sensitive in the case of VHR imagery, to the extent that the same object will be represented on the image by several pixels. It is therefore worthwhile carrying out joint processing on different pixels which correspond to the same object. This kind of processing requires an image segmentation.

Image segmentation divides an image into a group of homogeneous regions, which are disjoint from each other and their union forms the whole image. For segmentation, the parametrization choice can be critical, given that the aim is to have regions which are of a significant size and where the LC is consistent. Fine- or oversegmentation will contribute little in relation to classification carried out per pixel and will run the risk of leading to noisy classification, while undersegmentation, that is to say a segmentation that: is too coarse, containing regions of different classes, will also lead to mis-classification.

Various segmentation algorithms have been put forward, for example, [BAA 00, BEN 04, COM 02, GUI 06] and [ARB 11].

Regions created by segmentation can be used within the framework of a regularization process as for pixel-based classification. However, an object-based classification approach will use the segmented regions as input and derive specific object features [MAT 07, ZHA 10, MYI 11] and will immediately provide a less noisy classification. Examples of these objects features include:

– statistical features within the region (calculated from pixel-based features);

– geometric features describing the regions' shape;

– contextual features: these features take the local context into account. For example, the neighboring region of a shadowed area in the direction of the sunlight will most likely be above ground.

The use of segmentation therefore makes it possible to obtain a more smooth classification, but we have to keep in mind that there is a risk of error spreading from segmentation to classification.

Figures 1.12 and 1.13 show the regularization and improvement in classification through the object-based approach, especially for roads which are often confused with buildings. Object classification is more even but still leads to significant confusion when trying to distinguish between asphalt roads and asphalt roofs.

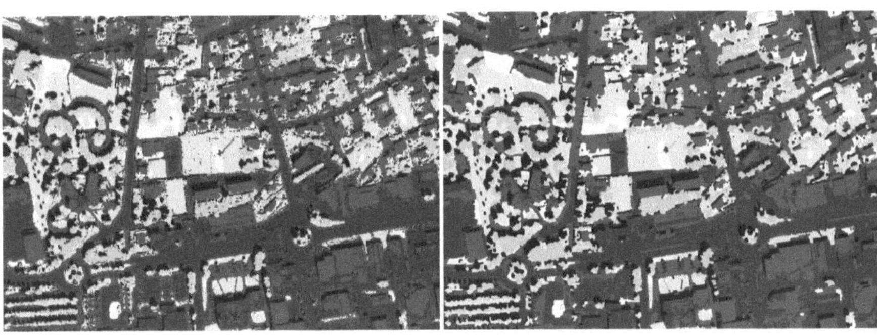

Figure 1.12. *Classification by maximum likelihood classifier by pixel a) and by region without using a Digital Surface Model b) (source: IGN). For a color version of this figure, see www.iste.co.uk/baghdadi/5.zip*

Figure 1.13. *Examples of classification by maximum likelihood classifier by pixel (center) and by region (right) taken from a Pléiades ortho-image (left) (source: IGN). For a color version of this figure, see www.iste.co.uk/baghdadi/5.zip*

1.2.1.5. *Contribution of digital surface models in urban classification*

The development of VHR satellite imaging in tandem with increased satellite agility (as is the case with the Pléiades sensor) has made it possible to acquire images in stereoscopy and even tri-stereoscopy without diachronism while. Aerial VHR acquisitions are almost always multi-view (more than three images). One of the advantages of multi-view VHR imaging is that it enables the calculation of DSMs with submetric resolution, i.e. a 2.5 D representation of the image, where the value of each pixel reflects its altitude on the terrain while taking into account objects above the ground (buildings, vegetation, etc.).

Such DSM are calculated using dense automatic correlation image matching techniques [PIE 06, GRA 11]. In urban environments, filtering techniques, for instance using an elastic grid algorithm [CHA 06], can then be used to delete above-ground objects making it possible to produce a digital terrain model (DTM), that is to say a grid of altitudes at ground level. It then enables us to derive from such data whether an object is at ground level or above for each pixel as a function of the difference between the DSM and the DTM. This difference is called normalized DSM (nDSM) and makes it easier to distinguish between classes which are spectrally similar such as roads and buildings with gray roofs on one hand and areas of bare ground areas and some red-tiled buildings on the other hand.

Depending on the type of classifier used, these data are either taken into account directly as new features based on the difference between the DSM and the DTM and provided as an input to the classifier or as *a priori* probabilities of classification given the difference between the DSM and the DTM. These probabilities are then merged retrospectively with posterior classification probabilities produced by the classifier through Bayesian [LEB 11] or fuzzy [ROT 07, POU 09] data fusion approaches. An example of ground versus above-ground probability classification described using a piecewise linear function is shown in Figure 1.14.

Figures 1.15 and 1.16 illustrate the benefit of using DSMs when classifying by pixel or by region. Ambiguities which were presented in Figure 1.12 have been removed by distinguishing between ground and above-ground objects.

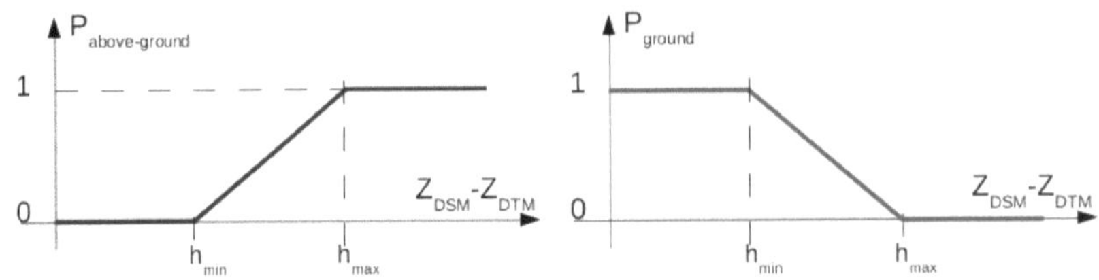

Figure 1.14. *Probability of being classified as ground level or above-ground level as a function of the difference in altitude between the DSM and the DTM (source: IGN).*

Figure 1.15. *Classification by maximum likelihood classifier using DSM, by pixel a) and by region b) (source: IGN).). For a color version of this figure, see www.iste.co.uk/baghdadi/5.zip*

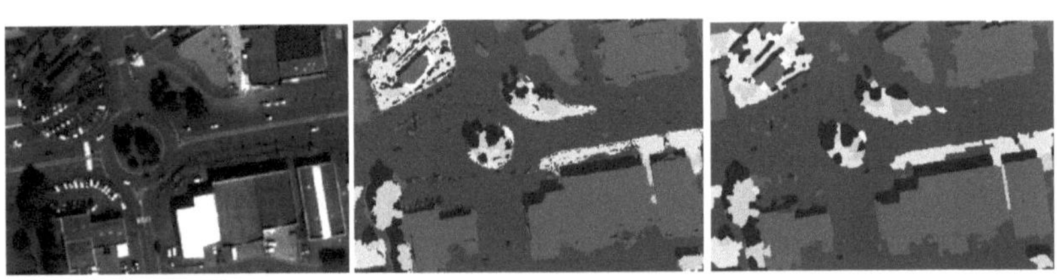

Figure 1.16. *Examples of classification by maximum likelihood classifier using DSM by pixel (center) and by region (right) on an extract from a Pléiades orthoimage (left) (source: IGN). For a color version of this figure, see www.iste.co.uk/baghdadi/5.zip*

1.2.2. *Biodiversity (blue belt and green belt) and vegetation detection in cities*

Despite a blurring of the concept [GUY 08], biodiversity, or at least the biological diversity of species found in cities [PRE 08], has taken on significant social, economic [MAR 06] and legal dimensions. The delicate balance between cities and the natural world is constantly changing, in part due to work carried out as part of the Millennium Ecological Assessment [MEA 05] which highlighted the challenges facing biological diversity and especially the strong ties between urban development and the health of ecosystems. In reality, "changes caused by anthropic activity have happened more quickly over the course of the last fifty years than during any other period of human history" [MEA 05]. Those transformations that have been observed are linked to the conversion of natural ecosystems into socio-ecosystems dominated by humans. Urbanization has had a negative effect on biological diversity and has caused the destruction of habitats by using these natural environments for various projects (construction, extraction, fragmentation and expansion), but it has also helped its development by creating specific zones – wildlife corridors, green belts, plants on roofs and walls, protected zones, etc. Green and blue belts can therefore be thought of as sites of both biodiversity and construction which spread across cities; they play an important role in the ecological functioning of species in this mostly mineral environment alongside existing vegetation (parks, gardens, rows of trees, etc.). In urban environments, biodiversity is localized depending on the different types of habitat (ground surfaces, vegetation, water and increasingly roofs and walls, trees or building), following the permeability gradient of the matrix and urban density. Vital for one or several species, these habitats are very diverse and are constrained by the urban matrix. They correspond to natural infrastructure that is necessary for the displacement of fauna and the dispersion of flora, but also for their reproduction, protection, feeding and rest.

Among this diverse range of infrastructures is where wildlife corridors are to be found. These natural or protected areas were set up following the application of the Habitats Framework Directive[16] and were extended in urban environments by clearly defining green and blue belts. In France, the implementation of green and blue belts since 2007 has been one of the major national projects to come from the Grenelle Environment Forum. This belt is made up of a network of biological corridors (or ecological corridors, existing or to be restored), "biodiversity reservoirs" and buffer zones or annexes ("natural relay spaces"). Documents provided by state services make it possible to identify these structures and to map them for town

16 Directive 92/43/CEE of the Committee on 21st May 1992 regarding the conservation of natural habitats as well as wild flora and fauna.

planning purposes using satellite data[17]. At the European level, protocols for evaluating and mapping ecosystem services have been studied and put forward by the European Commission, drawing on a range of data sources (Figure 1.17).

Mapping ecosystems	
Urban	**Land use land cover data, e.g.**
Cropland	Corine Land Cover
Grassland	Copernicus high resolution data
Woodland and forest	Elevation data
Heathland and shrub	Seabed maps
Sparsely vegetated land	National datasets
Wetlands	
Rivers and lakes	**Models** for spatially delineating wetlands or
Marine inlets and transitional waters	natural, unmanaged ecosystems
Coastal	
Shelf	
Open ocean	

Figure 1.17. *Mapping ecosystems (http://biodiversity.europa.eu/)*

Within the context of mapping habitats or the conservation of natural spaces, remote sensing makes it possible to identify, quantify and even model them. The data can also provide information on stress (in particular hydric stress, but also irradiance, constraints on the tree crown, etc.) and on the health and well-being of species. Spatial and spectral resolutions are key elements in these observations. Thus, satellite imaging has been used in order to establish continuity at a regional level (Regional schematic of ecological coherence (*schéma régional de cohérence écologique* – SRCE)): ecological continuity [HUB 12], wooded [VAN 11] or semi-natural environments [FAU 14].

In a similar manner, preserving ecological continuities and ensuring their quality incorporates a range of different activities in the sphere of environmental preservation, such as classified or unclassified wooded areas or even the development of waterproof surfaces or wetlands [RAP 12]. Such

17 Directed by the French Ministry for ecology, sustainable development and energy (MEDDE), the TVB Resource center is based on an umbrella organization which brings together Aten, the French federation of regional natural parks, the national institute for studies in science and technology for environment and agriculture (Irstea), the Museum of Natural History (MNHN) and the national office for water and aquatic environments (Onema).

environments facilitate hydric transfer and act as buffer zones against the risk of flooding [WEN 12, KAS 15].

In studies focused on urban biodiversity, particular attention has been paid to vegetation. It has been identified as a key element for quality of life in cities and reinforces the benefits of urban living [BON 06, VAN 03]. Urban vegetation is currently a key aspect of political agendas. Aside from monitoring ecological continuities or forests and parks, the aim is to be able to isolate and identify the relevant areas in order to characterize the evolution of these habitats within the urban belt [PUI 14, ROU 14, BOU 10, MAI 12] (Figure 1.18).

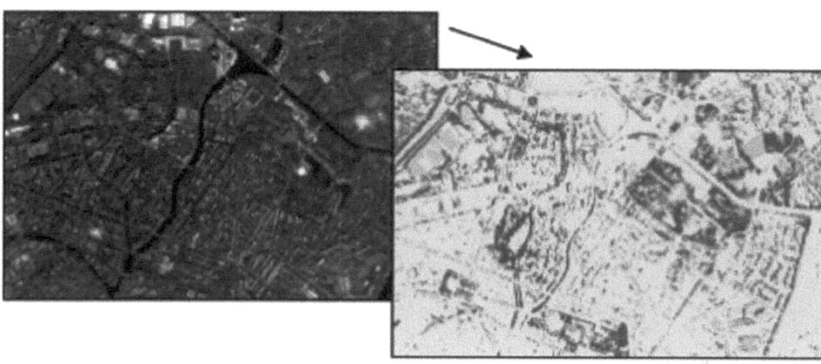

Figure 1.18. *Extraction of trees and shrubs using VHR images (Pléiades) [ROU 14]. For a color version of this figure, see www.iste.co.uk/baghdadi/5.zip*

Repeatable, generic methodological propositions for analysis using multi-source data are currently being developed in order to characterize urban spaces. Iovan *et al.* [IOV 08] and Ardila *et al.* [ARD 11, ARD 12] focused on the detection of tree crowns by using an object-oriented approach. Other developments based on training algorithms (Random Forest) currently make it possible to improve the detection and identification of trees in urban environments [PUI 14].

Over the course of the last decade, various initiatives have been introduced in order to promote green or open spaces [JOH 04]. Urban vegetation plays an essential role in the management of environmental characteristics such as wind, humidity and solar irradiance [GIV 98]. Thermal comfort is in correlation with the presence of vegetation [GOM 04, GUL 05] namely due to its capacity to reduce temperature through shadow (direct effect), heat islands and evapotranspiration (indirect effect). Large or dense areas are more effective than a few narrow or less dense areas [STU 90, DIM 03]. As a result, vast areas can be used in order to help cool down the air in the city, which can be felt far from the direct localizations of vegetation, namely, through cooling islands [SHA 00]. Through the quality of the information provided, hyperspectral imaging is an asset for analyzing the ecological diversity of vegetation [WAN 07b, GAN 08, BLA 06a, BLA 06b].

1.2.3. *Urban heat islands*

1.2.3.1. *Introduction to the concept of heat islands*

Studying urban areas through thermal infrared remote sensing (LWIR, long-wavelength infrared: 8–14 μm) dates back to the 1970s and studies carried out by Rao [RAO 72] and Carlson *et al.* [CAR 77]. There are currently three problems being studied using this technology [VOO 03]:

– the link between the spatial structure of urban thermal patterns and the characteristics of the urban surface;

– the energy balance of urban areas;

– urban heat islands (UHIs), a concept that will be explained here.

The following section will focus on this last point. Urbanization implies the replacement of natural surfaces by artificial materials, which possess different physical properties. This transformation causes changes in the area's climate characteristics and one of these is the UHI effect [OKE 81]. This phenomenon can be characterized as the warming of built-up zones in relation to surrounding, undeveloped areas. Oke [OKE 81, OKE 82] lists the main causes of UHI, which were then dealt with by Voogt [VOO 02]:

– geometry: solar irradiance is trapped in urban canyons which causes multiple reflections;

– the characteristics of urban materials: these materials have an increased thermal capacity and therefore store more heat than natural materials. Furthermore, these materials are insulating as well as being waterproof which reduces evaporation and thus warms the surrounding air;

– anthropogenic heat: human activity, particularly the energy output of buildings and vehicles, produces heat.

– contamination: urban areas are the main sources of greenhouse gas emissions.

The UHI effect can have negative as well as positive effects. In the first instance, one example is the direct influence it has on health and well-being. The effects of rising temperatures range from causing sleep problems to an increased mortality rate [BRU 05]. In the second instance, one example of the positive impact is with cold cities where rising air temperatures decreases the need to heat homes and, as a result, reduces energy consumption.

Characterizing UHI using weather stations is insufficient because they are not able to cover the entire urban area in a consistent manner. In contrast, remote sensing provides us with an instantaneous image of the entire city. Thus, UHI can either be defined at the atmospheric level or at the surface level. Atmospheric UHI depends on the position of the air temperature sensors: either in the urban canopy, i.e. using the air temperature measured between the road and the roofs, or in a boundary layer, situated above the preceding layer. Surface UHI on the other hand (referred to from now on as SUHI) is studied using remote sensing data and the parameter to be estimated is not the air temperature but the land surface temperature (LST):

$$SUHI = LST_{urban} - LST_{rural} \qquad [1.3]$$

where LST_{urban} refers to the surface temperature of the urban area and LST_{rural} refers to the surface temperature of the rural area. In order to estimate the SUHI, several stages of processing are necessary. The aim of the first stage is to estimate the materials' emissivity (section 1.2.3.2), followed by a second phase where the surface temperature is estimated (section 1.2.3.3). Finally, examples of SUHI studies are provided (section 1.2.3.4).

1.2.3.2. *The surface emissivity of urban areas*

In urban environments, an accurate estimation of the surface temperature[18] depends on surface characteristics and in particular its emissivity and geometry [VOO 03]. Thus, an uncertainty of 1% for the emissivity leads to an error of 0.78 K for temperature estimating [VAN 93]. In addition, due to its heterogeneity, an accurate estimation of emissivity is

18 The problem of estimating the surface temperature and emissivity is outlined in [BRI 16].

essential in order to deduce a temperature map in urban environments, which will be used in order to quantify the city's SUHI.

In [OLT 12], three methods for estimating emissivity are compared in order to create a map of surface temperature using data acquired by the hyperspectral scanner AHS (Airborne Hyperspectral Scanner) during the European Space Agency (ESA) mission DESIREX 2008 over the city of Madrid [SOB 13]. The results show that with these three methods, emissivity is estimated with a precision of 5.6% for the Normalized Difference Vegetation Index Threshold Method ($NDVI_{TM}$) [SOB 08], 3.0% for the TISI (Temperature Independent Spectral Indices [BRI 16, BEC 90] and 3.9% for TES (Temperature-Emissivity Separation [BRI 16, GIL 98]. However, there are operational disadvantages to all three methods. Even if the $NDVI_{TM}$ method is used in urban environments [RIG 06, STA 07], it is limited by the reduced number of classes of material that it takes into account (three classes: vegetation, bare ground and mixed pixels). The TISI method requires a channel in the mid-wavelength infrared (MWIR: 3–5 μm) as well as multitemporal acquisitions. Finally, the TES method developed for ASTER requires at least five bands in the low-wavelength infrared region (LWIR). Furthermore, the TES method is an empirical method and can lead to an error of 1.7% for emissivity [OLT 14] if urban materials are not taken into account during calibration.

A fourth method for mapping emissivity in urban areas is put forward by Sobrino *et al.* [SOB 12a]. A ground classification map (12 classes) is compiled using 80 AHS bands which cover the entire spectrum, from 0.4 to 13 μm. In order to associate with each class an emissivity spectrum, the ASTER database is used combined with field measurements. For devices with weak spectral resolution in the LWIR, this would appear to be the most promising method. The urban environments have slow temporal evolutions (apart from vegetation) and a map of materials (buildings, infrastructures, etc.) can be obtained from external high spatial resolution images in the reflective range. Then, the corresponding emissivity for each class will be chosen from a spectral library.

1.2.3.3. *Land surface temperature in urban areas*

Obtaining the land surface temperature, LST in urban environments is linked to the acquisition geometry of the 3D structure of the scene.

Furthermore, the temperature measured for an urban area shows significant daily directional effects, due to a city's high thermal heterogeneity (surfaces in shadow or in sunlight) and the directional properties of urban materials [LAG 04]. These phenomena are illustrated in Figure 1.19 which shows two temperature maps obtained using the TES algorithm [GIL 98] for two acquisitions of the AHS sensor during the ESA mission DESIREX 2008 [SOB 13], over the city of Madrid. The spatial resolution of the maps is 4 m and the AHS device has a field of view of 90°. Figure 1.19 corresponds to an acquisition carried out from the North West (NW) to the South East (SE) and Figure 1.19(b) corresponds to an acquisition across the same region but along the South (S) – North (N) axis. Both were acquired around midday local time with the second carried out at an interval of 20 min in relation to the first. There is a difference in temperature between the two maps, caused by the different surfaces viewed by the sensor depending on their position in relation to the sun. More shadows are observed when the city is covered in a NW to SE direction than when it is covered in an S to N direction. However, toward midnight, with the same acquisition configurations, any difference between the two flight lines disappears (Figure 1.20). This shows that it is easier to analyze urban environments at night in order to study urban heat islands [SOB 12b].

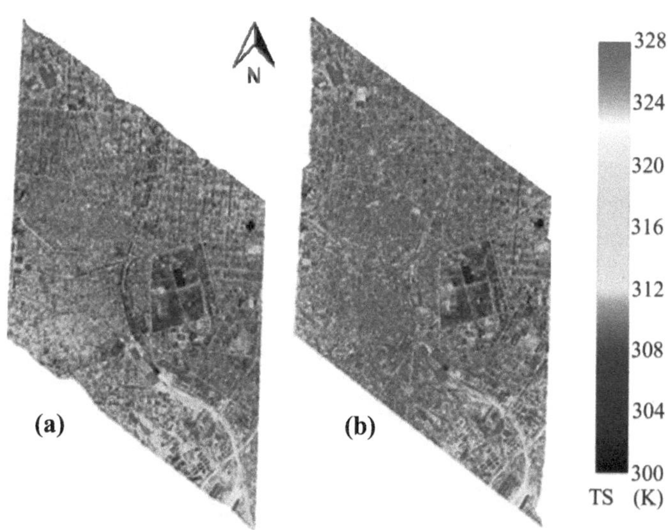

Figure 1.19. *Land surface temperature maps for the same area of Madrid at midday: a) transect from NW to SE; b) transect from S to N [OLT 13]. For a color version of this figure, see www.iste.co.uk/baghdadi/5.zip*

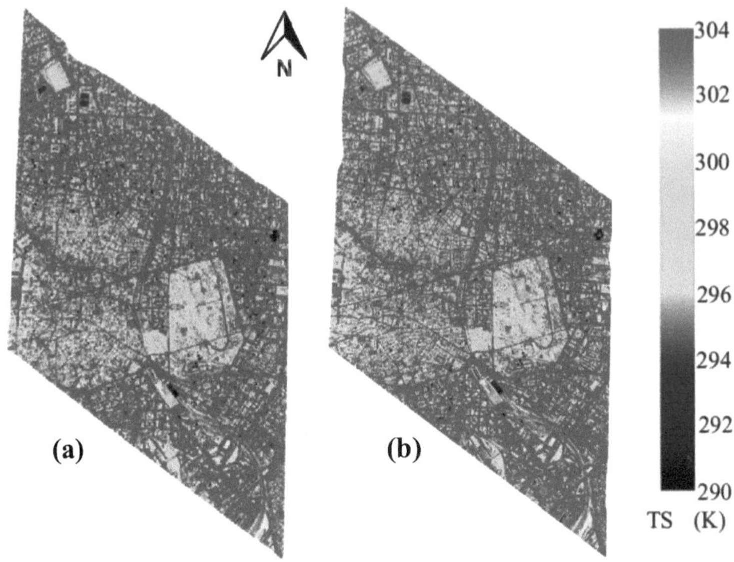

Figure 1.20. *Land surface temperature maps for the same area of Madrid at midnight: a) transect from NW to SE; b) transect from N to S [OLT 13]. For a color version of this figure, see www.iste.co.uk/baghdadi/5.zip*

The SUHI effect is studied in the bibliography using satellite and airborne sensors. One example of an airborne mission is the DESIREX mission, where the SUHI for Madrid is characterized during the summer with a maximum of 5 K [SOB 13].

The satellite devices currently in use do not possess a spectral or spatial resolution adapted to studying urban environments. However, they do offer a more cost-effective alternative to airborne acquisitions and are capable of providing temporal monitoring of the SUHI. Thus, the device MODIS, with a spatial resolution in the LWIR frequency of 1 km, was used by Bonafoni *et al.* [BON 15] in order to monitor the evolution of SUHI for the city of Milan (Italy) over the course of a day, using four MODIS acquisitions per day. A difference of more than 4 K can be observed between the diurnal and nocturnal SUHI. Various studies using ASTER images (a spatial resolution of 90 m and five spectral bands in the LWIR frequency) have also been carried out in order to characterize the SUHI, an example of which is shown in Figure 1.21. This image shows a temperature map for Madrid at night, where the city can be identified by its high temperatures (in red) compared to the surrounding area. In [TIA 08], LST maps obtained with

ASTER at night are compared to air temperature measurements for Grand Manilla (Philippines). The maximum SUHI registered there is 3 K and there is a linear correlation between LST and air temperature, with a correlation coefficient of 0.55.

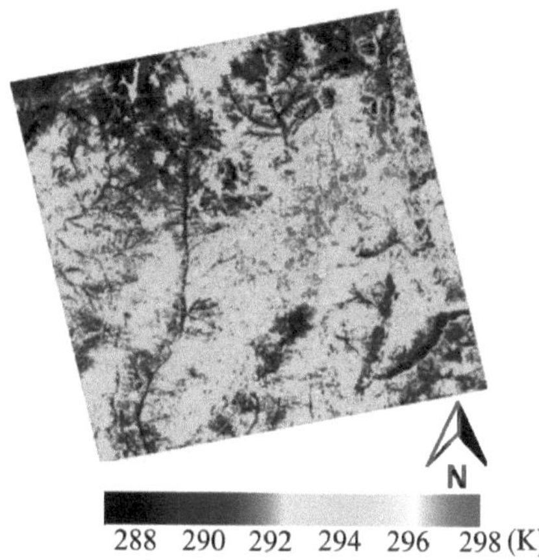

Figure 1.21. *Land surface temperature map for the night of 4 July 2008 obtained using the TES algorithm [GIL 98] using an ASTER image of the city of Madrid and its surrounding area. The SUHI effect can be observed [OLT 13]. For a color version of this figure, see www.iste.co.uk/baghdadi/5.zip*

Atmospheric UHI is characterized by higher levels at night and very weak, sometimes negative, levels (urban cooling islands) during the day, due to the shadows caused by buildings and urban vegetation. In the case of SUHI, we are able to note the same results for other cities such as Athens [STA 09] or Madrid [SOB 13]. However, SUHI does not always show this dynamic and it is possible to find examples where SUHI is more intense during the day than at night [BON 15]. Furthermore, this dynamic can change during the year, because the effect is linked to land cover. Even if there is no remarkable change to urban surfaces over a year, this is not the case for the surrounding areas of cities where agricultural activity, for example, can bring about seasonal changes in the surface cover. Figure 1.22 gives an example of this dynamic for the city of San Miguel de Tucuman (Northwestern Argentina), with the SUHI calculated using Landsat TM images at 2pm UTC [OLT 10]. We were able to observe a positive SUHI during the summer and a negative SUHI during winter and spring, with these results caused by crop conditions around the city, particularly sugar cane.

During September, October and November, when the SUHI is negative, the proportion of bare ground in the pixels which cover sugar cane crops is higher than during January, February and March, when sugar cane leaves are greener and more developed. As a result, for the first case, the rural surface at 2pm is warmer than the urban surface, while for the second case the inverse is true.

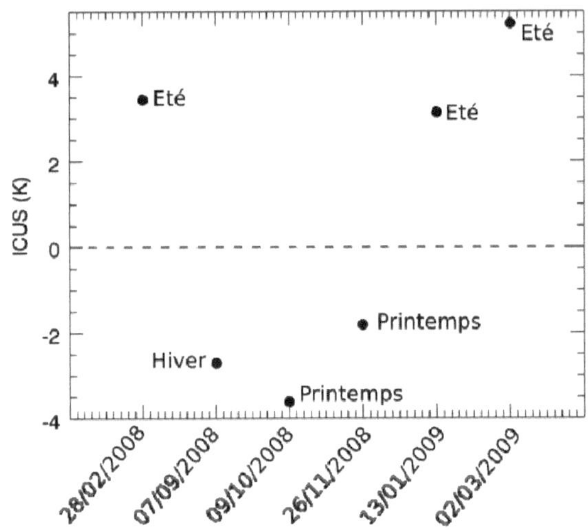

Figure 1.22. *SUHI measured using Landsat TM images (2pm UTC) for the city of San Miguel de Tucuman (Argentina) [OLT 10]*

1.3. Conclusions and prospects

Urban environments are characterized by a very high spatial heterogeneity in terms of materials, which requires remote sensing methods with high spatial resolution (lower than 5 m) and high spectral range in the optical domain.

From a methodological point of view, there will be need for research developments on the data processing level and the use of the available big data.

At the level of data preprocessing, studies which aim to improve atmospheric correction in the optical domain will need to be pursued. Developments based on the simultaneous use of data acquired in the reflective domain and the thermal infrared domain should be aided by the availability of new airborne devices covering the entire optical domain. Furthermore, linear and/or nonlinear separation techniques will need to be developed in order to identify materials at a subpixel level and to thus improve the quality of classification, in the reflective domain as well as the emissive domain.

For the production of LC maps in urban environments through classification, VHR imaging enables access to more detailed geometric data and thus improves the classification of homogeneous regions of LC taken from a segmentation by reducing the noise. Nevertheless, in urban environments, using the image alone leads to limited results caused by the confusion between similar classes. In the case of multi-view acquisitions, it is possible to improve these results by introducing 3D information regarding the height of objects (DSM and DTM).

However, the implementation of such approaches remains limited in the case of satellite imaging given the costly and difficult (in terms of programming) nature of acquiring multi-view scenes. In addition, approaches which combine pattern recognition and classification could make it possible to make better use of the geometric data accessible via VHR imaging in the case of monoscopic scenes [KAR 09, SID 14]. Another possible improvement could come through the use of deep learning methods which are starting to be introduced for geospatial images [MAR 15, MNI 12, PAI 15] after recent successes, particularly for image indexation. Such approaches have become possible due to the availability of significant volumes of learning data, in the case of existing database updates.

The next methodological developments will be based on the following two important points:

– the fusion of multi-source data (optic, radar, LiDAR and hyperspectral) [GUO 11] and multi-sensors (multiple spatial and temporal resolutions) in order to improve the classification of different usable urban objects;

– the use of temporal data for change detection and updating urban databases.

For studying biodiversity, hyperspectral imaging with high spatial resolution is necessary in order to be able to identify a vegetation species. Combining it with 3D data is necessary in order to be able to distinguish high vegetation from low vegetation.

Regarding the study of urban heat islands, there are a number of airborne methods which make it possible to study them at a local level. However, there is no satellite sensor with a sufficient spatial resolution for carrying out such a study. Merging spectral LWIR images with HR or VHR images in the reflective domain is one alternative for separating thermal infrared pixels and to gain a higher spatial resolution. These studies must be pursued with a view toward establishing a connection between the surface temperature estimated through remote sensing and the air temperature in order gain a better understanding of the temperature's directional activity and to establish a connection between remote sensing methods at different levels.

Sensors are constantly evolving. In the domain of satellite imaging, improvements in geometric resolution and the agility of satellites are in progress. At the same time, there has been an improvement in terms of spectral quality, with superspectral sensors such as Landsat-8 or Worldview-3, as well as in terms of the repetitivity of acquisitions over time using microsatellite constellations such as SkyBox. Greater spectral quality allows for a better quality of classification, as well as for a more detailed legend, including studying the state of vegetation or mapping materials, which are useful to model the urban system. High repetitivity of acquisitions over time also makes it possible to study phenomena more accurately, including the state of vegetation again, but also to meet urgent requirements in up-to-date data, in crisis situations or following a natural disaster, for example. In the aerial domain, Unmanned Aerial Vehicule (UAVs) acquisitions continue to be developed, and these have the advantage of a greater level of reactivity and agility, a higher spatial resolution (Ultra-High Resolution (UHR)) and lower costs (over small areas) compared to traditional airborne acquisitions. Their applications will be the same as for airborne acquisitions, but will feature higher resolution, increased accessibility and a wider viewing range. An example of this can be seen in the case of more detailed and more complete 3D architectural modeling which integrates building facades, in the monitoring of the condition of materials and measuring thermal flows. However, the use of UAVs in urban environments remains limited due to flight restrictions in populated areas as well as the significant weight of some of the imaging equipments.

While improvements in equipment and methods can be considered evidence of genuine progress, other barriers continue to stand in the way, including how best to use the wealth of data that are currently available. This massive amount of data leads us to consider current methods for using data and information. Within this context, there are possible avenues of research which should take into account the representation needs of data for several subjects or users. Metadata will be indispensable for describing selected data in order to meet the various challenges posed by urban ecosystems. Contextual, cognitive and expert information must be taken into account in order to strengthen the link between the image and reality. Finally, the uncertainty and the quality of available data will require to be specification and representation. In order to deal with applications linked to the use of satellite imaging, it will doubtless be necessary to develop a greater degree of harmony among producers, data processors and users while continuing to shape the process in order to satisfy the needs of final users within an intercognitive framework (elected officials, technicians, state representatives, NGO members or ordinary citizens). Formats adapted to users as well as to the different kinds of media used (digital or otherwise) will need to be developed and improved.

In a digital environment that is accessible to the majority of people living in urban environments, representations of urban space are an essential source of reference. Cartographic systems for identification, evaluation and planning are becoming increasingly diverse. Improvements in spatial resolution have made it possible to get closer to the geometric accuracy of equipment used to carry out measurements on the ground. As for needs, they are varied and often unspecified. It is clear that satellite imaging has a role to play, but often it must be accompanied by other sources of information (maps, on-ground data, historical records, etc.) These combinations can follow one of two paths: the relationship between the user and the product (the connection between the image and reality, the connection between the different classification used by users and providers of data) and the relationship between various sources produced using different intrinsic and extrinsic characteristics (between a cartographic element, information found online and the image itself, for example).

With the vast amount of information available on the Internet and in digital form, especially since data started to be considered "common property", that is shareable, modifiable and able to be used for various purposes and by a number of users, access to the different kinds of geographical data has widened, joining the flow of "open data" available since 2005 and especially since 2008 [PS1 05]. However, despite the increasing availability of information, questions regarding how accessible it is actually for urban populations continue to be asked.

The future is digital and the heterogeneous dataset available for characterizing urban systems is vast and will only continue to grow. Nevertheless, this does not guarantee the quality of information provided. The speed at which results are transmitted does not guarantee their quality nor does it mean that they will be evaluated in an equivalent manner. Masses of numerical, geographical and environmental data are already available and this will increasingly be the case taking into account current sociotechnical advances and those to come in the future. This will lead to new directions of research in more intelligent data models which will draw together the quality of the original information and the emergence of new knowledge while trying to objectify the relevant uncertainty.

1.4. Key points

In order to gain a proper understanding of the way in which the urban ecosystem functions at a local level, high spatial and spectral resolution optical remote sensing is relevant and necessary. It is relevant because it provides largely stable quality data for large areas with a level of repetitivity that is only going to increase. Furthermore, satellite imaging increasingly provides spatial resolution that is adapted for planning and organization carried out by local authorities.

Finally, the spectral resolution of satellites makes it possible to provide data on chemical compositions, changes and modifications in materials and urban shapes. As such, satellite imaging has become necessary in order to comprehensively cover the required areas, to monitor urban spaces and the effect of urbanization on the environment.

Methodological equipment exists for measuring surface properties: reflectance, emissivity and temperature. Urban remote sensing provides responses for a wide range of applications such as planning, detecting changes, heat islands, material dynamics and biodiversity. LC classification can be adapted to the needs of end users for representing, monitoring or prospective actions. Other products including maps of materials are also possible.

3D information continues to be necessary in order to be able to properly detect different above-ground classes.

1.5. Bibliography

[ADE 13] ADELINE K.R.M., CHEN M., BRIOTTET X. *et al.*, "Shadow detection in very high spatial resolution aerial images: a comparative study", *ISPRS Journal of Photogrammetry and Remote Sensing,* vol. 80, pp. 21–38, 2013.

[ARB 11] ARBALAEZ P., MAIRE M., FOWLKES C. *et al.*, "Contour detection and hierarchical image segmentation", *IEEE Transactions on Pattern Analysis and Machine Intelligence*, vol. 33, no. 5, pp. 898–916, 2011.

[ARD 11] ARDILA J.P., TOLPEKIN V.A., BIJKER W. *et al.*, "Markov-random-field-based super-resolution mapping for identification of urban trees in VHR images", *ISPRS Journal of Photogrammetry and Remote Sensing*, vol. 66, pp. 762–775, 2011.

[ARD 12] ARDILA J.P., BIJKER W., TOLPEKIN V.A. *et al.*, "Context-sensitive extraction of tree crown objects in urban areas using VHR satellite images", *International Journal of Applied Earth Observation and Geoinformation*, vol. 15, pp. 57–69, 2012.

[AVI 11] AVITABILE V., HEROLD M., HENRY M. *et al.*, "Mapping biomass with remote sensing: a comparison of methods for the case study of Uganda", *Carbon Balance and Management,* vol. 6, no. 7, 2011.

[AVI 12] AVITABLIE V., BACCINI A., FRIEDL M.A. *et al.*, "Capabilities and limitations of Landsat and land cover data for aboveground woody biomass estimation of Uganda", *Remote Sensing of Environment*, vol. 117, pp. 366–380, 2012.

[BAA 00] BAATZ M., SCHÄPE A., "Multiresolution segmentation – an optimization approach for high quality multi-scale image segmentation", in STROBL J., BLASCHKE T., GRIESEBNER G. (eds), *Angewandte Geographische Informationsverarbietung XII*, Wichmann, Heidelberg, pp. 12–23, 2000.

[BEC 90] BECKER F., LI Z.L., "Temperature-independent spectral indices in thermal infrared bands", *Remote Sensing of Environment*, vol. 32, pp. 17–33, 1990.

[BEN 01] BEN-DOR E., LEVIN N., SAARONI H., "A spectral based recognition of the urban environment using the visible and near-infrared spectral region (0.4-1.1 µm): a case study over Tel-Aviv", *International Journal of Remote Sensing*, vol. 22, no. 11, pp. 2193–2218, 2001.

[BEN 04] BENZ U.C., HOFMAN P., WILLHAUCK G. *et al.*, "Multi-resolution, object-oriented fuzzy analysis of remote sensing data for GIS-ready information", *ISPRS Journal of Photogrammetry and Remote Sensing*, vol. 58, no. 3, pp. 239–258, 2004.

[BEN 14] BEN-DOR E., VARACALLI G., "SHALOM: space borne hyperspectral applicative land and ocean mission: a joint project of ASI-ISA an update for 2014", *IGARSS'*14, Quebec, Canada, 2014.

[BLA 06a] BLANSCHÉ A., WANIA A., GANÇARSKI P., "Comparison of MACLAW with several attribute selection methods for classification in hyperspectral images", *ICDM Workshops 2006*, pp. 231–236. 2006.

[BLA 06b] BLANSCHE A., GANCARSKI P., KORCZAK J., "MACLAW: a modular approach for clustering with local attribute weighting", *Pattern Recognition Letters*, vol. 27, no. 11, pp. 1299–1306, 2006.

[BLA 10] BLASCHKE T., "Object-based image analysis for remote sensing", *ISPRS Journal of Photogrammetry and Remote Sensing*, vol. 65, no. 1, pp. 2–16. 2010.

[BON 06] BONAIUTO M., FORNARA F., BONNES M., "Indexes of perceived residential environment quality and neighbourhood attachment in urban environments: a confirmation study on the city of Rome", *Landscape and Urban Planning*, vol. 65, no. 2003, pp. 41–52, 2006.

[BON 15] BONAFONI S., ANNIBALLE R., PICHIERRI M., "Comparison between surface and canopy layer urban heat island using MODIS data", *Joint Urban Remote Sensing Event (JURSE)*, Lausanne, Switzerland, pp. 1–4, 2015.

[BOU 10] BOURGET E., LE DÛ-BLAYO L., "Cartographie des paysages : apport à l'analyse des trames vertes et bleues. Projets de paysage", available at: http://www.projetsdepaysage.fr/fr/cartographie_des_paysages_apport_a_l_analyse_des_trames_vertes_et_bleue, 2010.

[BRE 01] BREIMAN L., "Random forests", *Machine Learning*, vol. 45, no. 1, pp. 5–32, 2001.

[BRI 06] BRIOTTET X., LACHÉRADE S., PALLOTTA S. *et al.*, "Acquisition and analysis of a spectral and bidirectional database of urban materials over Toulouse (France)", *SPIE Defense and Security Symposium, Targets and Backgrounds XII: Characterization and Representation Conference - SPIE 6239-29,* Orlando (Kissimmee), Florida, USA, 17–21 April 2006.

[BRI 16] BRIOTTET X., "Radiometry in the optical domain", in BAGHDADI N., ZRIBI M. (eds), *Optical Remote Sensing of Land Surfaces*, ISTE Press Ltd, London and Elsevier Ltd, Oxford, 2016.

[BRU 05] BRÜCKER G., "Vulnerable populations: lessons learnt from the summer 2003 heat waves in Europe", *EuroSurveillance*, vol. 10, no. 7, p. 551, 2005.

[CAR 77] CARLSON T.N., AUGUSTINE J.A., BOLAND F.E., "Potential application of satellite temperature measurements in analysis of land use over urban areas", *Bulletin of the American Meteorological Society*, vol. 58, pp. 1301–1303, 1977.

[CAR 14] CARRÈRE V., BRIOTTET X., JACQUEMOUD S. *et al.*, "The French proposal for a high spatial resolution Hyperspectral mission", EGU General Assembly, *Geophysical Research Abstracts*, EGU2014-11791, Vienna, Austria, vol. 16, 27 April–2 May 2014.

[CAS 72] CASTELLS M., *La question urbaine*, François Maspero, Paris, p. 455, 1972.

[CAV 08] CAVALLI R.M., FUSILLI L., PASCUCCI S. *et al.*, "Hyperspectral sensor data capability for retrieving complex urban land cover in comparison with multispectral data: Venice City case study (Italy)", *Sensors*, vol. 8, pp. 3299–3320, 2008.

[CHA 06] CHAMPION N., BOLDO D., "A robust algorithm for estimating digital terrain models from digital surface models in dense urban areas", *International Archives of Photogrammetry, Remote Sensing and Spatial Information Sciences*, Bonn, Germany, vol. 36, no. 3, 2006.

[CHE 11] CHEN M., PANG S., CHAM T. *et al.*, "Visual tracking with generative template model based on Riemannian manifold of covariances", *Proceedings of 14th International Conference on Information Fusion (FUSION)*, pp. 1–8, 2011.

[CHE 13] CHEN M., BRIOTTET X., PANG S.K., "Efficient empirical reflectance retrieval in urban environments", *IEEE Journal of Selected Topics in Applied Earth Observations and Remote Sensing*, vol. 6, no. 3, pp. 1596–1602, 2013.

[CLA 80] CLARCK J., The effect of resolution in simulated satellite imagery on spectral characteristics and computer assisted land use classification, report jet Propulsion Laboratory Report, pp. 75–22, 1980.

[COM 02] COMANICIU D., MEER P., "Mean shift: a robust approach toward feature space analysis", *IEEE Transactions on Pattern Analysis and Machine Intelligence*, vol. 24, no. 5, pp. 603–619, 2002.

[DAL 05] DALAL N., TRIGGS B., "Histograms of oriented gradients for human detection", *International Conference on Computer Vision and Pattern Recognition (CVPR'05)*, San Diego, United States, IEEE Computer Society, vol. 1, pp. 886–893, June 2005.

[DAL 10] DALLA MURA M., BENEDIKTSSON J.A., WASKE B. *et al.*, "Morphological attribute profiles for the analysis of very high resolution images", *IEEE Transactions on Geoscience and Remote Sensing*, vol. 48, no. 10, pp. 3747–3762, 2010.

[DEL 08] DELM A.V., GULINCK H., "Classification and quantification of green in the expanding urban and semi-urban complex: application of detailed field data and IKONOS imagery", *Ecological Indicators*, vol. 11, no. 1, pp. 52–60, 2008.

[DIM 03] DIMOUDI, A., NIKOLOPOULOU M., "Vegetation in the urban environment: microclimatic analysis and benefits", *Energy and Buildings*, vol. 35, no. 1, pp. 69–76, 2003.

[DON 00] DONNAY J.P., COLLET C., WEBER C., *La télédétection en Francophonie: analyse critique et perspectives*, AUF, 2000.

[DUP 12] DUPUY S., BARBE E., BALESTRAT M., "An object-based image analysis method for monitoring land conversion by artificial sprawl use of RapidEye and IRS data", *Remote Sensing*, vol. 4, no. 2, pp. 402–423, 2012.

[DUR 08] DURIEUX L., LAGABRIELLE E., NELSON A., "A method for monitoring building construction in urban sprawl areas using object-based analysis of Spot 5 images and existing GIS data", *ISPRS Journal of Photogrammetry and Remote Sensing*, vol. 63, no. 4, pp. 399–408, 2008.

[EP 05] Directive 2003/98/EC of the European Parliament and of the Council of 17 November 2003 on the re-use of public sector information (OJ L 345, 31.12.2003 p. 90), 2005.

[FAU 14] FAUVEL M., PLANQUE C., SHEEREN D. *et al.*, "Télédétection des éléments semi-naturels: utilisation des données Pléiades pour la détection des haies", *Revue Française de Photogrammétrie et de Télédétection*, vol. 208, pp. 111–116, 2014.

[FLE 06] FLEYEH H., "Shadow and highlight invariant color segmentation algorithm for traffic signs", *Proceedings of IEEE Conference on Cybernetics and Intelligent Systems*, Bangkok, Thailand, pp. 1–7, 7–9 June 2006.

[GAN 08] GANÇARSKI P., BLANSCHÉ A., WANIA A., "Comparison between two coevolutionary feature weighting algorithms in clustering", *Pattern Recognition*, vol. 41, no. 3, pp. 983–994, 2008.

[GIL 98] GILLESPIE A., ROKUGAWA S., MATSUNAGA T. *et al.*, "A temperature and emissivity separation algorithm for Advanced Spaceborne Thermal Emission and Reflection Radiometer (ASTER) images", *IEEE Transactions on Geoscience and Remote Sensing,* vol. 36, pp. 1113–1126, 1998.

[GIV 98] GIVONI B., *Climate Considerations in Building and Urban Design*, John Wiley & Sons, New York, 1998.

[GOE 85] GOETZ A.F.H., VANE G., SOLOMON J.E. *et al.*, "Imaging spectrometry for Earth remote sensing", *Science*, vol. 228, no. 4704, pp. 1147–1153, 1985.

[GOM 04] GÔMEZ F., GLI L., JABALOYES J., "Experimental investigation on the thermal comfort in the city: relationship with the green areas, interaction with the urban microclimate", *Building and Environment*, vol. 39, pp. 1077–1086, 2004.

[GRA 11] GRABER G., POCK T., BISCHOF H., "Online 3D reconstruction using convex optimization", *ICCV Workshops*, pp. 708–711, 2011.

[GRI 02] GRIGORESCU S.E., PETKOV N., KRUIZINGA P., "Comparison of texture features based on Gabor filters", *IEEE Transactions on Image Processing*, vol. 11, no. 10, pp. 1160–1167, 2002.

[GUI 06] GUIGUES L., COCQUEREZ J.P., LE MEN H., "Scale sets analysis", *International Journal of Computer Vision*, vol. 68, no. 3, pp. 289–317, 2006.

[GUL 05] GULYAS A., "Differences in human comfort conditions within a complex urban environment: a case study", *Acta climatologica et chronologica*, Universitatis Szegediensis, vol. 38–39, pp. 71–84. 2005.

[GUO 11] GUO L., CHEHATA N., MALLET C. *et al.*, "Relevance of airborne LiDAR and multispectral image data for urban scene classification using Random Forests", *ISPRS Journal of Photogrammetry and Remote Sensing*, vol. 66, no. 1, pp. 56–66, January 2011.

[GUY 02] GUYON I., WESTON J., BARNHILL S. *et al.*, "Gene selection for cancer classification using support vector machines", *Machine Learning*, vol. 46, pp. 289–422, 2002.

[GUY 08] GUYADER A., "La biodiversité: un concept flou ou une réalité scientifique? ", *Courrier de l'environnement de l'INRA,* no. 55, February 2008.

[HAR 73] HARALICK R.M., SHANMUGAM K., DINSTEIN I., "Textural features for image classification", *IEEE Transactions on Systems, Man and Cybernetics*, vol. 3, no. 6, pp. 610–621, 1973.

[HEI 12] HEINRICH V., KRAUSS G., GÖTZE C. *et al.*, "IDB – www.indexdata base.de, Entwicklung einer Datenbank für Fernerkundungsindizes", *AK Fernerkundung*, vol. 10, pp. 4–5, 2012.

[HER 04] HEROLD M., ROBERTS D.A., GERDNER M. *et al.*, "Spectrometry for urban area remote sensing–development and analysis of a spectral library from 350 to 2400 nm", *Remote Sensing of Environment*, vol. 91, pp. 304–319, 2004.

[HER 05] HEROLD M., ROBERTS D., "Spectral characteristics of asphalt road aging and deterioration: implication for remote sensing applications", *Applied Optics*, vol. 44, no. 20, pp. 4327–4334, 2005.

[HER 11] HEROLD M., SKUTSCH M., "Monitoring, reporting and verification for national REDD+ programmes: two proposals", *Environmental Research Letters*, vol. 6, 2011.

[HUB 12] HUBERT-MOY L., NABUCET J., VANNIER C. *et al.*, "Cartographie des continuités écologiques: quelles données pour quelles échelles territoriales? Application à la sous-trame forestière", *Revue internationale de géomatique*, vol. 22, no. 4, pp. 619–640, 2012.

[ING 07] INGLADA J., "Automatic recognition of man-made objects in high resolution optical remote sensing images by SVM classification of geometric image features", *ISPRS Journal of Photogrammetry and Remote Sensing*, vol. 62, no. 3, pp. 236–248, 2007.

[IOV 08] IOVAN C., BOLDO D., CORD M., "Detection, characterization and modeling vegetation in urban areas from high resolution aerial imagery", *IEEE Journal of Selected Topics in Applied Earth Observations and Remote Sensing*, vol. 1, pp. 206–213, 2008.

[JAT 08] JAT M.K., GARG P.K., KHARE D., "Modelling of urban growth using spatial analysis techniques: a case study of Ajmer city (India)", *International Journal of Remote Sensing*, vol. 29, no. 2, pp. 543–567, 2008.

[JEN 12] JENSEN R.R., HARDIN P.J., HARDIN A.J., "Classification of urban tree species using hyperspectral data", *Geocarto International*, vol. 27, pp. 443–458, 2012.

[JOH 04] JOHNSTON J., NEWTON J., *Building Green: a Guide to Using Plants on Roofs, Walls and Pavements*, Ecology Unit, London, 2004.

[KAR 09] KARANTZALOS K., PARAGIOS N., "Recognition-driven two-dimensional competing priors toward automatic and accurate building detection", *IEEE Transactions on Geoscience and Remote Sensing*, vol. 47, no. 1, pp. 133–144, 2009.

[KAS 15] KASPERSEN P.S., FENSHOLT R., DREWS M., "Using Landsat vegetation indices to estimate impervious surface fractions for European cities", *Remote Sensing*, vol. 7, pp. 8224–8249, 2015.

[LAB 00] LABEN C., BROWER B., Process for enhancing the spatial resolution of multispectral imagery using pan-sharpening, U.S. Patent US 6 011 875, 2000.

[LAC 08] LACHÉRADE S., MIESCH C., BOLDO D. *et al.*, "ICARE: a physically-based model to correct atmospheric and geometrical effects from high spatial and spectral remote sensing images over 3D urban areas", *Meteorology and Atmospheric Physics,* vol. 102, no. 3, pp. 209–222, 2008.

[LAG 04] LAGOUARDE J.P., MOREAU P., IRVINE M. *et al.*, "Airborne experimental measurements of the angular variations in surface temperature over urban areas: case study of Marseille (France)", *Remote Sensing of Environment,* vol. 93, pp. 443–462, 2004.

[LAN 09] LANDRY S.M., CHAKRABORTY J., "Street trees and equity: evaluation of the spatial distribution of an urban amenity", *Environment and Planning,* vol. 41, pp. 2651–2670, 2009.

[LEB 96] LEBLON B., GALLANT L., GRANBERG H., "Effects of shadowing types on ground-measured visible and near-infrared shadow reflectances", *Remote Sensing of Environment,* vol. 58, no. 3, pp. 322–328, 1996.

[LEB 11] LE BRIS A., CHEHATA N., "Change detection in a topographic building database using submetric satellite images", *International Archives of Photogrammetry, Remote Sensing and Spatial Information Sciences (IAPRS)*, vol. 38, no. 3/W22, pp. 25–30, 2011.

[LON 15] LONCAN L., ALMEIDA L.B., BIOUCAS-DIAS J.M. *et al.*, "Hyperspectral pansharpening: a review", *IEEE Geosci. and Remote Sens. Mag.,* 2015.

[LU 06] LU D.S., WENG Q.H., "Use of impervious surface in urban land-use classification", *Remote Sensing of Environment*, vol. 102, no. 1–2, pp. 146–160. 2006.

[MAC 07] MACGRANAHAN G., "Urban environments, wealth and health: shifting burdens and possible responses in low and middle-income nations", Human Settlements Working Paper, Urban Environments No. 1, International Institute for Environment and Development (IIED), London, UK, p. 53, 2007.

[MAD 01] MADHAVAN B.B., KUBO S., KURISAKI N. *et al.*, "Appraising the anatomy and spatial growth of the Bangkok Metropolitan area using a vegetation-impervious soil model through remote sensing", *International Journal of Remote Sensing*, vol. 22, no. 5, pp. 789–806, 2001.

[MAI 12] MAIRE E., MARAIS-SICRE C., GUILLERME S. *et al.*, "High resolution remote sensing of trees and hedges using mathematical morphology", *International Journal of Geomatics and Spatial Analysis*, vol. 22/4, pp. 519–538, 2012.

[MAR 06] MARESCA B., RANVIER M., "Biodiversité : combien est-on prêt à payer ? Une méthode exploratoire appliquée au programme Natura 2000", *Centre de Recherche pour l'Etude et l'Observation des Conditions de vie (CREDOC)*, vol. 198, p. 4, 2006.

[MAR 15] MARMANIS D., DATCU M., ESCH T. *et al.*, "Deep learning Earth observation using ImageNet pretrained networks", *IEEE Geoscience and Remote Sensing Letters*, vol. 13, no. 1, pp. 105–109, 2015.

[MAT 06] MATHIEU N., "Pour un croisement transatlantique des recherches interdisciplinaires sur les socio-écosystèmes urbains", *Natures Sciences Sociétés*, NSS Dialogues, EDP Sciences, vol. 14, pp. 15–18, 2006.

[MAT 07] MATHIEU R., ARYAL J., CHONG A.K., "Object-based classification of Ikonos imagery for mapping large-scale vegetation communities in urban areas", *Sensors*, vol. 7, no. 11, pp. 2860–2880, 2007.

[MEA 05] MEA, Summary reports, available at: http://www.millennium assessment.org/fr/Synthesis.html, 2005.

[MNI 12] MNIH D., HINTON G.E., "Learning to label aerial images from noisy data", *Proceedings of the 29th International Conference on Machine Learning (ICML-12)*, pp. 567–574, 2012.

[MUR 13] MURATET A., LORRILLIERE R., CLERGEAU P. *et al.*, "Evaluation of landscape connectivity at community level using satellite-derived NDVI", *Landscape Ecology*, vol. 28, no. 1, pp. 95–105. 2013.

[MYI 11] MYINT S.W., GOBER P., BRAZEL A. *et al.*, "Per-pixel vs. object based classification of urban land cover extraction using high spatial resolution imagery", *Remote Sensing of Environment*, vol. 115, no. 5, pp. 1145–1161, 2011.

[OKE 81] OKE T.R., "Canyon geometry and the nocturnal Urban Heat Island: comparison of scale model and field observations", *Journal of Climatology*, vol. 1, pp. 237–254, 1981.

[OKE 82] OKE T.R., "The energetic basis of the urban heat island", *Quarterly Journal of the Royal Meteorological Society*, vol. 108, pp. 1–24, 1982.

[OLT 10] OLTRA-CARRIÓ R. SOBRINO J.A. GUTIERREZ-ANGONESE J. *et al.*, "Estudio del crecimiento urbano, de la estructura de la vegetación y de la temperatura de la supereficie del Gran San Miguel de Tucumán, Argentina", *Revista Nacional de Teledetección- Asociación Española de Teledetección*, vol. 34, pp. 69–76, 2010.

[OLT 12] OLTRA-CARRIÓ R., SOBRINO J.A., FRANCH B. *et al.*, "Land surface emissivity retrieval from airborne sensor over urban areas", *Remote Sensing of Environment*, vol. 123, pp. 298–305, 2012.

[OLT 13] OLTRA-CARRIÓ R., Thermal remote sensing of urban areas. The case study of the Urban Heat Island of Madrid, PhD Thesis, University of València, Spain, 2013.

[OLT 14] OLTRA-CARRIO R., CUBERO-CASTAN M., BRIOTTET X. *et al.*, "Analysis of the performance of the TES algorithm over urban areas", *IEEE Transactions on Geoscience and Remote Sensing,* vol. 52, pp. 6989–6998, 2014.

[PAI 15] PAISITKRIANGKRAI S., SHERRAH J., JANNEY P. *et al.*, "Effective semantic pixel labelling with convolutional networks and conditional random fields", *Proceedings of the IEEE International Conference on Computer Vision and Pattern Recognition Workshops (CVPRW),* pp. 36–43, 2015.

[PAL 05] PAL M., MATHER P.M., "Support vector machines for classification in remote sensing", *International Journal of Remote Sensing*, vol. 26, pp. 1007–1011, 2005.

[PEU 95] PEUQUET D., DUAN N., "An event-based spatio-temporal data model (ESTDM) for temporal analysis of geographical data", *International Journal of Geographical Information Systems,* vol. 9, no. 1, pp. 7–24. 1995.

[PHA 11] PHAM H.M., YAMAGUCHI Y., BUI T.Q., " A case study on the relation between city planning and urban growth using remote sensing and spatial metrics", *Landscape and Urban Planning*, vol. 100, no. 3, pp. 223–230, 2011.

[PHA 12] PHAM T.T.H., APPARICIO P., SÉGUIN A.M. *et al.*, "Spatial distribution of vegetation in Montreal: an uneven distribution or environmental inequity?", *Landscape and Urban Planning*, vol. 107, no. 3, pp. 214–224, 2012.

[PIE 06] PIERROT-DESEILLIGNY M., PAPARODITIS N., "A multiresolution and optimization-based image matching approach: an application to surface reconstruction from SPOT5-HRS stereo imagery", *IAPRS volXXXVI-1/W41 in ISPRS Workshop On Topographic Mapping From Space (With Special Emphasis on Small Satellites),* Ankara, Turkey, 2006.

[PLA 04] PLATT R.V., GOETZ A.F.H., "A comparison of AVIRIS and Landsat for land use classification at the urban fringe", *Photogrammetric Engineering and Remote Sensing*, vol. 70, pp. 813–819, 2004.

[POU 09] POULAIN V., INGLADA J., SPIGAI M. *et al.*, "Fusion of high resolution optical and SAR images with vector data bases for change detection", *IEEE International Geoscience and Remote Sensing Symposium (IGARSS'09)*, Cape Town, South Africa, 2009.

[PRE 08] PRÉVOT-JULLIARD A.C., SERVAIS V., "La ville, un dénominateur commun", in *Entre l'Homme et la nature, une démarche pour des relations durables*, Unesco – notes techniques, Paris, pp. 86–87, 2008.

[PUI 01] PUISSANT A., WEBER C., "Utilité des images haute résolution pour évaluer la place de la végétation dans l'aménagement urbain : Quelles résolutions pour quels besoins", *Vème rencontre Théo QUANT*, p. 10, 2001.

[PUI 03] PUISSANT A., WEBER C., "Les images à très haute résolution spatiale : une source d'information géographique en milieu urbain? État des lieux et perspectives", *L'Espace Géographique,* vol. 4, pp. 345–356, 2003.

[PUI 04] PUISSANT A., HIRSCH J., "Télédétection urbaine et résolution spatiale optimale : intérêt pour les utilisateurs et aide pour les classifications", *Revue Internationale de Géomatique*, vol. 14, no. 3–4, pp. 403–415, 2004.

[PUI 14] PUISSANT A., ROUGIER S., STUMPF A., "Object-oriented mapping of urban trees using Random Forest classifiers", *International Journal of Applied Earth Observation and Geoinformation,* vol. 26, pp. 235–245, 2014.

[RAO 72] RAO P.K., "Remote sensing of urban heat islands from an environmental satellite", *Bulletin of the American Meteorological Society*, vol. 53, pp. 647–648, 1972.

[RAP 12] RAPINEL S., Contribution de la télédétection à l'évaluation des fonctions des zones humides: de l'observation à la modélisation prospective, thesis, University of Rennes 2, 2012.

[RHE 14] RHEIN C., PALIBRK M., "Urban forms, land use and social mix in built up areas: the case of city of Paris", *Topics Images & Villes*, *European Journal of Geography*, vol. 685, 27 July 2014.

[RIC 03] RICHTER R., "Status of model ATCOR4 on atmospheric /yopographic correction for airborne hyperspectral imagery", *3nd EARSeL Workshop on Imaging Spectroscopy*, 13–16 May 2003.

[RIG 06] RIGO G., PARLOW E., OESCH D., "Validation of satellite observed thermal emission with in-situ measurements over an urban surface", *Remote Sensing of Environment,* vol. 104, pp. 201–210, 2006.

[ROU 14] ROUGIER S., PUISSANT A., "Improvement of urban vegetation segmentation and classification using multi-temporal Pléiades images", *Proceedings of the 5th Conference GEOBIA,* Thessaloniki, Greece, p. 409–414, 21–24 May 2014.

[ROU 15] ROUSSET-ROUVIÈRE L., "Sysiphe system: a state of the art airborne hyperspectral imaging system: initial results from the first airborne campaign", *EARSe*, Luxembourg, 14–16 April 2015.

[ROT 07] ROTTENSTEINER F., "Building change detection from digital surface models and multi-spectral images", *International Archives of the Photogrammetry, Remote Sensing and Spatial Information Sciences (IAPRS)*, vol. 36(3/W49B), Munich, Germany, 2007.

[SAI 08] SAINT-GÉRAND T., "Understanding to measure... or measuring to understand", in GUERMOND Y. (eds), *The Modeling Process in Geography*, ISTE, London and John Wiley & Sons, New York, 2008.

[SEB 10] SEBARI I., MORIN D., "Développement et défis de la télédétection urbaine", *Cahiers de géographie du Québec,* vol. 54, no. 151, pp. 117–132, 2010.

[SCH 12] SCHINDLER K., "An overview and comparison of smooth labeling methods for land-cover classification", *IEEE Transactions on Geoscience and Remote Sensing*, vol. 50, no. 11, pp. 4534–4545, 2012.

[SHA 00] SHASHUA-BAR L., HOFFMAN M.E., "Vegetation as a climatic component in the design of an urban street: an empirical model for predicting the cooling effect of urban green areas with trees", *Energy and Buildings*, vol. 31, no. 3, pp. 221–235, 2000.

[SHI 11] SHIMONI M., TOLT G., PERNEEL C. *et al.*, "Detection of vehicles in shadow areas", *Proceedings of 3rd Workshop on Hyperspectral Image and Signal Processing: Evolution in Remote Sensing, WHISPERS*, Lisbon, Portugal, pp. 1–4, 6–9 June 2011.

[SID 14] SIDI YOUSSEF M.M., MALLET C., CHEHATA N. *et al.*, "Détection de bâtiments à partir d'une image satellitaire par combinaison d'approches ascendante et descendante", *RFIA' 14 (19ème congrès national sur la Reconnaissance de Formes et l'Intelligence Artificielle)*, Rouen, France, June 2014.

[SIM 15] SIMOES M., BIOUCAS DIAS J., ALMEIDA L. *et al.*, "A convex formulation for hyperspectral image superresolution via subspace-based regularization", *IEEE Trans. Geosci. and Remote Sens.*, 2015.

[SMA 03] SMALL C., "High spatial resolution spectral mixture analysis of urban reflectance", *Remote Sensing of Environment*, vol. 88, pp. 170–186, 2003.

[SOB 08] SOBRINO J.A., JIMENEZ-MUNOZ J.C., SORIA G. *et al.*, "Land surface emissivity retrieval from different VNIR and TIR sensors", *IEEE Transactions on Geoscience and Remote Sensing*, vol. 46, pp. 316–327, 2008.

[SOB 12a] SOBRINO J.A., OLTRA-CARRIÓ R., JIMÉNEZ-MUÑOZ J.C. *et al.*, "Emissivity mapping over urban areas using a classification-based approach: application to the Dual-use European Security IR Experiment (DESIREX)", *International Journal of Applied Earth Observation and Geoinformation*, vol. 18, pp. 141–147, 2012.

[SOB 12b] SOBRINO J.A., OLTRA-CARRIÓ R., SÒRIA G. *et al.*, "Impact of spatial resolution and satellite overpass time on evaluation of the surface urban heat island effects", *Remote Sensing of Environment*, vol. 117, pp. 50–56. 2012.

[SOB 13] SOBRINO J.A., OLTRA-CARRIÓ R., SÒRIA G. *et al.*, "Evaluation of the surface urban heat island effect in the city of Madrid by thermal remote sensing", *International Journal of Remote Sensing*, vol. 34, pp. 3177–3192, 2013.

[STA 07] STATHOPOULOU M., CARTALIS C., PETRAKIS M., "Integrating Corine Land Cover data and Landsat TM for surface emissivity definition: application to the urban area of Athens, Greece", *International Journal of Remote Sensing*, vol. 28, pp. 3291–3304, 2007.

[STA 09] STATHOPOULOU M., SYNNEFA A., CARTALIS C. *et al.*, "A surface heat island study of Athens using high-resolution satellite imagery and measurements of the optical and thermal properties of commonly used building and paving materials", *International Journal of Sustainable Energy*, vol. 28, pp. 59–76, 2009.

[STE 06] STEWART I.D., OKE T.R., "Methodological concerns surrounding the classification of urban and rural climate stations to define urban heat island magnitude", Preprints, *6th International Conference on Urban Climate*, Goteborg, Sweden, 12–16 June 2006.

[SU 11] SU S., JIANG Z., ZHANG Q. *et al.*, "Transformation of agricultural landscapes under rapid urbanization: a threat to sustainability in Hang-Jia-Hu region, China", *Applied Geography*, vol. 31, pp. 439–449, 2011.

[STU 90] STULPNAGEL A. VON, HORBERT M., SUKOPP H., "The importance of vegetation for the urban climate", in SUKOPP H., HEJNY S., KOWARIK I. (eds), *Urban Ecology*, SPB Academic Publishing, The Hague, 1990.

[TIA 08] TIANGCO M., LAGMAY A.M.F., ARGETE J., "ASTER based study of the night time urban heat island effect in Metro Manila", *International Journal of Remote Sensing*, vol. 29, pp. 2799–2818, 2008.

[TOM 98] TOMASI C., ANDUCHI R.M., "Bilateral filtering for gray and color images", *Proceedings of IEEE International Conference on Computer Vision*, pp. 836–846, 1998.

[VAN 93] VANDEGRIEND A.A., OWE M., "On the relationship between thermal emissivity and the normalized difference vegetation index for natural surfaces", *International Journal of Remote Sensing*, vol. 14, pp. 1119–1131, 1993.

[VAN 03] VAN HERZELE A., WIEDEMANN T., "A monitoring tool for the provision of accessible and attractive urban green spaces", *Landsc. Urban Plan.*, vol. 966, pp. 1–18, 2003.

[VAN 11] VANNIER C., HUBERT-MOY L., NABUCET J., "Analyse spatiale de la dynamique de l'occupation du sol aux échelles de la parcelle et de l'îlot parcellaire, application en paysage bocager", *Revue Internationale de Géomatique*, vol. 21, no. 3, pp. 353–374, 2011.

[VOO 02] VOOGT J.A., "Urban Heat Island", in MUNN T. (ed.), *Encyclopedia of Global Environmental Change*, Wiley, Chichester, 2002.

[VOO 03] VOOGT J.A., OKE T.R., "Thermal remote sensing of urban climates", *Remote Sensing of Environment*, vol. 86, pp. 370–384, 2003.

[WAN 07a] WANIA A., WEBER C., "Hyperspectral imagery and urban green observation", *Proceedings of Urban Remote Sensing Joint Event*, pp. 1–8, 11–13 April 2007.

[WAN 07b] WANIA A., Urban vegetation: detection and function evaluation for air quality assessment, PhD thesis, University Louis Pasteur, 2007.

[WEB 15] WEBER C., "Formes urbaines et facteurs environnementaux: exposition et santé urbaine", *Environment, Risques & Santé*, vol. 14, no. 4, pp. 305–312, 2015.

[WEI 15] WEI Q., BIOUCAS DIAS J.M., DOBIGEON N. *et al.*, "Hyperspectral and multispectral image fusion based on a sparse representation", *IEEE Trans. Geosci. and Remote Sens.*, vol. 53, no. 7, pp. 3658–3668, 2015.

[WEL 82] WELCH R., "Spatial resolution requirements for urban studies", *International Journal of Remote Sensing*, vol. 3, no. 2, pp. 139–146, 1982.

[WEN 07] WENG Q., QUATTROCCHI D.A., *Urban Remote Sensing*, CRC Press, 2007.

[WEN 12] WENG Q., "Remote sensing of impervious surfaces in the urban areas: requirements, methods, and trends", *Remote Sensing of Environment*, vol. 117, 2012.

[WGI 14] WGIIARS, Intergovernmental panel on climate change, "Working Group II Contribution to the 5th assessment report of the intergovernmental panel on climate change", *Urban Areas*, 2014.

[WIL 11] WILSON E., TERMEER C., MEES H.P. *et al.*, "Adaptation to climate change in urban areas: climate-greening London, Rotterdam, and Toronto", *Climate Law*, vol. 2, no. 2, pp. 251–280, 2011.

[WU 03] WU C., MURRAY A.T., "Estimating impervious surface distribution by spectral mixture analysis", *Remote Sens. Environ.*, vol. 84, pp. 493–505, 2003.

[YOK 12] YOKOYA N., YAIRI T., IWASAKI A., "Coupled nonnegative matrix factorization unmixing for hyperspectral and multispectral data fusion", *IEEE Trans. Geosci. and Remote Sens.*, vol. 50, no. 2, pp. 528–537, 2012.

[ZHA 09] ZHANG X., ZHONG T., FENG X. *et al.*, "Estimation of the relationship between vegetation patches and urban land surface temperature with remote sensing", *International Journal of Remote Sensing*, vol. 30, no. 8, pp. 2105–2118, 2009.

[ZHA 10] ZHANG X., FENG X., JIANG H., "Object-oriented method for urban vegetation mapping using IKONOS imagery", *International Journal of Remote Sensing*, vol. 31, no. 1, pp. 177–196, 2010.

[ZHO 08] ZHOU W., TROY A., "An object-oriented approach for analysing and characterizing urban landscape at the parcel level", *International Journal of Remote Sensing*, vol. 11, no. 10, pp. 3119–3135, 2008.

[ZIN 15] ZINKO S., "Traitement de Données Hyperspectrales", Master Observation de la Terre et Géomatique, University of Strasbourg, p. 30, 2015.

2

Urban Scene Analysis with Mobile Mapping Technology

2.1. Introduction

Mobile mapping technology (MMT) involves the dynamic digitization at very high spatial resolution of complex environments (mainly urban environments) using a mobile terrestrial platform. The aim is to acquire very precise and very accurate georeferenced data of objects of interest as well as *in fine* data for applications which require this level of geometric precision (3D city modeling, pedestrian and vehicle detection for autonomous navigation, digitizing roadway property, archaeological excavations, etc.). This chapter will focus on mobile mapping from roads, but similar methods have also been developed for digitizing rail networks through the use of trains equipped with similar sensors.

Over the past 10 years, mobile mapping has undergone significant developments within remote sensing, photogrammetric, computer and robotic vision communities. The main reasons for this have been:

– previous limitations in the spatial resolution of airborne sensors made it necessary to move closer to objects being studied, in order to obtain a higher spatial resolution;

– increased capacity for storage and processing: advances in spatial resolution have led to a significant increase in the total amount of acquired data.

Chapter written by Bruno VALLET and Clément MALLET.

This development is heavily linked to in the introduction of sensors and devices designed for mobile terrestrial digitization:

– systems which enable very precise georeferencing for terrestrial moving vehicles;

– professional quality panoramic image sensors;

– laser sensors developed for the specific purpose of mobile mapping.

It can also be noted that the online service Google StreetView has led to a huge increase in the popularity of MMT and has brought it to the attention of the wider public [ANG 10]. There are now several companies carrying out operational mobile mapping for metrology, modeling and visualization of urban environments (Here, Cyclomedia, StreetMapper from 3DLaserMapping, TopCon and Apple) equipped with off-the-shell systems (Optech, RIEGL VMX-450, Trimble MX8 and Siteco). Similarly, research initiatives have been around for the last 20 years and have led to the emergence of increasingly complex and accurate innovative systems: VISAT [ELS 96], GPSVan [GRE 96], Lara3D [GOU 06], Roamer [KUK 07], Photobus [GON 07], RoadScanner [GAN 08], Stéréropolis [PAP 12] or the Kitti initiative [GEI 12].

The main issue with MMT is of a technical nature and relates specifically to equipping vehicles with sensors. This raises a number of mechanical, electrical and software-related questions and prompts a consideration of how best to produce terrestrial georeferenced data, i.e. precisely localized in a geographical reference system. There are also issues regarding the typology of data, because it is no longer possible to analyze data in 2D or 2.5D (2D plus depth data), unlike with geospatial data: the complex nature of the acquired scenes must be understood in three dimensions. These challenges will be discussed in the first part of this chapter which focuses on the state-of-the-art of the technology. The following two parts will deal with main issues which are not yet well managed by operational services and which are the focus of research studies.

The acquisition of MMT data poses a major problem for image registration because, as we will see, georeferencing systems remain limited and *a posteriori* image registration is necessary in order to guarantee a good quality of data georeferencing. This is critical for various applications of this type of data. These issues will be discussed in section 2.3.

The analysis and processing of MMT data will be the focus of section 2.4, where we will look specifically at the problem of scene analysis, i.e. in practice, segmentation and classification, object detection and reconstruction and possibly change detection in urban environments.

Finally, in section 2.5, we will detail research prospects for the forthcoming years by reviewing the main lines of research currently being carried out in this particular field.

2.2. Data acquisition

As is the case with any other data processing chain, describing the acquisition procedure is the first step in MMT. We will begin therefore by introducing the most commonly used sensors in MMT, before outlining how they are installed on a mobile mapping vehicle (MMV) in section 2.2.2. The calibration of these sensors for metrological purposes will be dealt with in section 2.2.3. Finally, the data produced and its characteristics will be discussed in section 2.2.4.

2.2.1. *Sensors onboard a mobile mapping system*

There are generally three main types of sensor which are installed onboard an MMV:

– image sensors;

– laser sensors;

– georeferencing sensors (Figure 2.1).

Figure 2.1. *The IGN Stéréopolis vehicle IGN [PAP 12]. See text for more details. Image sensor: blue: panoramic head; yellow: forward stereoscopic pair. Laser sensors: cyan: planar metrological laser; purple: robotic laser. Georeferencing sensors: green: GPS; red: odometer, the inertial central located down the mast is not visible. For a color version of this figure, see www.iste.co.uk/baghdadi/5.zip*

We will also see that other types of sensor or device can be useful for very particular applications [CAN 13].

2.2.1.1. Image sensors

The main type of data acquired by MMVs is images acquired by optical digital cameras. These can generally be divided into two types of configurations as Figure 2.2 illustrates:

– Panoramic: a group of cameras covers the majority of the sphere of visibility (apart from the vehicle itself), and therefore the majority of the vehicle environment. Imaging tends to be optical (visible+infra-red domains), but some MMVs also come equipped with thermal sensors. The most common configurations feature a camera pointed vertically and four or five others positioned horizontally in order to cover 360° with coverage ranging from a few pixels to a few dozen pixels. This creates a very large field. When four cameras are used, the device can be compared to a cube with the bottom side removed. Other configurations are possible, as seen with the IGN Stéréopolis vehicle [PAP 12] which uses more cameras (eight flat and two pointed upward, to the left and to the right at an angle of 45°), with narrower fields in order to improve the spatial resolution and to reduce the sky ratio in the panoramic view. The panoramic image is generated by resampling individual images in spherical geometry, which is only possible if all the optical centers of the cameras are colocated. Given that, in practice, this is impossible, the optical centers are either brought as close together as possible so as to be able to group the cameras together or they are aligned in the direction of the displacement and the trigger point depends on the MMV so that the centers are (almost) taken together at the trigger point.

– Stereoscopic pairs or triplets: a single panoramic image prevents from determining the 3D geometry of the scene. In order to be able to retrieve depth information, the MMV can be equipped with pairs or triplets of cameras pointing in the same direction but each shifted from a specific distance. These are known as stereoscopic pairs or triplets. This configuration makes it possible to acquire depth information (and therefore the 3D geometry) of the scene acquired through dense matching techniques. With pairs, the base (the section connecting the optical centers) can be horizontal or vertical depending on the system, with this choice having an impact on the reconstruction quality of linear elements parallel to the base.

Figure 2.2. *A Paris street viewed by an MMV with a panoramic head of eight cameras and two stereoscopic pairs pointing toward the front and the rear of the vehicle*

There are also systems with two panoramic views superimposed which combine the advantages of the two configurations (full coverage of the environment and access to depth information). "Pose" is the name given to the moment when all cameras onboard the MMV are used at once and the resulting set of images. These can be considered (almost) simultaneous. Poses are triggered over a temporal (necessary for a video camera) or spatial period, or a combination of the two (for example, every 3 m, but every 30 s in case of a stoppage). Sampling periods affect the density of the poses and have a significant impact on the volume of data acquired.

Images offer the advantage of being technically simple to handle (acquisition and processing). However, acquiring them onboard an MMV presents significant radiometric problems compared to aerial or satellite imaging: the scene displays a high dynamic range, between an area in shadow and a section of cloudy sky, for example. Cameras which take images both of building façades and the sky are subject to significant and highly variable backlighting. Two ways of dynamic adjustment (opening, time of exposure) are possible:

– a unique set-up for all cameras and the entire acquisition: the advantage is to facilitate texturizing (section 2.4.5) when it is necessary to "mix" different images;

– a set-up adapted automatically for each camera for each shot: this optimizes the dynamic of each image but creates splicing issues within panoramic views and successive panoramic views.

MMT cameras are still far simpler and a lot less expensive than their aerial counterparts. They are close enough to the scene that their spatial resolution is a less significant factor. MMT does, however, require sensors with far better dynamics.

2.2.1.2. Laser sensors

Once more, proximity to the scene puts very different constraints on MMT lasers compared to their airborne equivalents. They are closer to fixed terrestrial laser stations [PET 09, KUT 10]. Laser sensors onboard MMVs can be of the two types, illustrated in Figure 2.3:

– *planar metrological lasers:* the laser beam is emitted according to a fixed *A* axis in relation to the MMV and is reflected by a rotating mirror at 45° in an orthogonal direction to *A*, with the rays covering (approximately) an area orthogonal to A. The points captured by these lasers are positioned in the laser referential with a significant level of angular precision (one hundredth of a degree) and precision at a distance which ranges from a few cm to a few mm, depending on the quality of the sensor and the distance from the target. While the first sensors were only able to cover an angular sector somewhere between 70 and 90°, the latest sensors cover the full 360° and also have the capacity to record multiple returns (known as multi-echoes, similarly to airborne sensors) for an emitted pulse. From this point onward, sensors emit between 100,000 and 1,000,000 pulses per second with the mirror rotational speed roughly 5–20 turns per second, or an angular resolution in a direction orthogonal to *A* in the region of a tenth of a degree for average conditions. Finally, it is possible to identify two configurations as a function of the number of available lasers:

- a single laser: the *A* axis is angled toward the rear of the vehicle. This configuration is necessary for symmetry (both sides of the vehicle are of equal importance and one side must not be neglected in favor of the other) so that the axis is not vertical (in which case, a section at Z constant will be scanned) and the vehicle itself is not in the scanner's field. Ideally, the laser will be pointed upward to ensure that the sampling of vertical objects is consistent (a post can be "omitted" between two lines of vertical scans),

- two lasers: compared to the previous configuration, both axes are angled toward the outside of the vehicle (in a symmetrical manner). This configuration makes it possible to partly compensate for any occlusions, but requires the lasers to be very well calibrated;

– *robotic lasers*: these are also based on the principle of a rotation around an axis *A*, that is generally vertical, but the beam deviation mechanism which makes it possible to spatialize the measurement is replaced by a rigid system of "fibers" (from 16 to 64, a measurement is obtained by fiber) sweeping an angle of roughly 70°. These fibers rotate around *A* at a speed of between 6 and 15 turns per second, which makes it possible to obtain regular coverage of the immediate environment without requiring the displacement of the vehicle. However, they are less precise than metrological lasers (a few cm as opposed to a few mm) even if they acquire points at a similar rate (around 300,000 per second). The increased rotation speed of the head comes at the expense of point density and spatial resolution. They are well suited for monitoring mobile objects around an MMV, which explains why they are largely used in robotics.

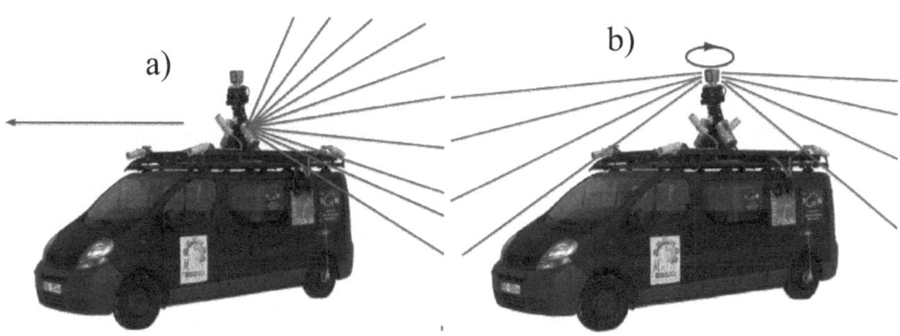

Figure 2.3. *Acquisition geometry for a metrological laser a) and a robotic laser b)*

Aside from giving the spatial position (x, y, z) of LiDAR points in their own reference frame (sensor reference), laser sensors allow us to obtain more information such as the exact time the pulse was emitted, the amplitude and the width of the peak of the backscattered signal, the number of the return (in case of multiple echoes), etc. The amplitude combined with the distance makes it possible to determine the reflectance of the target (following the hypothesis that surfaces exhibit a Lambertian behavior). This is a highly relevant attribute for the visualization and automatic interpretation of the scenes.

2.2.1.3. *Georeferencing sensors*

There are three types of sensors used onboard MMVs:

– satellite absolute positioning sensors (GNSS: GPS, Glonass and Galileo). These indicate the absolute position of a point of a vehicle in a cartographic reference frame, i.e. in relation to ground datum. Sometimes, two are used simultaneously in order to refine the measurement and in order to give an indication of the vehicle orientation. The quality of positioning in urban environments is affected by:

- masks (few visible satellites),

- the multipaths of GPS waves on building façades. GPS also gives a very precise measurement of the time which is highly useful for sensor integration (triggering poses and dating data);

– odometers: these sensors are placed on one of the vehicle's wheels in order to provide a precise measurement of its rotation. The press of the wheel must then be calibrated to link this measurement to the distance covered;

– accelerometers: there are various types of technology which enable the measurement of accelerations and these can be distinguished largely by their respective precision. In MMT, the tendency is to use inertial measurement devices based on the light pulse speed difference in optical fibers rolled up in the device depending on the axes being measured.

Georeferencing data collected in this way are combined by a geopositioning system that is (improperly) referred to as an inertial measurement unit (IMU) because this calculation is carried out in a device containing the computation electronics and the inertial measurement unit and is connected to the other georeferencing sensors (GPS/odometer). The IMU is thus able to estimate a "trajectory", i.e. the position and orientation of a reference frame connected to it (which will be referred to from now on as an IMU frame) at a frequency between 100 and 200 Hz. The an IMU often leads to acquisition restrictions, including the creation of a "static point" (remaining stationary for several minutes in an open space in order to position itself using as many satellites as possible) at the start and at the end of the period of acquisition. Calibrating the odometer may require more complicated procedures.

2.2.1.4. Other sensors

In addition, other sensors can be installed for specific purposes, including thermometers for thermal imaging acquisitions [HOE 09] or a sensor which measures ambient light for radiometric correction.

2.2.2. *Integrating sensors*

An MMV is made up of a subset of sensors, which must be integrated in an acquisition system which enables their power supply, their steering and the recording of data acquired. The sizeable flow of data can require the use of several computers:

– a master user in charge of acquisition (triggering the cameras, launching the initial center and laser sensors);

– slave computers which divide up the flow of data to be recorded.

The different elements communicate with each other via a local network or other methods (USB or PCI slot) depending on the sensor. The system is supplied by the vehicle's electrical energy, but it is also necessary to add batteries to improve the system's autonomy given that it can consume more electrical energy than the vehicle's alternator is able to produce. More recent systems tend to use components which consume less energy and it might not be too long before we see vehicles with power consumption lower than the energy an alternator is able to produce. The only constraint on the autonomy will then be the disk space and the vehicle storage capacity.

However, in order to ensure the continuous recording of the data acquired on disks, the system is restricted to the maximum number of poses that (ranging from 4 to 25 per second depending on the system). This will restrict the speed of the vehicle. Certain configurations mean that the vehicle is not able to surpass 10 km/h. However, by adapting the period to the type of road or to the environment (for example, 3 m in the city, 6 m on roads and 10 m on the motorways), more recent systems make it possible to meet the speed limit.

2.2.3. *Geometric calibration*

MMT sensors must be calibrated in order to be able to transmit the data acquired into geometry in a metrological manner, i.e. to express the coordinates of each laser point, of the optical center, and the axes of the cameras in a cartographic reference source. We separate:

– the internal calibration of a sensor: the aim is to precisely determine the position of a a 3D point or the 3D line corresponding to a point of an image in the sensor frame. The internal calibration of laser sensors is carried out by the manufacturer. For cameras, it is obtained through standard photogrammetric calibration procedures (laboratory measurements) using known targets [STU 04];

– the external calibration of a sensor involves precisely evaluating the position of the sensor reference points on the vehicle, and more precisely the IMU frame. For the images, this calibration can be achieved by acquiring a site with known targets which are detected in the images and matched in a semi-automatic way. There has been a recent trend to carry out this calibration "online", i.e. with data obtained during acquisitions and not on a site designed specifically for this purpose [CAN 10]. The problem must then be processed alongside the issue of repositioning [CAN 12] (section 2.3). There are a number of different methods for laser sensors. These are based on detecting objects with simple geometry and that are previously known (e.g. flat and spherical). A preliminary evaluation of the calibration can be carried out using topometry by measuring previously known targets positioned on the sensors themselves.

2.2.4. *Specificities of MMT data*

There are a number of differences between MMV data and airborne data:

– the complexity of the scene: in aerial imaging, it is possible to consider the scene along a plane or using a basic digital elevation model. This feature is often used to consider data processing as a 2D or a 2.5D standard problem: the altitude is only a defined attribute with regular 2D grid geometry (pixels). In MMT, the sensor is placed in the middle of the imaged scene which must be analyzed in 3D. This complexity also leads to anisotropy and variations in the spatial resolution which are considerably more significant than with aerial imaging, as is shown in Figure 2.4. Surface sampling using MMT LiDAR sensors is thus far more heterogeneous than with airborne devices;

– data volume: this will depend on the sensors used and how the data are compressed. We are able to estimate that roughly a terabyte of data is acquired every half day when using a system equipped with a panoramic head, two stereo pairs (with 3 m between poses) and a modern rotating laser (non-compressed) images correspond to 80% of the total volume;

– acquisition geometry: unlike an airplane, which will acquire data along parallel, rectilinear trajectories, an MMV is restricted by the layout of the road network. For a given area, the vehicle will have to cover every road, meaning going through the same junctions several times, and often going down the same streets, making each acquisition a unique configuration. Furthermore, unlike an airplane, the vehicle will often be stationary, which makes its trajectory even more complicated compared to the trajectory of a plane moving at a regular speed, and generating a large number of points for the same section (therefore unnecessary) with an MMT laser scan;

– dynamic: urban scenes are generally highly dynamic, with many vehicles and pedestrians and pieces of street furniture (benches and bins) which move between two acquisitions of the same street, even when they are several minutes apart. This dynamic makes it necessary to take the temporal dimension of the acquisition into account, especially in the registration procedure and for scene understanding.

These characteristics lead to tight restrictions on the data processing chain which must be highly efficient in order to cope with the volume and in order to take into account the complexity of the urban scenes. Furthermore, for laser acquisitions, the 3D point clouds generated are very anisotropic given that their density depends on the speed of the vehicle, the distance between the vehicle and the target object and the positioning of surfaces in relation to the laser beam.

In conclusion, while the MMT processing chain shares some similarities with the more traditional processing chain for aerial data, such as the complementarity between the image/laser and the acquisition system, its specifics (complexity and volume) mean that there are a number of particular challenges for setting up an efficient processing chain. This complexity also explains why objects of interest for MMT tend to be city centers where the challenges are more numerous and therefore justify a greater mass of data and so also higher costs.

Figure 2.4. *Terrestrial (orange/violet) and airborne (blue/green) laser cloud for the same area. For a color version of this figure, see www.iste.co.uk/baghdadi/5.zip*

2.3. Data registration and georeferencing

Geometrically aligning MMT data through the use of an inertial measurement unit and the calibration of each instrument is known as direct georeferencing. This includes geopositioning errors, related to the IMU, and calibration uncertainties. While we are often able to discount the latter, the position given by the IMU is more problematic. In ideal conditions (clear skies and few objects in the surrounding area), the precision of GPS and its integration in time coupled with the IMU guarantee subdecimetric geopositioning. However, the GPS signal can be significantly disturbed by masks (GPS no longer seeing enough satellites to give reliable positioning) and multi-paths (the GPS signal is reflected by building façades, which affects how the position is calculated) which is very common in dense urban areas. Accelerometer measurements of the IMU are only partly able to correct these errors, because the uncertainty relating to the measurement of acceleration, although very weak, is integrated twice: the inertial drift increases with the acquisition time. Thus, in urban canyons and at low speeds, the GPS signal can be lost for several minutes, in which case the drift (the difference between the real position and the position given by the inertial measurement unit) can be greater than 1 m. This is unacceptable for a large range of applications, particularly given the precision of the onboard sensors [CAN 13]. One of the major challenges in MMT is thus to guarantee subdecimetric localization, even in city centers. It is worth noting that the rotation precision is excellent in all conditions, with a standard deviation of less than 0.1°.

2.3.1. *Characteristics of the registration procedure*

The characteristics of the registration procedure in an MMV must take into account two factors:

– the data type (image or laser);

– the existence of external data for the acquisition area (topographic 2D or 3D database, aerial images or orthoimages, visual landmarks or known targets, etc.).

2.3.1.1. Image versus laser

For image-based registration, the process is similar to those employed for airborne imaging, with additional difficulties caused by the complexity of the non-static nature of scenes. Thus, even with robust keypoint detectors such as Scale Invariant Feature Transform (SIFT), it will generally not be possible to locate the majority of good correspondences between images [WEI 13]. A human operator must be put into the loop in order to validate all matches, which leads to higher costs and delays. Another specificity of MMV is that images from the same pose form a solid block and it is therefore necessary to only introduce one unknown pose per block in the photogrammetric system [CAN 13].

Laser-based registration is very different and it is therefore necessary to use other techniques. In an aerial context, acquisition is carried out by strips which highly overlap and can be matched all at the same time. For an MMV, laser acquisition is continuous and it is necessary to follow the road network: overlapping areas exist at road junctions between streets and on streets where acquisitions have been carried out several times. Furthermore, the vehicle will have stopped several times, which necessitates a non-rigid matching method. The aim is to evaluate the drift as a function of the time by re-estimating the position and the orientation of the inertial measurement unit, most of the time using external data (aerial images, known landmarks, etc.)

2.3.1.2. Relative versus absolute registration

Without external data, the sole aim will be to guarantee the consistency of data, i.e. data match at junctions or streets where several acquisitions have been carried out. If, however, external data are used, it will be referred to as absolute matching (assuming the external data are perfect) and surveys performed at various times are consistent, which opens up the field for change detection and monitoring applications.

2.3.2. *Deformation models*

In ideal circumstances, results from the registration process should make it possible to correct the trajectory given by the inertial measurement unit, i.e. the pose (position and angle) of the IMU frame at the frequency at which it is produced by the inertial measurement unit (100–200 Hz). However, an acquisition sequence can last up to 3 h between two fixed points, which corresponds to the shortest trajectory to be considered in its entirety for matching purposes. The resulting registration problem is too large (too many unknowns in the system) to be able to be processed efficiently. However, the trajectory of a vehicle is very regular owing to its limited turning radius and capacity for acceleration and braking. It is thus possible to subsample this trajectory in a certain number of control poses (which will be the real unknowns to be estimated) and to interpolate the trajectory between the control poses (Figure 2.5). When using images, these control poses are the poses of blocks of images acquired simultaneously. With lasers, where acquisition is continuous, these poses can be sampled regularly over time (1–5 Hz, for example) or using more precise rules taking into account the speed and the curve of the trajectory, in order to consider the specificities of the inertial measurement unit. Each control pose is connected to its control time, i.e. the moment during which the central marker had this pose.

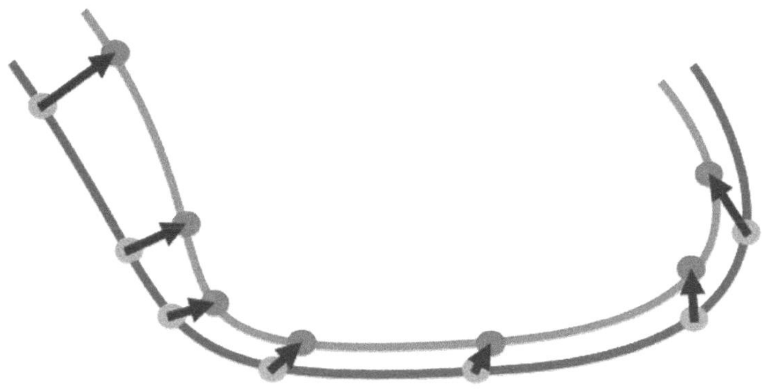

Figure 2.5. *Deformation trajectory expressed by a translation at certain points (control poses, blue points) of the trajectory. For a color version of this figure, see www.iste.co.uk/baghdadi/5.zip*

2.3.3. *Pairing methods*

Pairing is essential for image registration since it defines the restrictions that the data points will have to verify. For example, two 3D points or two pixels corresponding to the same point in the scene should coincide. Applying these restrictions will make it possible to re-estimate the trajectory. It should be noted that the inertial measurement unit generally provides good initial poses (no worse than 1 m in translation, 0.1° in rotation). Pairing methods depend on the type of data that is to be made coherent:

– *image/image*: matching between images is generally carried out using keypoints (SIFT or one of its variants). For MMV data, we can add specific constraints such as the distance from the 3D point corresponding to the center of the IMU frame. Generally, these matching points are not robust enough to rely too heavily on them. The intervention of a human operator is necessary in order to ensure their validity [CAN 12]. It is worth noting that certain studies suggest pairing shared objects taken from MMT and aerial images, where the geometry is better controlled [TOU 06];

– *laser/laser*: here, there are a wide range of possibilities. While attempts have been made in order to develop robust descriptors for 3D point clouds, these methods are still highly experimental. For lasers, we therefore tend to use more straightforward approaches such as plane-to-plane matching of building façades (see the section on building façade detection), linear primitives (corners of buildings, posts and tree trunks) or road markings, which are detectable using the laser intensity feature [HER 15]. Lasers also make it possible to carry out dense point-to-point or point-to-triangle matching [TAK 13], which is particularly useful for the ground surfaces. In this instance, local descriptors such as the deviation from the vertical direction are used in order to refine the matching process [GRE 13];

– *laser/image*: in general, it is preferable to pair the same type of data when it is available. However, laser/image data matching can be useful when fusing terrestrial laser data with aerial imaging, or in order to jointly calibrate the image and the laser data of the MMV. In this instance, the intensity of the laser is particularly useful and can be converted into image form using projection and interpolation in the geometry of the image. This gives us two images to match, which can be carried out sparsely, e.g. by detecting and matching points of interest, or dense points, by measuring the similarity between two images (by mutual information criterion for data from various sources);

– *laser/topographic database*: matching remote sensing sources with existing geodatabases is particularly helpful for navigation, change detection and geometrical refinement [LOT 10]. In an urban context, the most straightforward method is to use building façades derived from 3D city models [MON 13], as illustrated in Figure 2.6. Other primitives such as lamp posts or sidewalk edges are also relevant;

– *laser/DTM*: where a Digital Terrain Model (DTM, raster or triangulated) is available, it can be used in order to match the 3D laser cloud using height information. In this instance, matching will be carried out following a planarity criterion. It is then necessary to filter the 3D points from the terrestrial laser sensor in order to keep only those points lying on the ground. These are then matched to the pixel or to the triangle in which it projects itself.

These different kinds of possible matchings are known as "observations" and can be integrated in a process for estimating poses using the non-rigid deformation model outlined in section 2.3.2.

2.3.4. *Pose estimation*

Estimating the poses involves identifying the control poses which define the matched trajectory. Each observation listed above leads to a specific constraint: the two matched elements must coincide spatially. In general, there are more constraints than variables and therefore it is generally not possible to precisely verify coincidental restrictions. Instead, energy-based formulation for residual minimization (distances between the elements being matched) is used, often involving least squares techniques, in relation to the free variables (control poses). It is also possible to add observations which correspond to the initial trajectory: while this is not perfect, it is an invaluable source of information because the acceleration measured is very precise. In order to "smooth" the resulting trajectory residuals measuring the difference between the acceleration given by the inertial measurement unit, the acceleration of the matched trajectory can be inserted to the energetical formulation.

Figure 2.6. *Laser MMT point cloud (blue and violet) matched to a 3D city model of Toulouse (transparent brown) using the method described in [MON 13]. For a color version of this figure, see www.iste.co.uk/baghdadi/5.zip*

If the error in rotation is ignored, the resulting minimization problem is a least squares problem which can be formally solved by inverting a large sparse matrix (only those variables which correspond to the paired points are linked [CAN 13]). For automatic matching in a area confidence around the data, it is possible to refine these pairings by recalculating the pairings after estimation and iterating until convergence is reached. This very popular process in image registration is known as the Iterative Closest Point technique (ICP [GRE 13]).

If the error in rotation is not ignored, the minimization solution has no closed form and can be obtained by gradient descent (conjugated gradient, Levenberg–Marquart, etc.)

2.4. Analyzing urban scenes

Image and point cloud registration is only the preliminary stage for automatic data analysis. It should be noted at this stage that most registration procedures are based on a (rough) initial analysis of data. If the analysis takes into account the temporal aspect of the acquisition, i.e. it only takes into account a neighborhood in space and time, while multiple paths at the same junctions and streets become separated, then the analysis (detecting objects in particular) can be undertaken before image registration without being disturbed. In this instance, the same lamp post or the same building façade acquired from different points of view will be extracted twice and can be used for image registration. After image registration, it is possible to use the merged point cloud, in which case the different acquisitions enable a denser sampling of the scene.

The first stage of the analysis is low level: it often involves local point cloud (LiDARor image-based) analysis in order to extract the relevant descriptors. These descriptors can then be used for 3D scene analysis and reconstruction, these two processes being increasingly carried out jointly [HAE 13].

The following section will focus on methods for 3D point cloud scene analysis noting that there are already a wide range of publications on image-based approaches and these are often not specific to MMT. The majority of existing publications focus on the façades of buildings and make their hypothesis of orthogonal acquisition, without occlusion from objects in the street. Readers interested in state-of-the-art papers should refer to the following articles for up-to-date descriptions of studies carried out in this field [SIM 11, LIU 13, KOZ 15, MAR 15].

2.4.1. *Local descriptors*

We will gloss over the local descriptors of images (gradient, Laplacian, SIFT descriptors, etc.) for which abundant literature already exists, in order to focus more on point clouds which are specific to MMT [WEI 13]. At this stage, a point cloud is defined by a set of coordinates in a geographical reference frame, plus a few attributes (moment of acquisition, reflectance and echo number) per point. The first stage of knowledge extraction is to analyze the neighborhood of each point in order to analyze the local shape of the scene.

Unlike when using images, there is no trivial neighborhood (4 or 8 convexity) for photogrammetric or LiDAR point clouds. For an MMT point cloud, different types of neighborhood are used [DEM 11]:

– fixed spherical neighborhood: the neighbors of a point are points located at a set distance from this point below a certain threshold;

– k closest neighbors: the neighborhood is defined by the closest k points. This is a spherical neighborhood whose radius adapts to the density of the point cloud;

– cylindrical neighborhood: configured according to their radius and their height, they allow us to consider all 3D points in the same vertical axis;

– sensor neighborhood: the acquisition strategy will generally define a 2D topology (Figure 2.7). For the first dimension, this simply involves the following preceding points. For the second dimension, this involves the point acquired with the same angle in the turn before and after for a flat scanner and the point acquired simultaneously in the fiber before and after for a robotic laser. For a multi-echo laser, a third dimension can be added: the return before and after for the same laser pulse. The advantage of this topology is that it is very simple to calculate and that it is independent of any factor which might vary the resolution of the point cloud;

– dynamic neighborhoods: for each 3D point, the neighborhood adapts to the shape and the density of the cloud. Often, it is necessary to carry out a preliminary geometrical analysis of the point cloud (and therefore to use one of the neighborhoods listed above) in order to define them [DEM 11].

Once the neighborhood type is defined, it is generally used in order to perform a Principal Components Analysis (PCA) which gives the directions in which the point cloud locally extends and therefore makes it possible to evaluate its dimensionality: linear (wires, lamp-posts and thin tree trunks in urban environments), surface (ground and facades, see Figure 2.8) and volume (foliage). In a linear context, the PCA also gives the direction of this line. In a surface context, it gives the normal to the surface and thus makes it possible to evaluate its curvature. These three fundamental geometrical elements are the basis for other descriptors which are highly useful for classifying objects in urban environments [WEI 15]. They are used alongside other attributes which are linked to the altitude or the intensity of the 3D point as well as the presence of multiple backscatters in the neighborhood. More details can be found in [MAL 16].

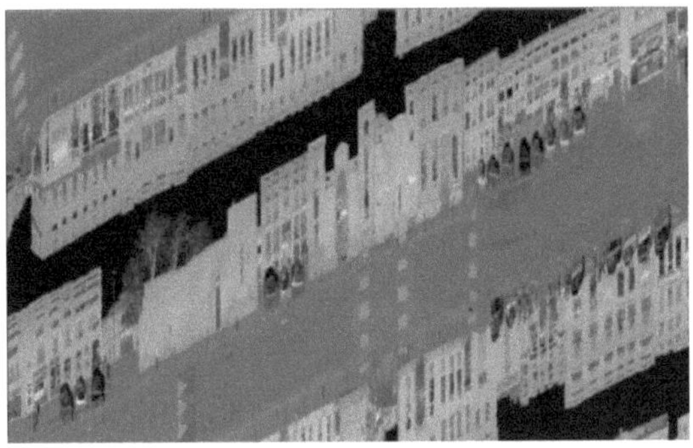

Figure 2.7. *Illustration of sensor topology: the columns are composed of points acquired in succession. A new column is generated each time the laser completes a rotation of 360° (the shift is due to the fact that there is not an integer number of points over 360°). Blue: no return; green: multiple returns. For simple returns, coloring uses the intensity and is orange-tinted for vertical surfaces. For a color version of this figure, see www.iste.co.uk/baghdadi/5.zip*

Figure 2.8. *Planarity descriptor on the façade of a building. Coloring: non-planar (black) to planar (white). Note the highly irregular sampling of a terrestrial scene with a mobile LiDARsensor as well as its ability to render fine objects like street lamps. For a color version of this figure, see www.iste.co.uk/baghdadi/5.zip*

2.4.2. *Segmentation and classification of 3D point clouds*

The segmentation and classification of terrestrial data is a vast and active field in computer vision for the purposes of reconstructing surfaces, detecting changes or visual landmarks (for example, traffic signs), navigation and even matching images from very large volume of data. We will not go into too much detail on the subject here and will instead direct the readers to the most prominent articles from the past few years [VAN 13, GEI 14, KUN 14, TAN 15, WAN 15]. We will also assume that non-relevant elements and different sources of noise were filtered from the 3D point cloud [DES 10].

The classification of 3D LiDAR MMT data (to attribute a class to each point) is a difficult problem because urban scenes are made up of various individual objects which are difficult to characterize at point level. One alternative is to subdivide the point cloud and to classify each segment instead of each individual point: this is referred to as object-oriented analysis. The main existing techniques for classifying point clouds are outlined in [MAL 16] and a comparison of the different methods for classification in urban environments is given in [WEI 15]. From this point forward, the current problems in the fields of classification and segmentation are grouped under the term *semantic segmentation*.

It is important to note that segmentation produced from pointwise classification will not necessarily be spatially consistent given that the simple connected component techniques do not make it possible to easily separate close objects labeled identically: this is particularly true for trees in the vegetal class. A method of segmentation *a posteriori* to classification is thus necessary [RUT 11, SER 14a].

Segmentation carried out prior to classification can have two objectives:

– to enable an object-oriented analysis and to facilitate the interpretation of complex urban scenes [XIO 11];

– to filter the background in advance of more specific methods for object detection (the ground and facades can be easily distinguished geometrically using the planarity descriptors outlined in section 2.4.1 [DEM 12]). Filtering them makes it possible to simplify the analysis and to exclude a high proportion of 3D points for more accurate classification.

The very high spatial precision of LiDAR data, the low noise level as well as the very high spatial resolution of the data makes it possible to differentiate between an increasingly large number of different classes (Figure 2.9): sidewalk edges, road signs, traffic signals, lamp posts, benches, letterboxes, different types of vehicle, pedestrians and trees [MON 12] as can be seen in the set of several annotated data made available to the research community in order to compare different methods of classification and segmentation [MUN 09, SER 14b, VAL 15b]. The latter is also essential for the learning stage of supervised classification methods.

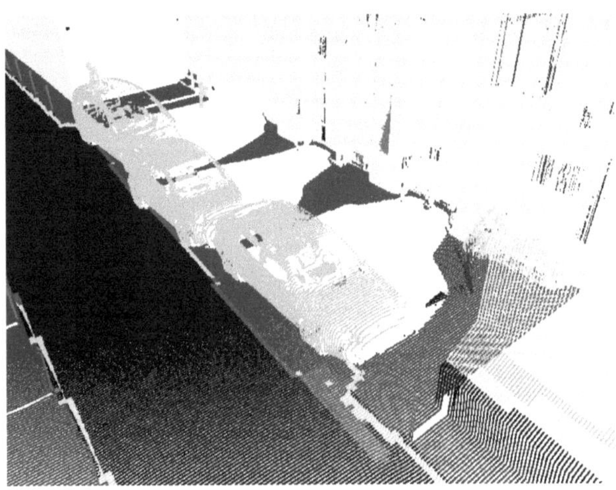

Figure 2.9. *Classification of an MMT point cloud: black=road, gray=pavement, beige=building façade, green=cars, cyan=two-wheeled vehicles and pink=posts. For a color version of this figure, see www.iste.co.uk/baghdadi/5.zip*

Finally, LiDAR data and images can be used together for classification since they each offer specific advantages: geometric data versus radiometric data and topology. One current trend is to develop mechanisms for joint analysis: e.g. the image will be used for segmentation while the supervised classification will be based on LiDAR descriptors [BAB 13]. On the other hand, in [LI 11], the geometric segmentation in the LiDAR data is disseminated throughout the available images, and completed using the different points of view available in order to manage the occlusions and to transfer it into 3D for refining models of building façades. One final approach involves classifying the two data sources simultaneously using labels and different levels of segmentation currently in use in LiDAR and image data [MUN 12].

2.4.3. *Object recognition*

Methods for recognizing (or detecting) objects are highly dependent on the shape of the object being detected. Objects with a well defined shape (road signs, road markings and registration plates) can be detected by pattern matching or by identifying simple geometric elements of the shape and using their color, even if this is subject to highly variable light conditions and requires excellent calibration of image and LiDAR sensors. Furthermore, the conditions for acquisition (angle and sampling) vary enormously within an urban scene: objects of the same class often have different geometric appearances, which, in practice, prevents the direct use of model-based methods. More advanced methods based on local descriptors have been able to show their capacity to handle these variations: this is particularly true for implicit shape models [VEL 12].

Thus, the methods used increasingly rely on a combination of segmentation and supervised classification using examples of objects of that class, including objects of known geometry such as posts or cars [LEH 10, LI 14]. The detection of objects therefore becomes a subproblem of classification/segmentation [BRE 13, SER 14a].

2.4.4. *Reconstruction*

Reconstruction is a term which encompasses various issues. Surface reconstruction aims to reconstruct a continuous surface of the objects of a given scene. Generalization aims for a more structured and more compact representation of these surfaces. Object reconstruction aims to instantiate parametric models of objects in the data. In the context of MMT, reconstruction is a particularly difficult problem because of the various occlusions, the incomplete nature of the data (only one side of a building can be seen) and the highly anisotropic sampling.

2.4.4.1. *Surface reconstruction*

We will not go into too much detail here on the methods for producing a point cloud from images using dense correlation. We will therefore assume that we have a point cloud as our input, whether it comes from photo or lasergrammety. The challenge with surface reconstruction involves reconnecting these points in order to make a continuous surface [DES 10]. It is possible to distinguish between several types of approach for addressing this issue:

– 2.5 D approaches: these are only applicable if the surface for reconstruction can be seen as a height map above a plane. Building façades [DEM 13] or the ground (after it has been determined) can be processed in this way, in which case it will be possible to use one of the DSM/DTM reconstruction methods discussed in [DEL 16, NIC 16];

– voxel-based approaches: the scene is subdivided into a grid of cubes (voxels) and the aim is to determine if the cube is inside or outside, with the final surface defined as the interface between the inside and the outside. The most popular approach, Poisson surface reconstruction, can be formulated as an optimization problem, based on a balance between its connection to the data and the surface's consistency [BER 14];

– tetrahedron-based approaches: a Delaunay tetrahedralization of a point cloud is calculated. If we are looking for a watertight surface, the problem is similar to the voxel-based approach in that it involves labeling tetrahedrons as either being inside or outside (see Figure 2.10). In this instance, we add restrictions on shape in order to exclude very flat tetrahedrons. Without the restriction of whether or not a surface is watertight, geometric criteria are used on triangles in order to decide whether or not to keep them. It should be noted that knowing the optical center (laser or image) can make it possible to fix angling ambiguities or to make the surface more consistent with the data by preventing it from intersecting a beam.

2.4.4.2. *Generalization*

Generalization involves data simplification in order to better enhance the underlying structure [VER 15]. It is then possible to distinguish between:

– mesh decimation: the edges of the mesh are contracted provided the induced error remains sufficiently low. This results in a simpler mesh;

– hybrid reconstruction: the sections of the mesh which correspond to simple geometric primitives (planes, cylinders, spheres and paraboloids) are replaced by these primitives. This can be applied to various artificial objects in urban scenes. For instance, paraboloids are useful for the surface reconstruction of a road where the shape is often that of a parabola;

– polyhedral reconstruction: here the aim is for an accurate polyhedral approximation of the surface using a minimum of polyhedrons. Widely used in aerial imaging for roof surfaces, it can also be applied to some urban objects such as building façades.

Figure 2.10. *Surface reconstruction using a mobile mapping point cloud based on a Delaunay tetrahedralization. For a color version of this figure, see www.iste.co.uk/baghdadi/5.zip*

2.4.4.3. *Object reconstruction*

Object reconstruction is highly dependent on the object in question and its specificities. These approaches require preliminary analysis of the scene as well as the preliminary classification and segmentation of the objects:

– building façade reconstruction: this is the most commonly studied object. The strong structure of building façades can be represented by a grammar for which we will aim to find the instantiation which best corresponds to the data [LI 11]. A simpler approach involves finding certain structures such as windows which can be added to the main plane;

– catalogs: catalogs of urban furniture can be used in order to find instances of these objects in 3D point clouds, which can then be resolved using registration techniques involving local descriptors. This detection enables the simultaneous reconstruction of the object by simple instantiation. This type of technique is also used for detecting and reconstructing road signs [ARL 09, SOH 13, VAL 14] or road marks [SOH 06, SOH 07];

– cars: parametric models of cars based on existing models making it possible to represent the majority of vehicles have recently been proposed [ZIA 14]. Reconstruction then involves the optimization of the parameters of the model;

– trees: trees are highly complicated objects to reconstruct. There are grammar-based methods which make it possible to generate 3D models of different species of trees, but their complexity makes them unsuitable for reconstruction for the time being. Depending on how it is to be used, the aim will be to estimate some of the tree parameters (height, diameter of the crown and the diameter at breast height) or to represent it using a billboard cloud.

2.4.5. *Texturing*

Once the reconstruction has made it possible to move from a sparse representation (point cloud) to a continuous representation (surface) of the scene, texturation can start, i.e. at least one image can be mapped onto it. This is mainly useful for visualization but it can also be relevant for automatically interpreting a scene [DES 10]. Compared to the texturing of city models using aerial images, terrestrial texturing is a lot more difficult due to the far greater number of occlusions. It is also necessary to mosaic these images, because the whole building façade is rarely seen in a single image (unlike with aerial imaging where this is always the case). Texturing relies on several techniques:

– visibility computation: for each geometry (triangle, plane, etc.) it is necessary to determine which images this geometry will "see" in good conditions (resolution, viewing angle, occlusions, etc.) [VAL 11];

– rectification: for each geometry/image pair, the image must be projected onto this geometry;

– mosaicking: it is necessary to optimize the connecting lines between images. The aim is to maximize the quality of the texture for each geometry while minimizing the number and visibility of jumps (changing between images in order to texture neighboring geometries);

– equalization: given that the dynamics of images can be different, the aim will be to optimize the dynamic in order to make the connections less visible;

– blending: images can be blended around discontinuities in order to mask them further.

Together, these steps make it possible to carry out a photorealistic reconstruction of the scene, which is even better if the current point of view is close to the point of view of the images used for texturing (Figure 2.11). There is thus a tendency to hide faults in texturing, and it is important not to use a textured surface in order to evaluate the quality of the surface reconstruction.

Figure 2.11. *Textured 3D model using terrestrial images plus images used for texturing. For a color version of this figure, see www.iste.co.uk/baghdadi/5.zip*

2.4.6. *3D change detection*

With multiple surveys in the same street, important elements of the scene can change, without taking into account objects that were moving at the point of acquisition. Analyzing urban scenes must take into account this dynamic aspect of scenes on different temporal levels:

– 0.1 s: enables the detection of mobile objects (cars and pedestrians) during acquisition, which makes it possible to extract images acquired simultaneously [VAL 15 b];

– minutes: enables the separation of fixed objects from mobile objects;

– hours: parking investigations;

– months: seasonal changes (for trees);

– years: permanent changes to building structures.

The problem is made more complicated by the fact that during different surveys, the sensor will operate at speeds and trajectories which are potentially different. A direct comparison between 3D point clouds makes it difficult to distinguish between changes in the scene, occlusion and changes in the acquisition geometry (variations in the sampling of surfaces). To solve such issues, two approaches have been proposed so far:

– semantic approach: the objects in the scene are directly classified as either being static or non-static objects [AIJ 13] using methods similar to those outlined in section 2.4.2 (an object is mobile because it is of a mobile type). This method can only be used for mobile objects and not for structural changes to the scene (constructions or destructions);

– geometric approach: there is a change because there has been a change to the geometry of the scene. This approach makes it possible to take into account all temporalities and will be described in the following section.

The information provided by the acquisition regarding the occupancy of space ranging from empty to occupied or no information is modeled (empty along the ray up to the target, occupied just after and unknown everywhere else) using the Dempster–Shafer theory [XIA 015] which defines the way in which these data are combined and conflicts involved in its combination. Thus, the information from each ray is combined for each survey, before combining the two surveys with each other, which makes it possible not only to define those parts of the scene which have changed (areas of conflict) but also those that have remained the same (same objects scanned) and those where there is no counterpart (object only visible in one of the surveys), as is shown in Figure 2.12.

Furthermore, the Dempster–Shafer theory makes it possible to combine an arbitrary number of acquisitions for the same area, and therefore to separate the unchanged part from the changing part for each acquisition using all of the information from the other acquisitions.

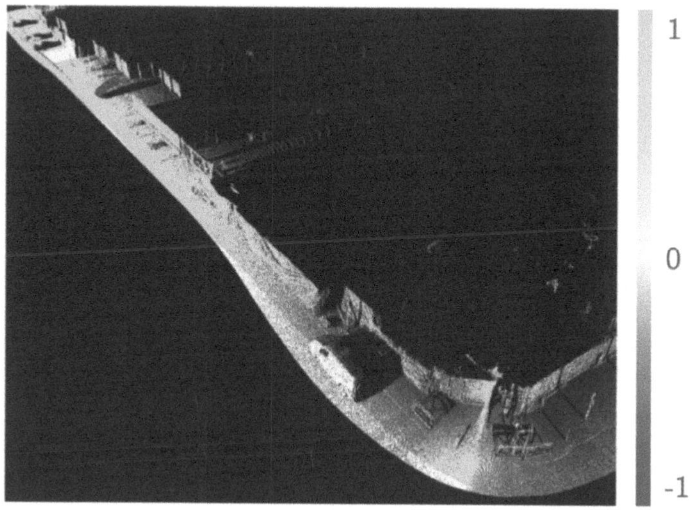

Figure 2.12. *Detecting changes in an urban area: green=consistent, yellow=no counterpart, red=change. Note here that the algorithm detects cars and pedestrians. For a color version of this figure, see www.iste.co.uk/baghdadi/5.zip*

2.5. Prospects

The wide range of MMT techniques listed above reflects the state of the art of research. While the acquisition of data is now quite well understood, the majority of techniques described in sections 2.3 and 2.4 are not yet fully operational today in companies where acquisitions for commercial purposes or for autonomous navigation are being carried out. These techniques have only recently been developed and they remain prospective. The most advanced studies for the time being tend to combine the techniques outlined above in favor of a more high-level application for navigating vehicles or people, such as the urban accessibility diagnosis for people with reduced mobility [SER 13]. From a methodological point of view, there are several trends for how MMT might develop in the future, and we will try to list these below.

2.5.1. *Uncertainty Management*

A key element for users of cartographic data is knowing that they can trust the data provided by a third party. Although the producer can give statistical elements relating to confidence using, for example, topometric surveys or GPS measurements, it is far more satisfactory to model and spread all of these uncertainties linked to the production of data in order to be able to give the user a reliable indication of the level of uncertainty for every area considered. The main source of geometric uncertainty in MMT is linked to the geopositioning of the vehicle by the inertial measurement unit which is several orders of magnitude higher than other uncertainties. The majority of recent inertial measurement units give an indication of this uncertainty, but given that their algorithms are industrial secrets, it is impossible to know how much trust we can place in these uncertainties. A good method of image registration should estimate all sources of uncertainties for input, uncertainties relating to the observations, and combine all of these in order to provide uncertainties for the image registration produced. This will also make it possible to raise alerts for an operator to come to refine the image registration at the points with the highest levels of uncertainty. This framework also raises the distinction between relative and absolute image registration since all of the data sources will come with their own uncertainties and will be processed in a similar manner. Thus, if all of the algorithms used in a processing chain take into account uncertainties, high level products will have their own uncertainties as is shown in [SOH 14] for road signs reconstructed in 3D. As a result, products created using these data (visual landmarks) would be able to take them into account, especially when used in localization and navigation (for example, road signs in [QU 15]).

2.5.2. *Image/laser fusion*

In this chapter, we have often looked at the different stages of the MMT processing chain independently depending on whether the data are image or laser data. However, MMT operators are increasingly using systems which incorporate both image and laser. The way in which these two types of data are able to work together means that we can hope their joint use will lead to a significant improvement in the quality of information produced (section 2.4.2).

2.5.3. *Semantization as segmentation/classification coupling*

A major part of scene analysis is based on the coupling of classification and segmentation, either in that order or vice versa. Issues relating to segmentation and classification must be dealt with jointly and integrated so that the geometric and semantic aspects will be better linked and processed in their entirety. Several recent studies have considered this to be a difficult but exciting challenge (section 2.4.2).

2.5.4. *Surface reconstruction and semantization coupling*

In much the same way, we have seen that reconstruction can be based on semantic information: it is possible to adapt a reconstruction algorithm to the reconstructed object. However, a textured surface is a more structured object than an image or a point cloud, and therefore classifying this type of data enables a more detailed analysis, which brings us back to the points outlined in sections 2.5.2 and 2.5.3.

2.5.5. *Fusion of aerial and terrestrial data*

Finally, MMT data are restricted to the part of the scene which is visible from the street, which means omitting a variety of elements such as the height of building façades, roofs, interior yards, etc. Optical and LiDARaerial data provide a coverage rate that is less precise geographically, but much more homogeneous and extensive spatially. These two points of view must be used together – for example, in order to reconstruct the lower parts of building façades using MMT and the upper parts using aerial data for modeling urban environments in their entirety. Such methods also require accurate data registration between terrestrial and aerial points of view.

2.6. Key points

– Mobile mapping produces data that are geolocalized and that have a high spatial resolution closer to the objects in question.

– The major challenges for data processing are related to volume and complexity.

– The difficult conditions in city centers lead to a drift in data positioning that is too high for direct georeferencing (sometimes greater than 1 m). Data registration of mobile mapping data is therefore a major challenge, and requires a more precise outside reference for comparison of several surveys.

– Analyzing data is often based on 3D local descriptors which are used to cluster and classify data in order to discriminate predefined types of objects (road, sidewalk, building façade, cars, pedestrians, lamp posts, trees, etc.).

– Surface reconstruction makes it possible to produce a continuous representation of the scene which can be textured to increase the level of realism.

– The dynamism of scenes prompts a consideration of the temporal dimension, and therefore mobile objects or changes to scenes between acquisitions.

2.7. Bibliography

[AIJ 13] AIJAZI A.K., CHECCHIN P., TRASSOUDAINE L., "Automatic removal of imperfections and change detection for accurate 3D urban cartography by classification and incremental updating", *Remote Sensing*, vol. *5, no.* 8, pp. 3701–3728, 2013.

[ANG 10] ANGUELOV D., DULONG C., FILIP D. *et al.*, "Google street view: capturing the world at street level", *Computer,* vol. 43, no. 6, pp. 32–38, 2010.

[ARL 09] ARLICOT A., SOHEILIAN B., PAPARODITIS N., "Circular road sign extraction from street level images using colour, shape and texture database maps", *International Archives of Photogrammetry, Remote Sensing and Spatial Information Sciences*, vol. 38, no 3/W4, pp. 205–210, 2009.

[BAB 13] BABAHAJIANI P., FAN L., GABBOUJ M., "Semantic parsing of street scene images using 3D LiDARpoint cloud", *ICCV Workshops*, Sydney, Australia, 2013.

[BER 14] BERGER M., TAGLIASACCHI A., SEVERSKY L. *et al.*, "State of the art in surface reconstruction from point clouds", *Eurographics*, vol. 1,no. 1, pp. 161–185, 2014.

[BRE 13] BREMER M., WICHMANN V., RUTZINGER M., "Eigenvalue and graph-based object extraction from mobile laser scanning point clouds", *ISPRS Annals of the Photogrammetry, Remote Sensing and Spatial Information Sciences*, vol. II-5/W2, pp. 55–60, 2013.

[CAN 10] CANNELLE B., PAPARODITIS N., TOURNAIRE O., "Panorama-based camera calibration", *International Archives of Photogrammetry, Remote Sensing and Spatial Information Sciences*, vol. 38, no. 3A, pp. 73–78, 2010.

[CAN 12] CANNELLE B., PAPARODITIS N., PIERROT-DESEILLIGNY M. *et al.*, "Off-line vs. on-line calibration of a panoramic-based mobile mapping system", *ISPRS Annals of the Photogrammetry, Remote Sensing and Spatial Information Sciences*, vol. I–3, pp. 31–36, 2012.

[CAN 13] CANNELLE B., Estimation de pose de grands blocs d'images panoramiques issues de systèmes de numérisation mobile, PhD thesis, University Paris-Est, 2013.

[DEL 16] DELVIT J-M., MICHEL J., "Digital terrain models from optical images", in BAGHDADI N., ZRIBI M., (eds), *Optical Remote Sensing of Land Surfaces*, ISTE Press Ltd, London and Elsevier Ltd, Oxford, 2016.

[DEM 11] DEMANTKÉ J., MALLET C., DAVID N. *et al.*, "Dimensionality based scale selection in 3D LiDARpoint cloud", *International Archives of Photogrammetry, Remote Sensing and Spatial Information Sciences*, vol. 38, no. 5/W12, pp. 97–102, 2011.

[DEM 12] DEMANTKÉ J., VALLET B., PAPARODITIS N., "Streamed vertical rectangle detection in terrestrial laser scans for facade database production", *ISPRS Annals of the Photogrammetry, Remote Sensing and Spatial Information Sciences*, vol. I–3, pp. 99–104, 2012.

[DEM 13] DEMANTKÉ J., VALLET B., PAPARODITIS N., "Facade reconstruction with generalized 2.5D grids", *ISPRS Annals of the Photogrammetry, Remote Sensing and Spatial Information Sciences*, vol. II-5/W2, pp. 67–72, 2013.

[DES 10] DESCHAUD J.E., Traitements de nuages de points denses et modélisation 3D d'environnements par système mobile LiDAR/Caméra, PhD thesis, École Nationale Supérieure des Mines de Paris, 2010.

[ELS 96] EL-SHEIMY N., "A mobile multi-sensor system for GIS applications in urban centers", *International Archives of Photogrammetry, Remote Sensing and Spatial Information Sciences*, vol. 31, no. B2, pp. 95–100, 1996.

[GAN 08] GANDOLFI S., BARBARELLA M., RONCI E. *et al.*, "Close photogrammetry and laser scanning using a mobile mapping system for the high detailed survey of a high density area", *International Archives of Photogrammetry, Remote Sensing and Spatial Information Sciences* vol. 35(Part B5), pp. 909–914, 2008.

[GEI 12] GEIGER A., LENZ P., URTASUN R., "Are we ready for autonomous driving? The KITTI vision benchmark suite", *IEEE Conference on Computer Vision and Pattern Recognition (CVPR)*, Providence, USA, pp. 3354–3361, 2012.

[GEI 14] GEIGER A., LAUER M., WOJEK C. *et al.*, "3D traffic scene understanding from movable platforms", *IEEE Transactions on Pattern Analysis and Machine Intelligence*, vol. 36, no. 5, pp. 1012–1025, 2014.

[GON 07] GONTRAN H., SKALOUD J., GUILLERON P.Y., "A mobile mapping system for road data capture via a single camera", in TAO C., LI D. (eds), *Advances in Mobile Mapping Technology*, Taylor & Francis Group, London, pp. 43–50, 2007.

[GOU 06] GOULETTE F., NASHASHIBI F., ABUHADROUS I. *et al.*, "An integrated on-board laser range sensing system for on-the-way city and road modelling", *International Archives of Photogrammetry, Remote Sensing and Spatial Information Sciences*, vol. 36, no. 1, pp. 3–5, 2006.

[GRE 96] GREJNER-BRZEZINSKA D.A., "Positioning accuracy of the GPSVan", *Proceedings of the 52nd Annual Meeting of The Institute of Navigation*, Cambridge, USA, pp. 657–665, 1996.

[GRE 13] GRESSIN A., MALLET C., DEMANTKÉ J. *et al.*, "Towards 3D LiDARpoint cloud registration improvement using optimal neighborhood knowledge", *ISPRS Journal of Photogrammetry and Remote Sensing*, vol. 79, pp. 240–251, 2013.

[HAE 13] HAENE C., ZACH C., COHEN A. *et al.*, "Joint 3D scene reconstruction and class segmentation", *IEEE Conference on Computer Vision and Pattern Recognition (CVPR)*, Portland, Oregon, pp. 97–104, 2013.

[HER 15] HERVIEU A., SOHEILIAN B., BRÉDIF M., "Road marking extraction using a Model&Data-driven RJ-MCMC", *ISPRS Annals of the Photogrammetry, Remote Sensing and Spatial Information Sciences*, vol. II-3/W4, pp. 47–54, 2015.

[HOE 09] HOEGNER L., STILLA U., "Thermal leakage detection on building facades using infrared textures generated by mobile mapping", *Joint Urban Remote Sensing Event*, Shanghai, China, pp. 1–6, 2009.

[KOZ 15] KOZIŃSKI M., GADDE R., ZAGORUYKO S. *et al.*, "A MRF shape prior for facade parsing with occlusions", *IEEE Conference on Computer Vision and Pattern Recognition (CVPR)*, Boston, USA, pp. 2820–2828, 2015.

[KUK 07] KUKKO A., ANDREI C.-O., SALMINEN V.-M. *et al.*, "Road environment mapping system of the Finnish Geodetic Institute–FGI Roamer", *International Archives of Photogrammetry, Remote Sensing and Spatial Information Sciences*, vol. 39, pp. 241–247, 2007.

[KUN 14] KUNDU A., LI Y., DAELLERT F. *et al.*, "Joint semantic segmentation and 3D reconstruction from monocular video", *European Conference on Computer Vision (ECCV)*, Zurich, Switzerland, pp. 703–718, 2014.

[KUT 10] KUTTERER H., "Mobile mapping", in VOSSELMAN G., MAAS H.-G. (eds), *Airborne and Terrestrial Laser Scanning*, CRC Press, pp. 293–311, 2010.

[LEH 10] LEHTOMÄKI M., JAAKKOLA A., HYYPPÄ J. *et al.*, "Detection of vertical pole-like objects in a road environment using vehicle-based laser scanning data", *Remote Sensing*, vol. 2, pp. 641–664, 2010.

[LES 13] LE SCOUARNEC R., TOUZÉ T., LACAMBRE J.B. *et al.*, "A positioning free calibration method for mobile laser scanning applications", *ISPRS Annals of the Photogrammetry, Remote Sensing and Spatial Information Sciences*, vol. II-5/W2, pp. 157–162, 2013.

[LI 11] LI Y., ZHENG Q., SHARF A. *et al.*, "2D-3D fusion for layer decomposition of urban facades", *International Conference on Computer Vision (ICCV)*, Barcelona, Spain, pp. 882–889, 2011.

[LI 14] LI G., Automatic detection of temporary objects in mobile LiDAR point clouds, Masters Thesis, University of Twente, Netherland, 2014.

[LIU 13] LIU J., PSARAKIS E., STAMOS I., "Automatic Kronecker product model based detection of repeated patterns in 2D urban images", *International Conference on Computer Vision (ICCV),* Sydney, Australia, pp. 401–408, 2013.

[LOT 10] LOTHE P., BOURGEOIS S., ROYER E. *et al.*, "Real-time vehicle global localisation with a single camera in dense urban areas: exploitation of coarse 3D city models", *IEEE Conference on Computer Vision and Pattern Recognition (CVPR)*, Anchorage, USA, pp. 863–870, 2010.

[MAL 16] MALLET C., CHEHATA N., J-S., "Airborne LiDAR data processing", in BAGHDADI N., ZRIBI M., (eds) *Optical Remote Sensing of Land Surfaces*, ISTE Press Ltd, London and Elsevier Ltd, Oxford, 2016.

[MAR 15] MARTINOVIC A., KNOPP J., RIEMENSCHNEIDER H. *et al.*, "3D all the way: semantic segmentation of urban scenes from start to end in 3D", *IEEE Conference on Computer Vision and Pattern Recognition (CVPR)*, Boston, USA, pp. 4456–4465, 2015.

[MON 12] MONNIER F., VALLET B., SOHEILIAN B., "Trees detection from LASER point clouds acquired in dense urban areas by a mobile mapping system", *ISPRS Annals of the Photogrammetry, Remote Sensing and Spatial Information Sciences*, vol. I–3, pp. 245–250, 2012.

[MON 13] MONNIER F., VALLET B., PAPARODITIS N. *et al.*, "Registration of terrestrial mobile laser data on 2D and 3D geographic database by use of a non-rigid ICP approach", *ISPRS Annals of the Photogrammetry, Remote Sensing and Spatial Information Sciences*, vol. II-5/W2, pp. 193–198, 2013.

[MUN 09] MUNOZ D., BAGNELL J.A., VANDAPEL N. *et al.*, "Contextual classification with functional max-margin Markov networks", *IEEE Conference on Computer Vision and Pattern Recognition (CVPR)*, Miami, USA, pp. 975–982, 2009.

[MUN 12] MUNOZ D., BAGNELL J.A., HEBERT M., "Co-inference for multi-modal scene analysis", *European Conference on Computer Vision*, Florence, Italy, pp. 668–824, 2012.

[NIC 16] NICOLAS J-M., TUPI F., "The principles of DTM reconstruction from SAR images", in BAGHDADI N., ZRIBI M., (eds) *Optical Remote Sensing of Land Surfaces*, ISTE Press Ltd, London and Elsevier Ltd, Oxford, 2016.

[PAP 12] PAPARODITIS N., PAPELARD J.P., CANNELLE B. *et al.*, "Stereopolis II: a multi-purpose and multi-sensor 3D mobile mapping system for street visualisation and 3D metrology", *Revue Française de Photogrammétrie et de Télédétection*, vol. 200, pp. 69–79, 2012.

[PET 09] PETRIE G., TOTH C.K., "Terrestrial laser scanners", in SHAN J., TOTH C.K. (eds), *Topographic Laser Ranging and Scanning*, Taylor & Francis, pp. 87–128, 2009.

[QU 15] QU X., SOHEILIAN B., PAPARODITIS N., "Vehicle localization using mono-camera and geo-referenced traffic signs", *IEEE Intelligent Vehicles Symposium*, Seoul, South Korea, pp. 605–610, 2015.

[RUT 11] RUTZINGER M., PRATIHAST A.K., OUDE ELBERINK S.J. *et al.*, "Tree modelling from mobile laser scanning data-sets", *The Photogrammetric Record*, vol. 26, pp. 361–372, 2011.

[SER 13] SERNA A., MARCOTEGUI B., "Urban accessibility diagnosis from mobile laser scanning data", *ISPRS Journal of Photogrammetry and Remote Sensing*, vol. 84, pp. 23–32, 2013.

[SER 14a] SERNA A., MARCOTEGUI B., "Detection, segmentation and classification of 3D urban objects using mathematical morphology and supervised learning", *ISPRS Journal of Photogrammetry and Remote Sensing*, vol. 93, pp. 243–255, 2014.

[SER 14b] SERNA A., MARCOTEGUI B., GOULETTE F. *et al.*, "Paris-rue-Madame database: a 3D mobile laser scanner dataset for benchmarking urban detection, segmentation and classification methods", *4th International Conference on Pattern Recognition, Applications and Methods*, Angers, France, pp. 819–824, 2014.

[SIM 11] SIMON L., TEBOUL O., KOUTSOURAKIS P. *et al.*, "Random exploration of the procedural space for single-view 3D modeling of buildings", *International Journal of Computer Vision,* vol. 93, no. 2, pp. 253–271, 2011.

[SOH 06] SOHEILIAN B., PAPARODITIS N., BOLDO D. *et al.,* "3D zebra-crossing reconstruction from stereo rig images of a ground-based mobile mapping system", *International Archives of Photogrammetry, Remote Sensing and Spatial Information Sciences*, vol. 36, no. 5, pp. 291–296, 2006.

[SOH 07] SOHEILIAN B., PAPARODITIS N., BOLDO D. *et al.*, "Automatic 3D extraction of rectangular roadmarks with centimeter accuracy from stereo-pairs of a ground-based mobile mapping system", *International Archives of the Photogrammetry, Remote Sensing and Spatial Information Sciences*, vol. 36, no. 5/C55, pp. 22–27, 2007.

[SOH 13] SOHEILIAN B., PAPARODITIS N., VALLET B., "Detection and 3D reconstruction of traffic signs from multiple view color images", *ISPRS Journal of Photogrammetry and Remote Sensing*, vol. 77, pp. 1–20, 2013.

[SOH 14] SOHEILIAN B., BRÉDIF M., "Multi-view 3D circular target reconstruction with uncertainty analysis", *ISPRS Annals of the Photogrammetry, Remote Sensing and Spatial Information Sciences*, vol. II–3, pp. 143–148, 2014.

[STU 04] STURM P., RAMALINGAM S., "A generic concept for camera calibration", *European Conference on Computer Vision (ECCV)*, Prague, Czech Republic, vol. 3022, pp. 1–13, 2004.

[TAK 13] TAKAI S., DATE H., KANAI S. *et al.*, "Accurate registration of MMS point clouds of urban areas using trajectory", *ISPRS Annals of the Photogrammetry, Remote Sensing and Spatial Information Sciences*, vol. II(5/W2), pp. 277–282, 2013.

[TAN 15] TANEJA A., BALLAN L., POLLEFEYS M., "Geometric change detection in urban environments using images", *IEEE Transactions on Pattern Analysis and Machine Intelligence*, vol. 37, no. 11, pp. 2193–2206, 2015.

[TOU 06] TOURNAIRE O., SOHEILIAN B., PAPARODITIS N., "Towards a sub-decimeter georeferencing of ground-based mobile mapping systems in urban areas: matching ground-based and aerial-based imagery using roadmarks", *International Archives of Photogrammetry, Remote Sensing and Spatial Information Sciences*, vol. 185, pp. 84–89, 2006.

[VAL 11] VALLET B., HOUZAY E., "Fast and accurate visibility computation in urban scenes", *International Archives of Photogrammetry, Remote Sensing and Spatial Information Sciences*, vol. 38, no. 3/W22, pp. 77–82, 2011.

[VAL 14] VALLET B., SOHEILIAN B., BRÉDIF M., "Combinatorial clustering and its application to 3D polygonal traffic sign reconstruction from multiple images", *ISPRS Annals of the Photogrammetry, Remote Sensing and Spatial Information Sciences*, vol. II–3, pp. 165–172, 2014.

[VAL 15a] VALLET B., BRÉDIF M., SERNA A. *et al.*, "TerraMobilita/iQmulus urban point cloud analysis benchmark", *Computers & Graphics,* vol. 49, pp. 126–133, 2015.

[VAL 15b] VALLET B., XIAO W., BRÉDIF M., "Extracting mobile objects in images using a Velodyne LiDARpoint cloud", *ISPRS Annals of the Photogrammetry, Remote Sensing and Spatial Information Sciences*, vol. II-3/W4, pp. 247–253, 2015.

[VAN 13] VAN GOOL L., MARTINOVIC A., MATHIAS M., "Towards semantic city models", *Proceedings of the 54th Photogrammetric Week*, Stuttgart, Germany, pp. 11–15, 2013.

[VEL 12] VELIZHEV A., SHAPOVALOV R., SCHINDLER K., "Implicit shape models for object detection in 3D point clouds", *ISPRS Annals of the Photogrammetry, Remote Sensing and Spatial Information Sciences*, vol. I–3, pp. 179–184, 2012.

[VER 15] VERDIÉ Y., LAFARGE F., ALLIEZ P., "LOD generation for urban scenes", *ACM Transactions on Graphics*, vol. 34, no. 3, pp. 30:1–30:14, 2015.

[WAN 15] WANG S., FIDLER S., URTASUN R., "Holistic 3D scene understanding from a single geo-tagged image", *Conference on Computer Vision and Pattern Recognition (CVPR)*, Boston, USA, pp. 3964–3972, 2015.

[WEI 13] WEINMANN M., "Visual features - from early concepts to modern computer vision", in FARINELLA G.M., BATTIATO S., CIPOLLA R. (eds), *Advanced Topics in Computer Vision. Advances in Computer Vision and Pattern Recognition*, Springer, London, pp. 1–34, 2013.

[WEI 15] WEINMANN M., JUTZI B., HINZ S. *et al.*, "Semantic point cloud interpretation based on optimal neighborhoods, relevant features and efficient classifiers", *ISPRS Journal of Photogrammetry and Remote Sensing*, vol. 105, pp. 286–304, 2015.

[XIA 15] XIAO W., VALLET B., BRÉDIF M. *et al.*, "Street environment change detection from mobile laser scanning point clouds", *ISPRS Journal of Photogrammetry and Remote Sensing*, vol. 107, pp. 38–49, 2015.

[XIO 11] XIONG X., MUNOZ D., BAGNELL J.A. *et al.*, "3-D scene analysis via sequenced predictions over points and regions", *IEEE International Conference on Robotics and Automation*, Shanghai, China, pp. 2609–2616, 2011.

[ZIA 14] ZIA M.Z., STARK M., SCHINDLER K., "Are cars just 3D boxes? - Jointly estimating the 3D shape of multiple objects", *IEEE Conference on Computer Vision and Pattern Recognition (CVPR)*, Columbus, Ohio, USA, pp. 3678–3685, 2014.

3

Satellite Imagery: a Tool for Territorial Development

3.1. Introduction

Spatial information linked to the observation of the Earth is increasingly present in natural resource management. Since the American Landsat program launched in 1972, many nations have set out to observe the Earth. More than 100 operational satellites are currently in orbit around the Earth; they allow access to geographical information from anywhere in the world, with increasing regularity. These satellite systems allow the monitoring of continental surfaces from local (submetric resolution sensors) to global (hectometric and decametric resolution) scales, with frequencies of observation going from daily to multiannual, and with different wavelengths going from the visible spectrum to microwaves. The growing expansion and complexity of the systems allow the development of increasingly specialized applications. The arrival of very high spatial resolution (VHSR), in the 2000s, expanded the scope of mapping applications that were previously reserved for aerial photography. In the near future, the Sentinel (S1 and S2) family of satellites from the European Copernicus program, providing free access to weekly or decadal decametric data, will allow the development of new services, in particular concerning real-time monitoring of the evolution of the landscape and use of resources.

Chapter written by Jean Philippe TONNEAU and Pierre MAUREL.

Today, images are used for various applications, such as monitoring the evolution of cultivated land, monitoring clear cuts and forested areas, checking farmers' statements for the CAP (European policy concerning agricultural subsidies), early warning systems for food security, evidence for insurance claims, land registration for rural, taxation or legal purposes, or services for precision agriculture.

The applications supporting spatial planning and regional development policies are not as significant. In this chapter, we define the potential contribution of geomatics and, particularly, remote sensing data to regional development policies. To define this contribution, we will determine the information needs for the development, implementation and evaluation of spatial planning and regional development policies. We hypothesize that this information depends on the nature of the policies (objectives, philosophy of action, etc.) and on the approaches implemented. To this end, we will specify the decision processes linked to existing regional development designs. We will infer from these the information needs the approaches require. Then, we will compare satellite imagery specificities with other types of geographic data. This will allow us to characterize the potential of remote sensing in response to these needs. In the literature, we identify the different functions which spatial information can provide, by showing its various uses, including social, relational or cognitive points of view, which are rarely considered in scientific literature. We will use these functions to analyze the use of spatial information in case studies in France and in Madagascar. In the conclusion, the initiatives currently developed to facilitate the use of satellite remote sensing and the new research problems raised by it will be discussed.

3.2. Sustainable territorial development, decision-making and information

3.2.1. *Regional policies*

The development models designed by the North are used to determine models in the South. Demands for a more "endogenous" design, especially for the programs supported by the United Nations in the 1970s (small is beautiful), have not been successful. Territorial action was often identical in the North and the South.

Policies were mainly concerned with "structural" investments (electricity, transport, service infrastructure etc.), that were relatively localized (urban policy, low mountain range policy). Essentially, these investments were made by the national state, and after decentralization made in association with local authorities in planning contracts. In this period, known as economic growth, the challenge is to mobilize existing resources as well as possible. These policies are questioned at the end of the economic growth. Investors find it hard to create economic activity with a direct impact and the gap between those who win and those who lose is widening.

Since the 1980s, sustainable development and resource management have become a concern for the international community. The United Nations development program and organizations (World Bank, FAO, etc.) promote methods for using natural resources which allow the needs of the population to be met and keep "ecological capital in a good condition".

The Brundtland Commission [WOR 87] defines sustainable development as "a development that meets the needs of the current generations without compromising the capacity of future generations to meet their needs". This definition, with which most of people agree, underestimates the contradictions between economic and market (competitiveness, attractiveness and performance of the productive system in the broad sense), social (solidarity, expansion of general interest and human progress) and environmental logic (sustainability of the global system, taking into account its limited resources). Sustainable development "is one of these words that allow a tactical political consensus between people with different ideas, actually based on their lack of precision. This polysemy allows a consensus or at least a productive misunderstanding. It is however difficult to stabilize or accumulate knowledge with it" [HUF 07].

The difficult implementation of sustainable development is causing debates, which show "significant differences concerning the concept of what should be sustainable development" [OWE 02]. These debates have had an impact on territorial public policies. The objectives (economic growth vs. sustainable development) and the implementation (bottom-up approaches of emerging projects vs. top-down approach of guidance through directives) of those policies have both evolved.

The bottom-up approaches have allowed us to develop projects for territories, created on the basis of a cohesive social structure and the endogenous will for a shared future, embodied in a charter with a symbolic value stronger than the simple contract. Article 22.-I of the French Act of February 4 1995 on guidance for spatial planning and territorial development (LOADDT) institutionalizes endogenous and participatory approaches. "The territory expresses the community of economic, cultural and social interests of its members. It represents the framework for the development of a common sustainable development project intended to develop the territorial assets concerned and to reinforce the mutual solidarities between urban and rural spaces. This project takes the form of a charter for the development of the country." These approaches have mainly been used in rural areas.

At the beginning of the 21st Century, more binding policies appeared. Through directives they define objectives that are often quantitative and predetermined. In order to access funding they impose tools and procedures on local territories. The urban power of institutional and planning communities (communities municipalities, cities and urban) is reinforced. Planning instruments such as the SCoT (Coherent Territorial Planning Scheme) are promoted. The SCoT, a highly technocratic tool with a strong urban connotation, replaced the country charter which primarily represented a shared political will.

In the South, funding from landlords in the 1980s and 1990s has been replaced by private and public bodies that implement integrated development programs.

3.2.2. *Territorial development process*

The endogenous bottom-up approach used more local knowledge/skills and was validated by the general population. The top-down approach through directives emphasizes general governance, project output and impact indicators. The two options have always coexisted. Today, the endogenous approach has not disappeared with the appearance of regulatory top-down territorial planning procedures. These procedures often impose a form of "mandatory communication" between local parties [THO 96]. The injunction to create a "territorial project", despite all the remaining ambiguities [MAU 12], also helps local actors to develop and express a strategic vision of the territory's future.

There has always been the need to reconcile directives with the aspirations of the population. For some people, the implementation of territorial policies must support the aspirations of local society, symbolized by a unique political project, and must be reflected in the normative reference models imposed by regional, national, and bureaucratic European organizations. For others, it is about adapting normative directives to the needs of the populations and territories.

A hybrid concept between classic planning (planning, programming, etc.) and facilitation of development (promoting synergies, organizing consistency, facilitating initiatives, etc.) is increasingly being implemented. "Territorial engineering"[1] [LAN 07] not only contributes to the understanding of territorial phenomena and the development of technical projects, but also to organizational, institutional, social and individual changes allowing the territory to gain reflexivity, autonomy, adapt its capabilities and assert its own identity.

Informational, communicational and mediation dimensions have major roles in territorial engineering. They are about producing inherently spatial information on the territory but also about sharing and using information. The information exchange is mainly via formal representations of reality, as is highlighted in quantification sociology [DES 00] and illustrated by the management of flood risk [LEB 07]. Representations can take the shape of:

– measurement, figures and statistics;

– databases with conceptual models and semantic description;

– maps with legends;

– indicators;

– model results.

They can also be based on physical or digital media (maps, aerial or satellite images, field photos, graphics, 3D models, films, etc.) [LAR 01].

1 The DIACT (Inter-ministerial Agency for Spatial Planning and Competitiveness) defined territorial engineering as "all the concepts, tools and devices made available to the territory actors to support the design, execution and evaluation of their territory projects".

Spatial representations of territorial reality can correspond to mental representations of reality, i.e. to the way an individual or a group of actors cognitively see larger locations, local territories or geographic spaces.

Information is necessary for the design, implementation, management and evaluation of the project(s) of (on) the territory, during knowledge and skill acquisition. The formulation and resolution of territorial problems impose collective learning, which is both cognitive and relational – a form of "common conceptualization" [THO 96]. All the actors must be mobilized to use their knowledge and agree to ensure efficient participation in the implementation of projects. Representations play an essential role for the bearer/promoter of public policy, because they allow him, in particular through mapping, to propose interpretive guides of a complex reality in order to come up with shared representations of both the object to be managed and the community of actors concerned. Representations also reinforce the authority of this policy promoter, legitimize public action and regulate the practices of territory actors. Local mediation mechanisms aim to facilitate communication between actors, in order to set up first meetings, to promote existing skills, and to develop capacities necessary for the development and/or implementation of the project [BER 04].

3.2.3. *Territorial socio-technical dispositive/apparatus of information and communication*

In order to formalize the informational and communicational dimensions at work in these processes, we rely on socio-technical information and communication arrangements (STICA), which has been developed by the I3M Laboratory. A "STICA" is defined as "a place of mediation, made up of multiple semiotic, aesthetic and technical factors in interaction, which link up social actors through sensory and mediated means"[2]. A STICA is composed of three inseparable entities:

– the "media product(s)" the information produced, mainly spatial representations (text, speech, film, graphic, table, map, image, hypermedia, etc.), requiring some language proficiency (e.g. mapping language and infrared imagery language). Media products are disseminated by means of

2 I3M Laboratory source: Information Milieux Médias Médiations, Université Nice Sophia Antipolis - Université de Toulon, France, http://i3m.univ-tln.fr/.

agents of mediation, either technical (channel, display devices, etc.) or human (expert, facilitator, etc.);

– the "area of social cooperation for production" characterized by the intentionality of the designers of the STICA and the media products;

– the "area of social cooperation for reception", in which the actors are not just receivers or message decoders, but autonomous and reflexive social subjects equipped with multiple resources, who can alter the interpretations and the use of media products initially planned by the designers [BER 99][3].

Figure 3.1 shows how the STICA concept and the territorial decision process with multi-actors driven by a project bearer are organized. We have differentiated here temporary STICA which are used to support punctual events (studies, workshops, public meetings, etc.) and STICA that have a longer lifespan (newspapers, Websites, steering committees, community councils, etc.), comparable with "*trace leavers*". We insist in this figure on two permanent STICA, which we called "*Heritage-type STICA*" and "*Observatory-type STICA*". We specify these concepts in Box 3.1.

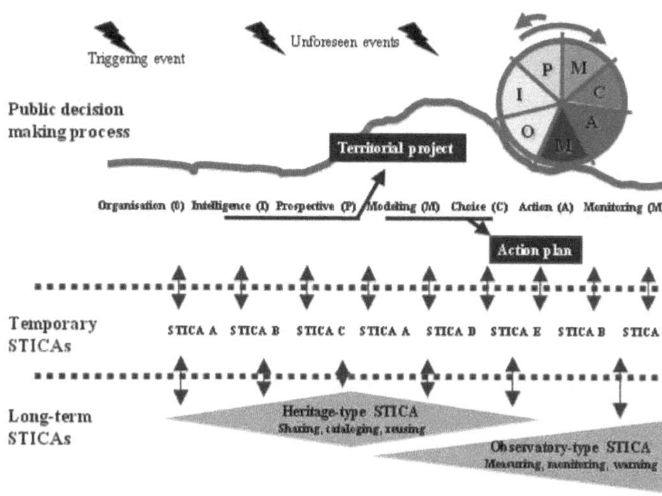

Figure 3.1. *Permanent STICA and temporary STICA in the territorial decision cycle*

3 Further analysis into spatial data coconstruction and their use, as well as the related network reconfigurations of actors, is currently being conducted in the framework of research in information and communication sciences [MAR 13], based on the network actor theory [AKR 88].

The "Heritage-type STICA" provides the various temporary STICAs with cartographic supports (maps and aerial or satellite imagery used as background layers) and with informational content (data, reports, thematic maps, ...) from the first phases of the decision cycle up to the choice of a project and the implementation of an action plan. In return, that STICA receives, indexes (cataloguing by metadata), archives, and makes available, with regulated access if necessary, media products made in the temporary or durable STICAs of the territorial partners.

The "observatory-type STICA" complements the "heritage-type STICA" when the territory has established an object of management to follow or, more broadly, a strategic development plan, with a holistic meaning, legitimized by local political power and implemented by the actors concerned with action plans. Based on monitoring and warning indicators which make sense for the territorial actors, this STICA allows them to assess changes made at the physical level of the territory. Such an arrangement requires preliminary analysis and design work with all stakeholders in order to determine relevant media products [LEM 11]. In the "area of social cooperation for reception", pooling and interpreting this information and the related signs allows us to assess the progress of the territory towards the desired endpoint. This STICA can also help to detect unplanned gaps and allows the territorial community to adapt its actions [HER 05].

Box 3.1. *"Heritage-type STICA" and "observatory-type STICA"*

3.2.4. *Functions supporting the territorial decision process*

The STICA allows the production and use of information and representations by meeting the needs of the territorial development process. Analysis of the scientific literature [MAU 12, pp. 223–224] highlights six main categories of function, each concerning a specific register:

– analytical (understanding);

– creative (imagining);

– cognitive (learning);

– relational – between individuals, social groups, institutions (linking);

– decisional (choosing);

– operational (acting).

According to the actors, the dynamics and the situations, these different functions are going to be developed to varying degrees. In conflict and crisis, or even in participative approaches, the "relational" and "creative" functions will probably be decisive. During economic growth, the "operational" functions will be promoted.

Figure 3.2 summarizes these six categories. We will return to these functions after discussing the contribution of remote sensing to the construction of spatial representations.

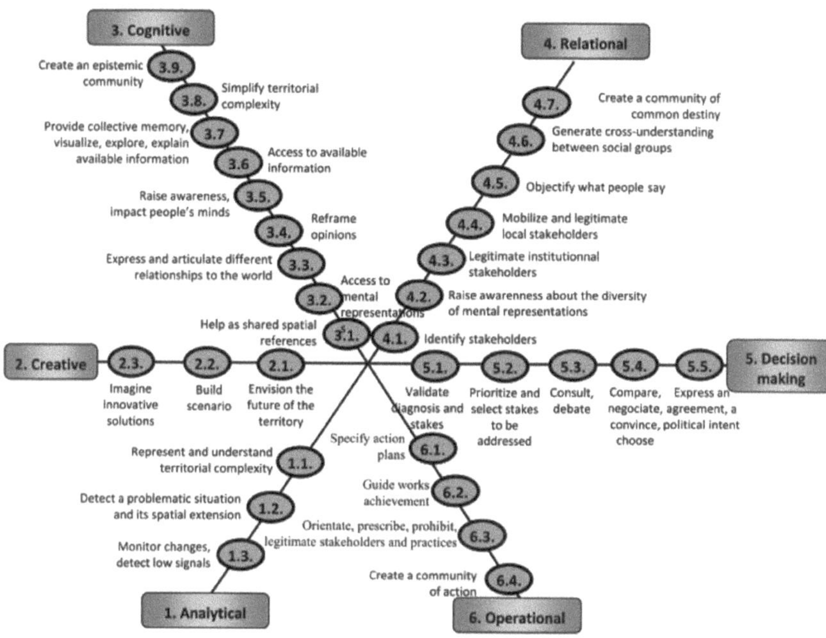

Figure 3.2. *Categorization of the functions of spatial representations in STICAs [MAU 12]*

3.3. Spatial representations derived from remote sensing

Remote sensing enables us to produce a great variety of spatial representations. Figure 3.3 shows the main ones.

The production of spatial representations derived from remote sensing is highly automated with very little human intervention (raw images, not interpreted). The remote sensing images have specific intrinsic properties that differentiate them from other representations (maps, field photographs, graphics, etc.). From a semiotic point of view[4], remote sensing images can contain iconic, indexical and symbolic signs. The degree of iconicity, the opposite of the degree of abstraction, of a satellite image will be even greater as it will graphically represent objects in a similar way to that perceived by people, i.e. in natural colors and with a very high spatial resolution. Thus, the vision provided by very high resolution satellite images is similar to the human eye in terms of the objects seen, which facilitates their reading, even for non-experts [FRE 00]. However, the degree of abstraction increases with the pixel size. In terms of radiometry, images acquired in the visible spectrum in black and white or natural color can easily be "read". Compositions that incorporate channels other than the visible ones (near, middle and thermal infrared, microwave) in contrast are difficult to interpret by untrained individuals.

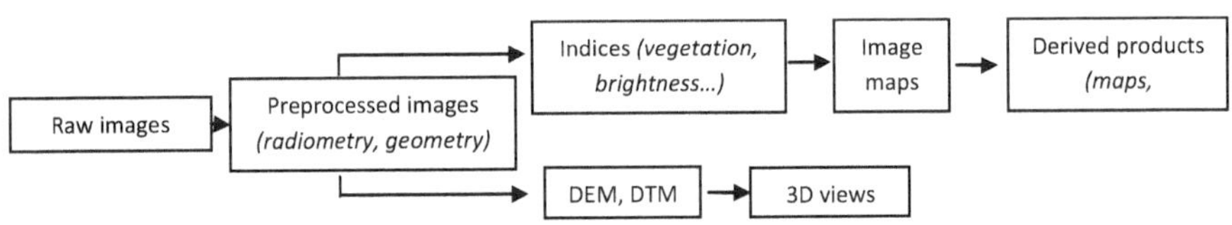

Figure 3.3. *Spatial representations derived from remote sensing data*

An image can also contain indexical signs, i.e. signs that maintain a physical continuity, an existential link with the object represented, because of the interaction during the acquisition process between the radiation measured in a given wavelength and the material of the object observed. The various indices often used in remote sensing (vegetation, brightness, etc.) thus represent indexical signs linked to the physical quantities of the object measured. Their interpretation requires prior training on the meaning of their values.

Finally, the image and its derived products also contain symbolic signs, or "intellectual signs", which are linked to the objects represented by social

4 Semiotics is the science of linguistic and non-linguistic signs and their meanings. The spatial representations were studied with Charles Sanders Peirce semiotics, Bertinand's graphic semiotics and μ group's visual semiotics.

convention, law or habit. They lose their meaning as signs if no one can interpret them. Thus, a satellite image represented in infrared false color requires knowledge of this coding system to give meaning to the signs contained in the image. It is the same for land-use maps, the interpretation of which requires knowledge and understanding of the related nomenclature.

These semiotic properties combined with remote sensing sensors and a largely automated and digital image-making process, gives images a "short hand power" [SOU 08] as well as a "truth effect" [LUS 03]. This is important because at the European and national level through the Territorial Agenda of the European Union (2007–2013) and the DATAR[5] promote territorial evidence-based planning, by recreating a privileged relationship between politics and science [FAL 09]. This raises the question of whether the remote sensing image is socially acceptable as a substitute for the object represented and provides an irrefutable truth [CHA 04].

The analysis of the 13 most frequently used methodological guides [ATG 00, DAT 01, DES 07, GRA 06, HEN 01, IAA 05, KIL 07, RAR 98, RAY 14, SAG 95, SCO 10, SOU 09, UNA 03] by territorial engineering professionals shows that half make no reference to spatial representations [MAU 12]. In the other guides, the map is the most quoted representation. The images (satellite image, aerial photo and field pictures) are absent, except for their analytical function. Their potential as media, as common reference, is never mentioned. Several functions identified in Figure 3.2 are absent from these guides: decentering or taking different points of view, explanation of mental representations, commitment and legitimization of actors, objectification of the statements of actors, generation of interknowledge between the actors, creation and cultivation of a shared vision. The co-construction of territorial knowledge between heterogeneous actors and the importance of local knowledge are not discussed either.

These results indirectly reveal that the model of expertise and technical rationality, dominated by "elected officials/technicians", is still shaping methodological guide content, even though they deal with the participation of territorial actors. They also demonstrate that the cognitive and relational functions of spatial representations, including remote sensing images, are not

5 DATAR: Interministerial Delegation for Territorial Planning and Regional Attractiveness.

taken into account and are not translated into methodological recommendations in territorial mediation practices, thus confirming a finding already established by territorial development specialists [DEB 03]. These findings reinforce the need for support and for a learning process intended for potential users, hence the importance of the product usage range in the STICAs. Such support and learning creates diverse functions that can be applied to the images and their derived products.

From two examples, one in Languedoc-Roussillon France (territorial planning) and one in Southern Madagascar (land management), we are going to show how spatial information derived from remote sensing can meet the needs of territorial decision processes, by ensuring some of the functions described above. We focus on high and very high spatial resolution images, including some products derived from digital aerial photography. These seem to offer the most potential regarding the diversity of functions and use, and their quantities are going to significantly increase with the arrival of new satellite sensors and the boom in images acquired by drones. We also give priority to operational examples rather than work within the scientific community. We refer to the functions of spatial representations described in Figure 3.2 by using their index and a brief description (for example, "F4.1. Identifying the actors concerned").

3.4. STICA based on spatial representations at the service of integrated land management

3.4.1. *Thau territory and the challenge of urban sprawl*

Thau territory is located on the Mediterranean coast in the department of Hérault (France, 43.46°N and 3.67°E), approximately 20 km west of Montpellier, the regional capital. Sète, Mèze and Frontignan are the main cities. With a population of 127,000 inhabitants in 2015 (figure estimated from the last National Institute of Statistics and Economic Study census of 2010), Thaubasin is the third center of Hérault department. It is a territory dominated by water:

– 30 km of coastline;

– several lagoons including the 7,500 ha Thauone, which is the unifying element of the territory;

– a watershed of 44,000 ha (distributed over 22 municipalities);

– 27% of the surface of the Thau SCoT area is classified as water (the highest proportion for the French coastal territories).

This territory is characterized as being extremely rich in biodiversity and landscapes, with maritime plain and wooded reliefs, lagoons, wetlands and agricultural plains. Economic activities are diverse:

– industries linked to Sète's harbor;

– shellfish farming, and fishing;

– wine growing;

– spa industry, tourism and recreational activities.

Due to its geographic situation, this highly attractive territory is subject to considerable demographic pressures and rapid urban sprawl. These dynamics are challenging the balance of the environmental system, creating tensions between traditional economic activities and residential economy, and leading to social inequity linked to steep real estate price increases. Industrial and wine productions are in crisis, even though the areas benefiting from the AOC (Protected Designation of Origin[6]) are more viable. More global changes, including climate change, also affect this territory, with impacts such as coastline erosion and marine submersion hazard.

3.4.1.1. *Organization phase: initial investments in geomatics and remote sensing*

The Syndicat Mixte du Bassin de Thau (SMBT) is a territorial engineering structure created by two intermunicipal authorities in order to better deal with the challenges linked to the territorial dynamics. The SMBT is responsible for coordinating the development and implementation of several management and planning instruments (successive lagoon contracts, a SCoT, a water planning and management scheme, and a Natura 2000 procedure for the whole lagoon of Thau).

Several governance bodies were created: the trade-union committee, the mayor's commission, as well as *ad-hoc* technical groups on topics requiring

---

6 The "Appellation d'Origine Contrôlée" is a French official label guaranteeing the quality of an agricultural product, with the manufacturing (production and processing) steps being executed within the same geographical area and according to a recognized expertise (adapted from Wikipedia).

the production of new knowledge on the territory or the networking of stakeholders. From the beginning, the elected officials and the SMBT technicians chose to encourage the active participation of local actors and civil society with public officials (state services, local and regional authorities) that must be present during land planning procedures. The SMBT also developed numerous collaborations with the scientific community in order to promote a research partnership and to produce the necessary knowledge to meet challenges. Geomatics was one of the disciplines mobilized with the creation of a geographical information system (GIS) and the acquisition of 20 cm resolution aerial color orthoimagery (2005 and 2012).

3.4.1.2. *Diagnostics phase: the challenge of urban sprawl*

One of the main challenges in the Thau territory is urban sprawl linked to the population growth which is causing various problems: environmental degradation and landscape destruction through urban sprawl, risks for lagoon professions due to a decrease in water quality, soaring land prices penalizing low-income inhabitants, consumption of agricultural land, increase in vehicle traffic, etc. It is also a national challenge, as that the Agriculture Modernization Act of July 27 2010 aims (the pace of urban sprawl to helve in France by 2020. In this respect, an Observatory of natural, agricultural and forestry areas ("Observatoire des Espaces Naturels, Agricoles et Forestiers") was created by the Decree of June 29 2015. However, it is a phenomenon spread out in time and space. It is particularly complex due to the multiplicity of explicative factors and the impacts generated. At the beginning of the diagnosis phase of the SCoT in 2006, the phenomenon was not very well known. It was considered unmanageable, even unstoppable, and generated numerous controversies.

3.4.1.2.1. *Informal use of images*

Aerial photographs and satellite images, as well as derived representations, were mobilized in formal and informal STICAs concerning urban sprawl. The images enabled us to create spatial representations of areas at variable scales characterized by a low level of abstraction. According to the "zoom", the objects represented ranged from global objects at a territorial level (the lagoon, the coastline, the wide limestone massifs, and the outline of the cities and villages) to the types of objects that individuals and social groups experience on a daily basis (houses, paths, public places, land parcels, trees, woods, hedges, vehicles, boats, etc.). Thus, the images offer a very close relation to the real world. The similarities between the signified (physical space of the territory) and the signifier (aerial photograph or very high spatial resolution satellite image) require little specific skills for cognitive operations in scouting tasks and for the production of meanings. They facilitate informal communication between the actors of the territory.

For example, the 2005 ortho-photograph was used in 2006 during an initial meeting between the newly created SMBT and representatives of the shellfish farmers. Seen on the SMBT's laptop, the ortho-photograph allowed the lagoon professionals to voice their concerns, in particular regarding urban sprawl and its impact on the quality of the lagoon water. The ortho-photo was used as (1) a spatial reference to support speeches (F3.1. Acting as shared spatial reference), (2) a concrete support in order to become acquainted with the representations of the territorial actors' reality (F3.2. Accessing mental representations) and to ascertain their veracity (F4.5. Objectifying what people say). In a derived manner, the ortho-photo showed the SMBT engineers the wealth of local knowledge held by the fishermen and the shellfish farmers, even if this knowledge remained in a tacit form (F4.2. Discovering the diversity of representations). This meeting represented one of the first interactions between the SMBT and this socio-professional community, interactions which helped to gradually build trust (F4.6. Generating cross-understanding). Globally, remote sensing images played the role of boundary objects, in the sociological meaning of the word [STA 89], i.e. objects capable of linking socio-heterogeneous worlds, without standardizing them or making them invisible to each other.

3.4.1.2.2. Formal use of images to objectify the urban sprawl phenomenon

In order to objectify the urban sprawl phenomenon, which was still not very well understood, an initial remote sensing and geomatics study intended to create quantitative and spatial representations of the evolution between 1944 and 2005 was conducted from old topographic maps over several periods and a 20 cm ortho-photograph of 2005 [CEM 07]. The objective was to provide unbiased information for the diagnostics phase of the SCoT.

In the "area of social cooperation for production" of the initial STICA, the methodology used to represent this phenomenon was cobuilt in the framework of a technical committee on urban sprawl, which brought together the SMBT engineer responsible for the SCoT, the SMBT's GIS engineer, researchers and engineers[7] from Irstea, the state services, the CCNBT[8] and the general board of the Hérault department.

The development of the nomenclature to be used to map the urban sprawl phenomenon provided an opportunity to challenge various points of view on the categorizations of geographic space, and to agree on the classes and subclasses to be selected and their meaning (F3.3. Expressing and articulating different relationships to the world). An embedded nomenclature on three levels was finally chosen. This nomenclature played a normative role of boundary object (F3.1. Acting as shared reference) in order to stabilize an agreement on the definition of the object to be measured.

In this collective work of reality codification, the ortho-photograph was used as a semantic reference (F3.1. Acting as shared spatial reference) by allowing us to illustrate some classes of the nomenclature with representative ortho-photo thumbnails (Figure 3.4). From a semiotic point of view, they correspond to "iconic types" (μ Group 1992).

A geographical database was then created from topographic maps from the archives (in 1944, 1971, 1981 and 1995) and from the 2005 ortho-photograph. This latter enabled us to characterize through photo-interpretation the urban sprawl situation in 2005, at the beginning of the SCoT (F1.1. Representing and understanding territorial complexity). It was also used as a geometric and semantic reference to normalize the representations of the previous periods (F3.1. Acting as shared spatial reference).

A mapping simulation of the urban sprawl in 2020 was also created with a cellular automata spatial modeling tool using the rate of demographic growth observed during the 1995–2005 period [CHE 08]. Figure 3.5 compares the situations in 1944 and 2005. Figure 3.6 shows the simulated sprawl in 2020.

---

7 Research Institute of Science and Technology for the Environment and Agriculture.

8 Communauté de Communes du Nord Bassin de Thau – Community of Municipalities of the North Thau Basin.

Theme level 1	*Class level 1*	*Class level 2*	
1-Built area	1.1- Old housing	1.1.1- Village center 1.1.2- Suburb 1.1.3- Hamlet 1.1.4- Individual house 1.1.5- Collective housing 1.1.6- Farm house	
	1.2- Collective housing	1.2.1- Big collective housing (2 floors and more) 1.2.2- Small collective housing (1 floor)	
	1.3- Individual housing	1.3.1- Ordered residential housing composed of terraced houses. 1.3.2- High density structured residential housing (average land <1,000 m^2). 1.3.3- Low density of structured residential housing (average land >1,000 m^2). 1.3.4- High density of scattered individual housing (average land<1,000 m^2). 1.3.5- Low density scattered individual housing (average land>1,000 m^2).	

Figure 3.4. *Extract from the urban sprawl nomenclature and illustrative ortho-photo thumbnail for iconic type "1.3.1. Ordered residential housing composed of terraced houses" (source: Irstea© Gaia Mapping, 2005)*

3.4.1.2.3. Derived representations in order to facilitate the ownership of the diagnosis on urban sprawl

Derived representations of this urban sprawl map were executed in order to provide media products to different STICAs mobilizing a variety of actors from the territorial governance. Please find their descriptions below.

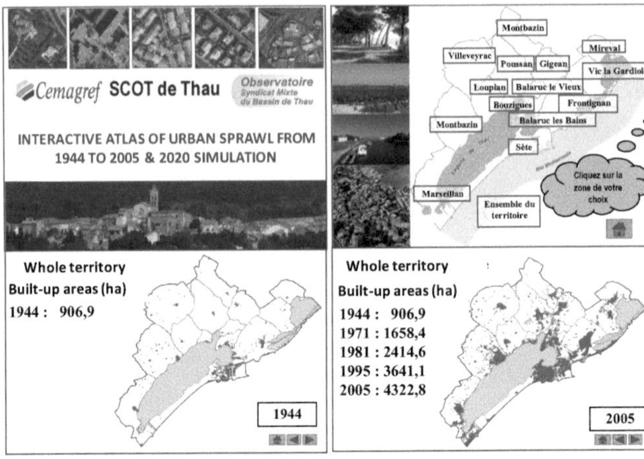

Figure 3.5. *Extract from the Thau urban sprawl atlas (in red) derived from remote sensing images. For a color version of this figure, see www.iste.co.uk/baghdadi/5.zip*

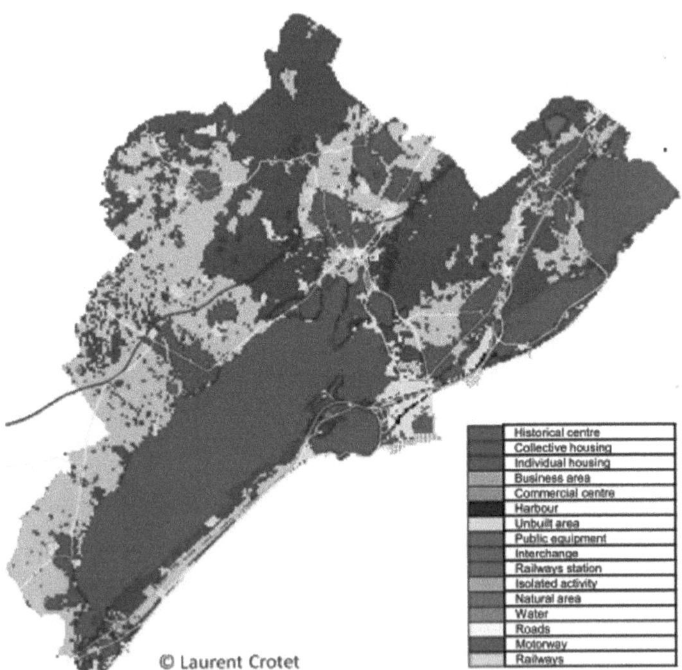

Figure 3.6. *Projection of urban sprawl dynamics in the Thau territory in 2020. For a color version of this figure, see www.iste.co.uk/baghdadi/5.zip*

Various indicators were created per municipality and for the whole territory (e.g. evolution of the developed surface between two periods (ha and %); average urbanized surface per new inhabitant between two periods (m^2)). One estimated the number of years remaining before complete saturation of the space suitable for development in each municipality and the whole territory. It was calculated from the potentially remaining surface suitable for development (on the basis of biogeographical and regulatory criteria) and the annual consumption of land based on the urban growth rate measured during the 1995–2005 period.

This indicator enabled us to express the urban sprawl phenomenon in a space–time reference compatible with the frames of reference of the elected officials (the number of electoral mandates of a mayor and the perimeter of his/her constituency), as well as those of the inhabitants (the number of generations and their local place of residence) (F3.3. Express and articulate different relationships to the world). At the urbanization rate observed during the 1995–2005 period, this indicator revealed that some municipalities will fill all of land suitable for development in less than 20 years, equivalent to three mayor mandates and one generation. By thus highlighting the magnitude and speed of the phenomenon, the indicator acted as a "wake-up call" (F3.5. Raise awareness). The objective of this indicator was to project oneself into the future, rather than simply report past dynamics (F2.1. Envision the future of the territory). It was about providing a sign with a potential meaning in order to recognize the concept of sustainability, responsibility and solidarity toward those who "will come after" (future elected officials, next generations of children and grandchildren, etc.) (F4.7. Create a community of common destiny).

An interactive and dynamic atlas was created in the form of an animated Powerpoint slide show. The user could see the progression of urban sprawl in a given area (the whole Thau territory or by municipality) due to the perfect superposition of the mono-date maps with each other (Figure 3.5). Synthetic indicators complemented the mapping representations: developed surface, average annual expansion of the developed area compared to the previous period (F1.3. Monitor changes, detect low signals).

Two 3D physical models of the Thau territory at a scale of 1:25,000 were created. One of the models was painted in white and was used as a support to project the mapping atlas on, in order to show spatial planning constraints due to geomorphology (mountains and water). On the second model, the reference mapping elements (roads and water area) were painted, as well as the urban sprawl in 2005. The two models were used in a mobile exhibit reproducing the SCoT diagnostics, which took place over 2 months in 2007 in several Thau municipalities (Figure 3.7). Associated with posters summing up the different challenges threatening the territory, including the one of the urban sprawl, these media products facilitated mediation between the SMBT and the visitors in order to develop an awareness of the magnitude of urbanization and its consequences (F3.5. Raising awareness). They were also used during a meeting to get the SCoT diagnostics validated by the decision makers of Thau (mayors and their technicians) (F5.1. Validate diagnosis and stakes), in addition to other representations based on factual evidence regarding the state of the territory.

Figure 3.7. *3D models as mediation supports representing urban sprawl phenomena*

To enhance understanding of the urban sprawl phenomenon and to identify solutions, the SMBT asked two local associations, dealing with environmental education and urbanism, to provide a playful STICA mediation called the "*jeu du petit urbaniste*" (the little urban planner game) (Figure 3.8). It included an ortho-photograph representing a site to be developed on the fringes of a village in Thau (F3.1.Help as shared spatial references) and a series of items in wood and paper (building elements, land parcels and roads).

Participants (schools, general public and exhibit visitors) were invited to devise individually or in groups, a development plan for the site in order for it to welcome a given number of new inhabitants (F2.3. Imagine innovative solutions). The exercise, which was supervised by a mediator, required a reflection on urban forms that consumed less space than the peri-urban detached house, while addressing issues such as livability, sociability, links between private spaces and public spaces, traffic, connection with the urban center and peripheral spaces, etc. At the end of each exercise, a photo of the solution created was taken by the mediator, and then printed and displayed on a board. Comparing the solutions enhanced reflection by cross-examining the points of view (F2.3. Imagine innovative solutions, F3.3. Express different relationships to the world) and encouraging debates (F5.3. Consult, debate). This game was used in the SCoT mobile exhibit, but also during festive events and extracurricular activities.

Analysis of the answers given to questionnaires by participants during the mobile exhibition, as well as discussions held in the framework of research work [MAU 12], indicates that the different representations of the urban sprawl and the mediation exercises implemented captured people's imagination, whether it was elected officials, the general public or even state service representatives who visited the exhibit.

Figure 3.8. *The little urban planner game*

3.4.1.2.4. Cross-referencing the agronomic potential of the land

During the mayors' commission, spatial representations other than the ones previously presented were used for cognitive purposes in order to strengthen the responsiveness of elected officials and to challenge their thought patterns. The objective was to produce information changing their representation of the magnitude urban sprawl of and its challenges.

Among those representations, a map of the agronomic potential of the agricultural land of the Languedoc-Roussillon region created by the LISAH[9] was used. Superposition with the urban sprawl map shows that urbanization is preferentially developing in the plain areas, where agricultural land with high agronomic potential is located. These representations crystallized "epiphany" operations, within the chemical meaning of the word. They helped elected officials and technicians to become aware of a reality, which had until then been partly overshadowed (F3.4. Reframe opinions). Thus, they helped to reinforce the importance of taking into account the rural dimension in the SCoT, meeting expectations expressed during the diagnostics phase by several elected officials, state services and representatives of the farming profession.

Once the SCoT diagnostics were validated by the elected officials and the mobile exhibit for the general public had ended, the next step was spatial planning, in particular regarding urbanization regulations, which became a priority recognized by all.

Three examples of the contribution of spatial representations reflecting fundamental choices for the Thau development are given the first example concerns a decision, during the mayor's commission, related to a request from one of the municipalities to expand its built-up area with a residential housing project. To justify its request, which was contrary to the vision announced by the territory to reduce urban sprawl, the mayor argued that the town had developed very little during the last 10 years. In this closed arena of social cooperation between elected officials and technicians, the use of the urban sprawl database, the related indicators and the ortho-photograph objectified the actual situation of this municipality, which had in fact had very significant recent urbanization. The defenders of strict control of urbanization were able to play on "*the truth effect*" created by superposition of the urban sprawl map onto the ortho-photo background to reinforce the credibility of their arguments.

---

9 LISAH *Laboratoire d'étude des Interactions Sol - Agrosystème - Hydrosystème* (Ground Interactions study laboratory–Agro-system–Hydro-system), France.

The negotiation conducted internally within the mayor's commission finally resulted in an unfavorable opinion of the urban expansion project (F5.4. Compare, negotiate, convince, choose). More generally, the elected officials understood that it was now possible to refer to ortho-photography and urban sprawl maps in order to obtain undisputed evidence on the actual situation be dealt with (F5.1. Validate diagnosis and stakes). One external consequence was to reinforce in the eyes of the state and the prefect, the credibility that the Thau territory (F4.3. Legitimate institutional stakeholders) could itself decide and regulate the complex phenomenon of urban sprawl, which is particularly significant in the coastal areas and often blamed on the laxity of the local authorities.

The second example concerns the (SMVM) work performed by the SMBT by means of the same data sets to assess the marine scheme (Schéma de Mise en Valeur de la Mer), a land zoning document established and controlled by the state since 1995, which is supposed to protect lagoon water quality in order to maintain shellfish farming and fishing activities. The SMBT managed to support state services in their analysis on the limits of this regulation instrument by supporting its arguments with evidence (F4.5. Objectify what people say F5.4. Compare, negotiate, convince, choose).

Ultimately, these arguments convinced the Prefect not to renew the SMVM (which would have remained under state control), and to replace it with an additional maritime section in the SCoT placed under the responsibility of Thau-elected officials. This deal represented a turning point in the development of the Thau territory, because it proved to the Prefect the expertise and capacity for autonomy of the territory (F4.3. Legitimate institutional stakeholders).

The third and last example concerns the strategic choices made by the elected officials once the SCoT diagnostics was correctly completed and shared, concerning urbanization over the next 10 years. These choices were written down in the SCoT spatial planning project and were the subject of in-depth debates and long negotiations during the nine meetings of the mayor's commission between the end of 2007 and the beginning of 2009 with the support of consultants. Maps were used to present spatial planning scenarios on the residential growth of the various municipalities (F2.2. Build scenarios). Finally, the process resulted in a decision to reduce the global demographic growth in the territory and to concentrate it in a particular area (the triangle between Sète, Balaruc les Bains and Frontignan) in order to preserve the upstream part of the watershed and so limit the impact on the lagoon water quality. The SCoT map demonstrates that choice and represents a symbolic assertion of a shared political vision for the future (F5.5 Express an agreement, a political intent).

3.4.1.3. *Implementing phase of the territory project*

The SCoT approval in 2014 and SAGE (water planning and management scheme) approval in 2015 ended the long planning period and its registration in regulatory documents. The territory then entered the operational implemention phase of its strategic vision by means of an action program (2014–2017) entitled "Integrated management agreement of the Thau territory".

To support this new phase, the SMBT developed its information system by complementing it with an "observatory-type" at the end of 2015. The objective of this observatory was to accurately assess the territory evolution, in particular under the impact of the action program, with the help of monitoring and warning indicators. That assessment was supplied to the technical and political authorities in charge of the project in order to adjust the action program as they go along if the evolutions observed were not compliant with the directions of the SCoT and the SAGE.

In the context of the fourth action plan, land use mapping, which initially focused on the developed areas, was extended to the whole territory by adding the natural and agricultural areas, and by categorizing them by means of a more detailed four-level nomenclature. The methodology combined the photo interpretation on the urban areas of an ortho-photo of 2012 with an object-orientated classification of multidate Pleiades satellite (50 cm) images for mapping the agricultural and nature areas [DUP 14]. This mapping should be updated in 2017. The comparison of the two maps (2012 and the recent one) will give information on some of the planned state and evolution indicators (F1.1. Represent and understand the territorial complexity F1.3. Monitor changes, detect low signals).

3.4.1.4. *Summary*

This case study showed how different spatial representations derived in part from aerospace remote sensing resulted in a long territorial decision process concerning the control of urban sprawl by supplying a wide spectrum of functions. The spatial representations played the role of boundary objects between different groups of actors within the territorial governance system (elected officials, state services, technical services of the intermunicipal and municipal authorities, economic actors, journalists, associations, schools and citizens).

The emergence and consolidation of this network spread over approximately 10 years, going through the successive phases of : (1) an epistemic community (F3.9) (coproduction of new knowledge on the complex phenomenon of urban sprawl during the diagnostics phase of the SCoT); (2) then a common destiny (F4.7) (common vision of a shared future for the territory expressed in the spatial planning and sustainable development project of the SCoT validated in 2014 and relying in particular on better control of urban sprawl); (3) a common community of action (F6.3) (collective commitment in an action plan in order to reach the vision shared by a common destiny – here the Thau SCoT and the municipalities local development plan that will supervise urban development of the territory in the decade to come).

The neutral point of view that we have chosen to adopt in order to describe the spatial representations and their use must not overshadow the moments of social, economic, environmental and political tension that this territory has gone through in the last 10 years. In addition, our functionalist approach of the representations can lead to belief in a form of technological determinism. Even if this approach is useful in helping the territorial engineers to better formalize the STICAs design according to specific objectives, contexts and public, it will never prevent unplanned use and misuse by users. Opening the black box of the uses of these representations and STICAs, and their multiple, positive or negative, direct or indirect, impacts represents a complete field of research in terms of theoretical foundations and analytical methods.

3.4.2. *Use of spatial information for land management in Madagascar*

3.4.2.1. *Land issue in Africa and Madagascar*

The land situation in Africa is complex. The demographic growth (2.5% per year during the last decade) is putting significant pressure on resources, resulting in competition between multiple uses and forcing an intensification of production systems. Land generates tensions arising from use (agriculture, urban growth, conservation areas, tourism, etc.) and actors (native population and immigrants, rural producers and urban elites, family agriculture and agribusiness). Tensions also arise due to rights systems with laws and administration methods defined by the state, and rules and management methods based on custom.

Many land policies in Africa have tried to reduce this duality between common, often informal, and statute law. The programs of land titling tried to give a written legal form to local rights according to the standards defined by the state. From just after independence until the 1980s, the reforms imposed a system based on private property and on the issuance of land titles. They relied on complex procedures and rarely took into account local land right systems, especially their collective dimensions. Thus, they offered tools that were too costly, made no sense for the population and created some exclusion [DEI 09, COL 09].

In the 1990s, a new wave of land reforms were created in the southern countries [BRU 94]. These reforms aim to decentralize land management in order to introduce, at the local level, innovative processes for recognizing existing land rights and thus trying to reconcile the legality and legitimacy of local practices [LAV 98]. These reforms are part of a paradigm supporting the local evolution of rights (for example, formalizing customary rights) and are no longer of a paradigm replacing existing rights with private property [BRU 94, COL 09][10].

For all these policies, one of the factors of success, or one of the constraints, is associated with the tools used. Geomatic technologies, which are "guarantees of modernity", are often mentioned and help to give an image of the ease of registering rights in the mind of promoters, operators, professionals, landlords, etc. But the progress of imaging and, more generally, of geomatic tools does not solve issues concerning the identification of rights (in their diversity) and their holders. In particular, technology cannot replace social validation of the nature of rights and the holders' legitimacy [TOU 08]. "New technologies allow to increase productivity for the surveying of land parcels (GPS) or information processing (GIS). But, they cannot ignore the time spent mediating and

10 Nevertheless, since the 2000s, registration-orientated policies and systematic rights privatization have reappeared, as they are relegitimized by technological evolutions and theses that present rights formalization as a means of fighting poverty and the pressure of private investments on the land.

negotiating before that, or the time spent on social investigation concerning lands allowing to identify in a reliable manner the rights holders" [COM 05]. It is very difficult to update rights. Transfer registration – purchase/sale, inheritance, etc – is rarely performed. The institutions do not have the necessary human or material means and are not motivated enough – promoting confusion leading to an informal source of income derived from corruption [MAT 96, COL 09].

3.4.2.2. *Malagasy land reform*

3.4.2.2.1. Main institutional innovations: legal recognition of local rights and decentralization of land management

The Malagasy land reform belongs to the second generation of reforms based on the legal recognition of local rights and the decentralization of land management [TEY 09]. Since 2005 its objective has been to contribute, using the words of the land policy letter [LET 05], to "a massive land tenure security accessible to most and at low cost". The main measures of the land reform consist of recasting the legislative framework, modernizing land services and decentralizing land management.

Since 2005, suitable untitled land no longer falls within the scope of state property, but within a new legal status: untitled private property (UPP)[11]. The municipalities have increasing powers. They can, through the creation of a land office, legally recognize user land rights within the scope of the UPP. On request they then issue the owners, with an individual or collective certificate.

Thus, two types of legal authorities are responsible for legal land management and two arrangements to register land property are available: state land services, which issue titles, and land offices, which issue certificates. Certificates, as well as titles, legally validate private property and have the same attributes (transferable, transmissible and mortgageable). Unlike the titles issued by the state on land considered unoccupied, the certificates are only granted after a local social validation process of the existing land rights.

11 There is no longer a presumption of public domain, but a presumption of private property.

3.4.2.2.2. Main technical innovation: unique information system

Noting the obsolescence of land services (premises with archives: maps, land registries) and the inadequacy of citizens service (few requests for titles, complex and costly procedures, high corruption, etc.), the 2005 national land program has included a highly technical dimension and modernization actions, but they are used for institutional changes. The image use and the establishment of a common information system were the compromise that allowed the land administration to accept the loss of its land management monopoly and the transfer of its competence to the municipalities responsible for certification. The tool based on images allows us to "compare, negotiate, convince, choose" (F5.4). Land administration works from maps, land parcel limits and registries of the names of the owners. It can only consider a different right registration system providing that there is no overlap between certificates and titles. The local land tenure plan (PLOF), which is a tool created by the reform, allows information transfer from the central state to the municipalities (specifying where the titles are located), and so information circulates between state services and municipalities (specifying the location of the new certificates and titles). Management seen as more professional than a simple registry, has given legitimacy to the municipalities (F4.3. Legitimizing institutional actors).

3.4.2.2.3. Role of images in the PLOF

The PLOF, which is a map registering the legal status of sites, has no legal value but is necessary for the opening of the municipal land office.

Most of the PLOFs were executed from ortho-photoplans, using aerial shots. There are a few PLOFs with no background image and some PLOFs with a background satellite image. In this last case, the images have a very high resolution, are ortho-rectified and geo-referenced (satellites Quickbird and Ikonos) [BUR 14]. Their resolution is between 0.5 and 1 m. They are projected in the Madagascar Laborde system and are used at a scale between 1/2,500 and 1/10,000. These images were used to create PLOFs of the municipalities in the Alaotra region. The project was funded by the AFD on watershed in the AloatraBvLac lake region.

The need to create PLOFs led the land services to do some modernizing: reclassification and digitization of land information (maps and registries). In particular, geo-referenced images were used to help locate the outlines of the titled land parcels, as the original paper maps sometimes gave very brief indications concerning the location of parcels and finally offered little security to the title holders (Figure 3.9). The creation of the PLOFs naturally allowed the municipalities to obtain information which was previously kept by the land services and was very hard to access.

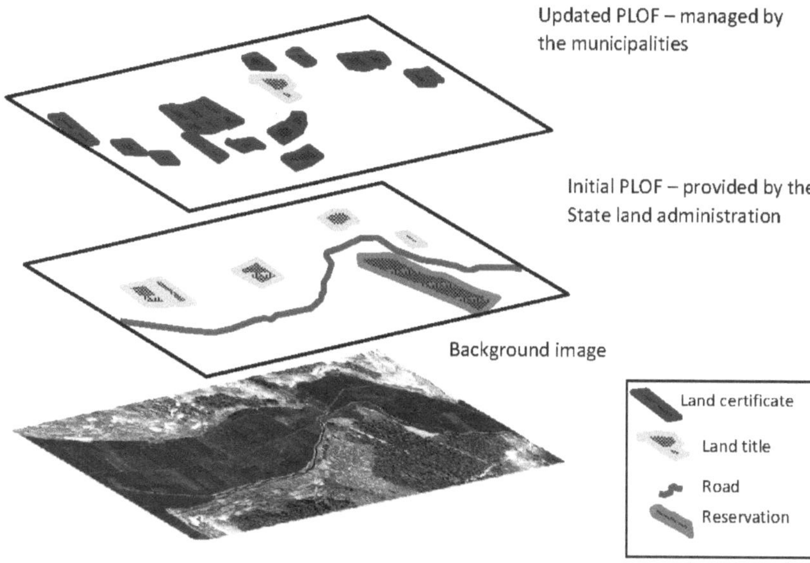

Figure 3.9. *The three layers of the local land tenure plan (PLOF) (source: National land program)*

In particular, the use of the PLOFs and the presence of images often facilitate the certification process. For each land certificate application, a Local Recognition Commission (CRL) gathers on site (Figure 3.10). It is composed of the neighbors of the applicant, elected representatives at the local level (usually older people who know the local land history) and members of the municipal team (the land office agent and a member of the municipal council). The office agent brings with him/her an extract of the PLOF in paper form, or the image of the land parcels with the limits of the possible parcels already titled or certified in color. During these CRLs, the certificate applicants and their neighbors trace the limits of the parcels to be certified on the basis of "land markers" visible on the image (irrigation canal, bund, track, noticeable tree, development, etc.). The image is thus used to support a local participative land tenure mapping (F3.1. Shared references). Subject to an agreement of the members of that CRL, minutes register the rights claimed and the parcel limits are indicated on the extract of the PLOF (F4.5. Objectifying the statements of actors; and F5.1. Validating the situation diagnostics). After this procedure and if no opposition is made, the office agent prepares a certificate that will be signed by the mayor. The paper edition of each land certificate (Figure 3.11) is composed of an extract of the PLOF drawn on the image, allowing the owners to visualize the outlines of their own parcel and their neighbors' parcels.

Figure 3.10. *Local recognition commission in action (source: Madagascar land observatory)*

Figure 3.11. *Land certificates. (source: Madagascar land observatory)*

3.4.2.3. *A STICA at the service of a territorial reform*

This recognition procedure can be considered as a STICA that mobilizes spatial information. Ortho-rectified and geo-referenced remote sensing images with a very high resolution are used for several functions (Figures 3.9–3.11).

The first of these functions concerns *analytics* (F1). Spatial information helps to understand the complexity of the land phenomena. The physical observation of the area consists of characterizing its physical use (forest canopy, development, arable land, etc.) and the function performed by that use (for example, for development, habitat or activities; for arable land, food crop or industrial use; etc). The remote sensing, due to the synoptic vision that it brings, allows us to obtain a comprehensive and global mapping of the land use, derived from a classification or a photo-interpretation of satellite or aerial images. The cross-referencing of the of land use maps obtained at different dates allows us to characterize the evolution of land use and thus to provide information on the ongoing dynamics in the territory. The land use study and its evolution in time increases our knowledge on the ecosystem and territory dynamics, as well as on the strategies of natural resource management and spatial planning of the territories. Land use maps also allow us to create zoning. Zoning divides a given area (space, territory, etc.) into smaller units according to differentiation criteria. The basic zoning unit presents a set of homogeneous problems with a minimum variability according to scale. In the case of land reform, zoning sets the types of rights and the beneficiaries. GISs provide information and helps out in locating additional information, such as information concerning economic values (sale price, rents and taxes).

A second function falls within the scope of the *cognitive* (F3). VHSR images are shared spatial references (F3.1), which give access to various mental representations (F3.2), formulating and voicing multiple relations to the world (F3.3) within the framework of the CRL, mobilizing available information from villagers (F3.6), raising the actors' awareness about land issues (F3.5) and creating common memories (F3.6). Images also create a social link ("relational" functions, F4) by objectifying the statements of the actors (F4.5), mobilizing local actors (F4.4) and legitimizing institutional actors (F4.2). An interknowledge between social groups is developed (F4.6). The CRL being recognized, another function is associated with the decisional one (F5). The PLOF turns into a legally recognized agreement (F5.5), but, through its cooperative construction, it validates diagnostics (F5.1) and allows us to consult and debate (F5.3), and compare, negotiate, convince and choose (F5.4).

Finally, the procedure also contributes to the "operational" function (F6). The PLOF can be used as a planning (land use planning, location of investments, etc.) and tax (taxation on transactions, for the establishment of land certificates and land taxes) tool. In that sense, it allows us to direct, require, prohibit and legitimize actors and practices (F6.3). In some cases, which are still very few, it has facilitated territorial spatial planning actions (location of water supply service in the context of the development project and identification of locations for infrastructures). In a more usual way, the PLOFs have facilitated the participative and collective identification of land parcels in order to perform a more comprehensive tax inventory. The number of parcels recorded – which was previously done through statements – increased by more than 100% in some municipalities. However, in a context of political crisis and because of the prospect of the municipal elections, mayors did not proceed with any tax recovery.

In some situations, in the context of the project supporting land reform, the PLOFs had a "creative" function (F2). Thus, consolidation and reforestation experiences were implemented.

3.4.2.4. *Tools that must be supported by political will*

The use of images was a crucial element for the decentralization of the land tenure security administrative procedures at the municipal level, with the creation of 400 land offices responsible for certifying the existence of "untitled private properties". However the tool cannot solve all the issues. For example, linking titles and certificates, as well as the central administration and decentralized offices, remains difficult. The land reform responded to individual applications for land tenure security, which was possible through a quasi-market approach. But the reform did not allow the set up of a local public service to ensure the provision of land data and user information. Such a provision is, however, necessary in order for land tenure and its security to actually provide sustainable development. It depends on a reinforced political will.

3.5. Conclusions

To varying degrees, depending on their nature, territorial public policies are seeking the participation and mobilization of actors. The commitment of those actors requires mechanisms of co-construction and the common implementation of projects. Spatial information plays a decisive role by promoting reflection, decision-making and by helping to legitimize the action of the decision makers, and so the acceptance of those decisions.

In order to reach those objectives, information is produced, shared, discussed, assimilated and used. The examples, in France and in Madagascar, seem to have illustrated the whole point of collective learning approaches within the "STICAs". These devices allow us, in a complex learning process, to express needs, formalize local and expert knowledge, exchange, decide, operationally implement, etc. We have described in detail the functions to which information can contribute in support of the different phases of the project cycle: diagnosing (analytical), programming (imagining), learning together (cognitive), organizing oneself (relational), choosing and evaluating (decisional), implementing (operational), etc. In order to be really efficient, its use must be supported by facilitation processes and support must be provided by high-quality professionals. In spite of this, there will still remain unexpected uses and impacts, encouraging us not to be tricked by the illusion of a technological determinism, which would overshadow the social dimension in action in the design and use of the STICAs.

The nature of spatial information derived from remote sensing makes the processes even more complex. The spatial representations obtained can sometimes be remote from the object observed. Thus, there is a great distance between the product and the user. All the more so as the conclusions of the report ("Observation of the Earth for agricultural information in Africa" [BEG 15]) can easily be generalized for other continents and fields, such as territorial development. "*The entire satellite experience and its diversity remain hard to understand for non-specialists... The working time and the technical level necessary to create relevant information are often under-estimated...The so-called "decision support" products are less operational than the so-called "biophysical" products... They also often have an insufficient accuracy level (spatial, temporal or thematic) compared to the needs... The small number of operational services expresses those difficulties, which are recognized. They are at the origin of initiatives, such as the GEOSUD project http://www.geosud.teledetection.fr/) [MAU 15] and the THEIA center (https://www.theia-land.fr/) [LER 13] in France, GEOGLAM*[12]*- Countries at risk [WHI 15] or AFRIGEOSS in the South".*

These projects all want to:

– guarantee access to images and software at low cost;

– develop methods. Methodological innovation requires the adoption of an integrated approach mixing knowledge from different disciplines (social sciences, agronomy and environmental sciences), allow us to find indirect indicators and, through multi-source approaches, allow us to jointly process multi-resolution satellite and optical/radar data, but also other types of data (field and environmental data), expertise and modeling;

– develop automated processes;

– train process specialists, but also mainly users;

– raise the awareness of potential users to geo-spatial technologies, by insisting on potentials, but also on constraints in order to avoid false expectations;

– prompt and recommend the use of spatial observation, by planning *ad hoc* funding, in particular that granted by major funders;

– structure and support the final users in order to create a critical mass of technicians, professionals and researchers able to control the technology and, in particular, to develop applications appropriate for the environment. In the creation of that core, emphasis should be placed on design offices and on professional NGOs. These structures do not have the means to invest in the capitalization of the methods they develop. The data they process and the intermediary products that they produce are undervalued and rarely used.

These projects are very recent. Their impacts are still limited. However, it is due to this type of effort, which is being reinforced, that spatial information derived from remote sensing will become usual and common practice for spatial planning and territorial development, and public investments in the spatial field will be written off.

12 GEOGLAM: Group on Earth Observations Global Agricultural Monitoring.

3.6. Key points

Satellite remote sensing represents a source of original information for territory management, whatever the geographic location and the challenges to be dealt with. Its use is limited to certain functions, mainly those linked to the identification of land use. However, a satellite image can be used for numerous functions in support of the decision-making process in territorial projects: diagnosis, planning, implementation, monitoring, etc.

The two examples in France and Madagascar illustrate the possibilities with image uses. But, in the two cases, these uses depend on the territorial governance and engineering process, which guarantee both an institutional inclusion, giving them the necessary legitimacy, and training in order for the actors to make use and mobilize those tools to develop, implement and evaluate territorial projects.

3.7. Bibliography

[AKR 88] AKRICH M., CALLON M., LATOUR B., "A quoi tient le succès des innovations? 1 : L'art de l'intéressement", Gérer et comprendre, *Annales des Mines*, vol. 11, pp. 4–17, 1988.

[ATG 00] ATGER C., GALLETY J.C., ROUSSET M., "La concertation en aménagement: Eléments méthodologiques", *Dossiers du CERTU*, vol. 104, Lyon, 2000.

[BEG 15] BEGUE A., LEROUX L., LO SEEN D. *et al.*, "L'observation de la Terre pour le renseignement agricole en Afrique", *Notes et Techniques Agence Française de Développement,* 2015.

[BER 99] BERTEN A., Dispositif, médiation, créativité : petite généalogie, *Hermès*, vol. 25, pp. 33–47, 1999.

[BER 04] BERTACHINI Y., "Entre information & processus de communication: l'intelligence territoriale", *Les Cahiers du Centre d'études et de Recherche, Revue Humanisme et Entreprise*, vol. 267, pp. 13–29, 2004.

[BRU 94] BRUCE J., MIGOT-ADHOLLA S., ATHERTON J., "Institutional adaptation or replacement", in BRUCE J., MIGOT-ADHOLLA S. (eds), *Searching for Land Tenure Security in Africa*, Kendall/Hunt, Dubuque, 1994.

[BUR 14] BURNOD P., ANDRIANIRINA-RATSIALONANA R., ANDRIAMANALINA B.S., "Une réforme foncière fragilisée par la crise mais qui parvient à maintenir le cap", *Revue Afrique contemporaine*, vol. 251, pp. 152–154, 2014.

[CEM 07] CEMAGREF, Scot du territoire de Thau. Réalisation d'une base de données géographique des dynamiques du bâti sur le territoire entre 1944 et 2005, Montpellier, 2007.

[CHA 04] CHATEAU RAYNAUD F., "L'épreuve du tangible. Expériences de l'enquête et surgissements de la preuve", in KARSENI B., QUERE L. (eds), La croyance et l'enquête, *Raisons Pratiques,* vol. XV, EHESS, Paris, 2004.

[CHE 08] CHERY J.P., MAUREL P., "Cellular automata simulation of urban sprawl in the Mediterranean zone: support for participatory planning of the Thau area (France)", *Spatial Landscape Modeling: from Dynamic Approaches to Functional Evaluations*, Toulouse, p. 2, 3–5 June, 2008.

[COL 09] COLIN J.PH., LE MEUR P.Y., LEONARD E. (eds), *Les politiques d'enregistrement des droits fonciers : du cadre légal aux pratiques locales*, Karthala, Hommes et Sociétés, Paris, 2009.

[COM 05] COMITÉ TECHNIQUE "FONCIER ET DÉVELOPPEMENT", La formalisation des droits sur la terre dans les pays du Sud: dépasser les controverses et alimenter les stratégies , Ministère des Affaires étrangères et du Développement international (Maedi)/Agence française de développement (AFD), Paris, 2005.

[DAT 01] DATAR, Repères sur la Charte des pays, Délégation à l'aménagement du territoire et à l'action régionale (DATAR), 2001.

[DEB 03] DEBARBIEUX B., LARDON S., *Les figures du projet territorial. La Tour d'Aigues*, Editions de l'Aube, 2003.

[DEI 09] DEININGER K., FEDER G., "Land registration, governance, and development: evidence and implications for policy", *World Bank Research Observer,* vol. 24, pp. 233–66, 2009.

[DES 00] DESROSIÈRES A., *La politique des grands nombres. Histoire de la raison statistique*, La Découverte-Poche, Paris, 2000.

[DES 07] DESMARAIS C., *La concertation dans la conduite de projet : Guide méthodologique*, Territorial Editions, 2007.

[DUP 14] DUPAQUIER C., DESBROSSES A., MAUREL P. *et al.*, "Apports de l'imagerie Pléiades à la gestion intégrée des zones côtières", *Revue Française de Photogrammétrie et de Télédétection*, vol. 208, pp. 45–50, 2014.

[FAL 09] FALUDI A., "A turning point in the development of European spatial planning? The Territorial Agenda of the European Union and the First Action Program", *Progress in Planning*, vol. 71, no. 1, p.1–42, 2009.

[FRE 00] FREUNDSCHUH S., "Micro- and macro-scale environments", in KITCHIN R., FREUNDSCHUH S. (eds), *Future, Routledge Frontiers of Cognitive Science,* Routledge Inc, New York, 2000.

[GRA 06] GRAND L., Guide méthodologique : la concertation dans la conduite de projet, DPSA (Grand Lyon), 2006.

[HEN 01] HENOCQUE Y., DENIS J., "Des outils et des hommes pour une gestion intégrée des zones côtières : Guide Méthodologique", Commission Océanographique Intergouvernementale, Paris, UNESCO-COI, vol. 2, no. 42, p. 64, 2001.

[HER 05] HERBAUX P., BERTACCHINI Y., "La rupture, un des corollaires de la vitesse", L'intelligence territoriale, une posture d'anticipation, in *Colloque TIC et Territoires, Quels développements ? Revue ISDM,* île Rousse, p. 10, 2005.

[HUF 07] HUFTY M., DORMEIER FREIRE A., PLAGNAT P. *et al.*, *Jeux de gouvernance: regards et réflexions sur un concept*, Karthala, Paris, pp. 117–122, 2007.

[IAA 05] IAAT Poitou-Charentes, Guide méthodologique du travail en commun, Institut Atlantique d'Aménagement du Territoire (IAAT), 2005.

[KIL 07] KILLI A., Manager un projet territorial, Association Rhône-Alpes des responsables du développement économique local (ARADEL), 2007.

[LAN 07] LANDEL P.A., "Entre politique publique et action publique: l'ingénierie territoriale", in FAURE A., NÉGRIER E. (eds), *Les politiques publiques à l'épreuve de l'action locale. Critiques de la territorialisation*, L'Harmattan, Paris, pp. 117–122, 2007.

[LAR 01] LARDON S., MAUREL P., PIVETEAU V., *Représentations spatiales et développement territorial*, Hermès, Paris, 2001.

[LAV 98] LAVIGNE DELVILLE PH., "La sécurisation de l'accès aux ressources par le titre ou l'inscription dans la communauté? ", in LAVIGNE DELVILLE Ph. (ed.), *Quelles politiques foncières en Afrique noire rurale? Réconcilier pratiques, légitimité et légalité*, Karthala, Paris, pp. 76–86, 1998.

[LEB 07] LE BOURHIS J.P., "Du savoir cartographique au pouvoir bureaucratique. La politique des risques en France 1970-2000", *Genèses*, vol. 68, no. 3, pp. 75–96, 2007.

[LEM 11] LEMOISSON P., MAUREL P., TONNEAU J.P., "Observatoires et gouvernance territoriale: une approche par la co-construction de modèles", *Fonder les sciences du territoire,* Paris, France, 23–25 November, 2011.

[LER 13] LEROY M., BAGHDADI N., CHERCHALI S. *et al.*, "First achievements of the French National Land Data Centre", *ESA Living Planet Symposium*, Edinburgh, UK, p. 9, September 2013.

[LET 05] Lettre de politique foncière, République de Madagascar. Ministère de l'Agriculture, de l'Elevage et de la Pêche, Secrétariat Général, 2005.

[LUS 03] LUSSAULT M., "L'espace *avec* les images", in DEBARBIEUX B., LARDON S. (eds), *Figures du projet territorial, La Tour d'Aigues*, l'Aube, pp. 39–60, 2003.

[MAR 13] MARTIN-SCHOLTZ A., MAYERE A., BARBE E. *et al.*, "Quand l'information échappe à ses créateurs. Le cas de l'artificialisation des terres agricoles en Languedoc-Roussillon", *Etudes de Communication,* vol. 40, pp. 145–165, 2013.

[MAT 96] MATHIEU P., "Pratiques informelles, gestion de la confusion et invention du foncier en Afrique", in DE VILLERS G. (ed.), *Phénomènes informels et dynamiques culturelles en Afrique*, Cedaf, Paris, pp. 64–87, 1996.

[MAU 12] MAUREL P., Signes, Données, Représentations Spatiales: des éléments de sens dans l'élaboration d'un projet de territoire intercommunal. Application au territoire de Thau, PhD thesis, University of South Toulon/Var, Toulon, 2012.

[MAU 15] MAUREL P., FAURE J.F., CANTOU J.P. *et al.*, "The GEOSUD remote sensing data infrastructure", *RSDI Seminar, Geospatial Week,* La Grande Motte, France, 28 September –2 October 2015.

[OWE 02] OWENS S., COWELL R., *Land and Limits: Interpreting Sustainability in the Planning Process*, Routledge, London/New York, 2002.

[RAR 98] RARE, Le développement durable : une autre politique pour les territories? Définition pratique/mise en oeuvre, Réseau des Agences Régionales de l'Energie et de l'Environnement (RARE), 1998.

[RAY 14] RAYSSAC G.L., DE LA GUÉRONNIÈRE C., *Guide de la concertation locale. Pour construire le vivre ensemble*, Territorial Editions, p. 122, 2014.

[SAG 95] SAGE, Guide cartographique des Sage, vol. 2: Informations pour la réalisation des cartes, 1995.

[SCO 10] Schéma de Cohérence Territoriale & Développement Durable, un nouvel avenir pour les territoires : Guide méthodologique, 2010.

[SOU 08] SOUCHIER E., ROBERT P., "La carte, un média entre sémiotique et politique", *Communication & Language,* vol. 158, pp. 25–106, 2008.

[SOU 09] SOUHEIL H., BOIVIN D., DOUILLET R. *et al.*, Guide méthodologique d'élaboration des Documents d'objectifs Natura 2000, Atelier Technique des Espaces Naturels (ATEN), Montpellier, 2009.

[STA 89] STAR S.L., GRIESEMER J., "Institutionnal ecology, 'Translations' and Boundary objects: amateurs and professionals on Berkeley's museum of vertrebate zoology", *Social Studies of Science*, vol. 19, pp. 387–420, 1989.

[TEY 09] TEYSSIER A., ANDRIANIRIANA-RATSIALONANA R., RAZAFINDRALAMBO R. *et al.*, "Décentralisation de la gestion des terres à Madagascar: processus, enjeux et perspectives d'une nouvelle politique foncière", in COLIN J.P. *et al.* (eds), *Les politiques de reconnaissance des droits fonciers. Du cadre légal aux pratiques locales,* IRD/Karthala, Paris, pp. 273–297, 2009.

[THO 96] THOENIG J.C., DURAN P., "L'État et la gestion publique territoriale", *Revue française de science politique,* vol. 46, no. 4, pp. 580–623, 1996.

[TOU 08] TOULMIN C., "Securing land and property rights in sub-saharan Africa: the role of local institutions", *Land Use Policy*, vol. 26, pp. 10–19, 2008.

[UNA 03] UNADEL, "Les cahiers de l'expérimentation. La mobilisation", Union nationale des acteurs et des structures de développement local (UNADEL), 2003.

[WHI 15] WHIRCRAFT A.K., BZCKEZR-RESHEF I., JUSTICE C.O., "A Framework for defining spatially explicit earth observation requirements for a global agricultural monitoring initiative (GEOGLAM)", *Remote Sensing*, vol. 7, pp. 1461–1481, 2015.

[WOR 87] WORLD COMMISSION ON ENVIRONMENT AND DEVELOPMENT, *Our Common Future*, Oxford University Press, Oxford, 1987.

4

Remote Sensing and Ocean Color

4.1. Introduction

In the Earth's remote sensing field, the term "water color" is used when analyzing the spectral distribution of visible radiation exiting an aquatic environment (ocean, lakes and rivers) to reach a satellite or airborne spectroradiometer sensor. In particular, when studying the determination and estimation methods of suspended matter concentration in the oceans, such as phytoplanktonic organisms, matters of mineral origin or even detrital and dissolved organic matters, the term "ocean color" is used. The main and historical interest in using ocean color satellite technologies to observe aquatic environment relies on the quantification of the phytoplanktonic biomass on a global scale. Indeed, phytoplanktonic organisms, which are microscopic unicellular algae often present in low concentrations in a given region, represent approximately 50% of the vegetation on the planet [LON 95] because of the very large surface of the world's oceans. Consequently, these organisms have a significant impact both on marine life, since they form the first link in the marine food chain, and on the climate, especially through the carbon cycle. Those organisms in fact help to capture the atmospheric carbon dioxide, one of the main greenhouse gases, through the process of the photosynthesis and primary production. A wide variety of phytoplankton species exist in seawater, categorized into different types according to their size and the photosynthetic pigments they contain. For example, we can mention the type of the diatoms, dinoflagellates

---

Chapter written by Malik CHAMI and Tristan HARMEL.

or cyanobacteria. Other common applications of ocean color satellite technologies concern the monitoring of toxic algae (for example, red tide), the study of nutrient dynamics, the study of marine ecosystem ecology, the monitoring of river discharge in coastal areas and their impact on the health of coastal ecosystems.

The geographic distribution of the suspended matters present in the oceans (phytoplankton, mineral matters and dissolved organic matters) is very different according to ocean regions and hence regular monitoring of this spatial distribution on a global scale is only possible from space. The development of satellite sensors dedicated to the observation of the ocean color started at the end of the 1970s with the launch of the first sensor by NASA the *Coastal Zone Color Scanner* (CZCS) sensor in 1978 [HOV 80].

Ocean color variability is linked to two fundamental optical processes: the absorption and the scattering of the solar light penetrating the water column by the suspended matters present. The absorption can simply be defined as the transformation of radiant energy into heat. Thus, it concerns a transformation mechanism of radiative energy caused by photons. The scattering is an angular redistribution of incident energy on a suspended particle. As such, scattering is different from absorption, since photons remain "visible", but their direction is modified after the interaction with the particle. Figure 4.1 schematically represents the two processes of light absorption and scattering, when it interacts with a particle. It should be noted that photons scattered in the direction of the source define the backscattering.

The absorption and scattering processes, called "optical properties", of oceanic light radiation, as will be measured by a satellite sensor, can be used in the study of the biological and mineral matters that we are interested in. Consequently, it is very important to accurately know the optical response of the different types of suspended matters, when they are exposed to solar radiation. The term "bio-optical" is commonly used to describe the link between solar radiation and the optical properties of suspended matters. Thereafter, we will refer to "marine constituents" or "hydrosols" as the suspended matter able to interact with solar radiation according to the two processes of absorption and scattering. These marine constituents are typically phytoplankton, mineral matters, detrital and dissolved organic matters. In order to determine the bio-optical concentrations and characteristics of the marine constituents from space, it is above all necessary to understand the different components of radiation that will be captured by an ocean color satellite sensor (section 4.2).

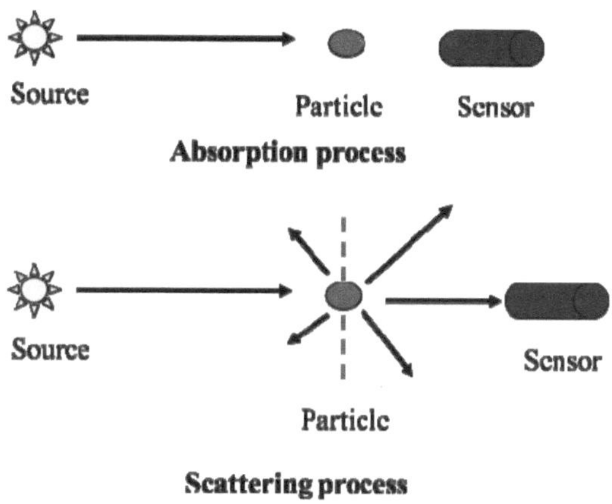

Figure 4.1. *Schematic representation of the light absorption and scattering processes by a particle*

4.2. Radiation components received by an observation satellite of the ocean color

The interaction of marine constituents with light radiation takes place mainly across the visible and near infrared wavelengths of the solar spectrum. Figure 4.2 depicts the variety of wavelengths, as well as the colors associated with the ocean.

The radiation (also called radiance) received by an ocean color satellite sensor goes through several interactions with all the elements (matters or surfaces) it meets along its journey from ocean to space. The radiation that composes the light reaching the satellite comes from three sources in the atmosphere–ocean coupled system:

– the water leaving radiance, written as L_w, which represents the ascending radiation L_w is the radiometric value that takes into account the absorption and scattering mechanisms, which take place in the water column between the solar light and the hydrosols. Consequently, the water leaving radiance. L_w is the most important component of the radiation sensed by a satellite, which defines the ocean content in hydrosols. This component is essential to estimate ocean color through remote sensing;

– the radiance as a result of the direct reflection of the solar radiation on the sea surface, called *sunglint* radiance (or also *glitter*), written as L_G.. This component results in the glare of the satellite sensor by the reflection of the sun on the sea, as shown in Figure 4.3;

– the atmospheric radiance, written as L_{atm}, which represents the interactions of the light with the constituents present in the atmosphere able to scatter radiation. These constituents are of two types: air molecules and suspended particles in the atmosphere, called aerosols. The scattering by air molecules is commonly called "Rayleigh scattering". Thus, the atmospheric radiance is the sum of the radiance linked to aerosols, written as $L_{aerosol}$, and of the radiance linked to air molecules, written as $L_{Rayleigh}$.

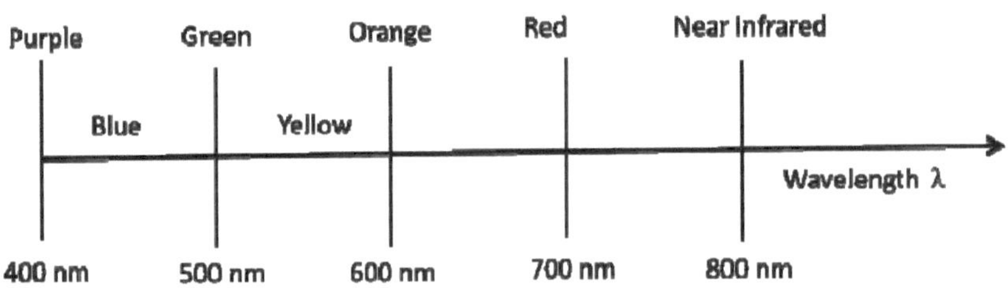

Figure 4.2. *Wavelengths and colors associated with the ocean color research field. For a color version of this figure, see www.iste.co.uk/baghdadi/5.zip*

Figure 4.3. *Illustration of the component of the sunglint radiance L_G that reaches a satellite sensor: a) photo of a marine surface; b) satellite image of the MODIS sensor (NASA), in which the white area covering the left side of the image is caused by the reflection of the sun on the sea surface, or glare*

All the components contributing to the solar radiation sensed by an ocean color observation satellite are represented in Figure 4.4, in which the processes of absorption and scattering are also mentioned. The propagation of solar radiation through the atmosphere–ocean system and the interactions of this radiation with atmospheric and marine constituents define the concept of "radiative transfer".

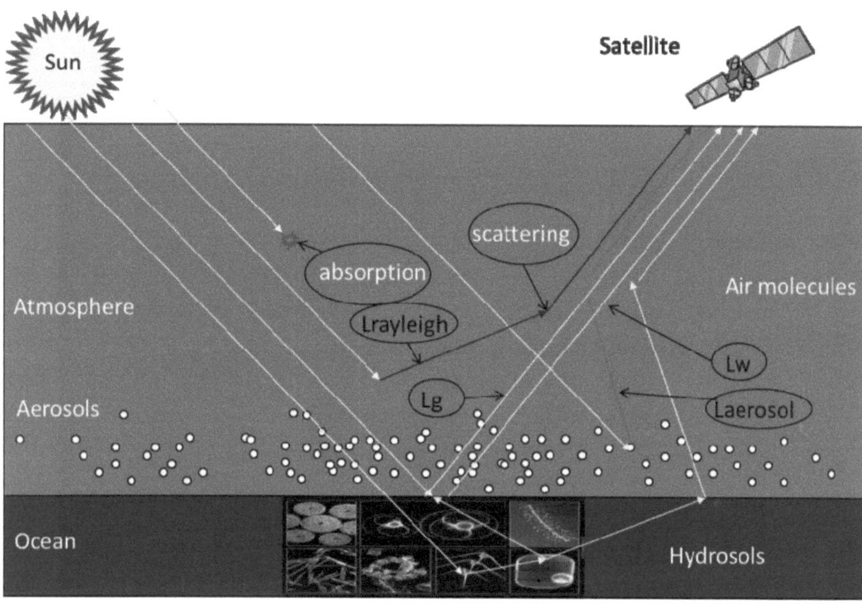

Figure 4.4. *Radiation sources sensed by an ocean color satellite sensor. L_w, L_G, $L_{aerosol}$, $L_{Rayleigh}$ are, respectively, the water-leaving radiance, the* sunglint *radiance, the radiance linked to aerosols, the radiance linked to air molecules. We will note that the sum of $L_{aerosol}$ + $L_{Rayleigh}$ makes up the atmospheric radiance L_{atm}*

The total radiance received at the top of the atmosphere by a satellite, $L_{satellite}$, is mathematically formulated as follows:

$$L_{satellite} = [L_{atm} + T_{atm}.L_W + T_{atm}.L_G]\ T_{gas} \quad [4.1]$$

where:

– T_{gas} is the transmittance linked to the radiation absorption by the gases present in the atmosphere. Atmospheric gases, which are mainly involved in solar radiation absorption in the visible and the near infrared (spectral field of interest for ocean color), are the ozone and water vapor.

– T_{atm} is the atmospheric transmittance, which is determined from the Beer–Lambert law as follows:

$$T_{atm} = \exp(-\tau/\cos\theta_v) \quad [4.2]$$

In equation [4.2], τ represents the total optical thickness of the atmosphere, and θ_v is the viewing angle of the satellite toward the Earth surface.

4.3. Correction of atmospheric effects from satellite images

As shown in Figure 4.4, the total radiance reaching the satellite, $L_{satellite}$, is significantly influenced by the scattering processes occurring in the atmosphere [BRI 16]. Consequently, the determination of the water-leaving radiance L_W, the only one carrying information on the nature of the hydrosols and on the ocean color, requires that the influence of the atmosphere on $L_{satellite}$ be corrected. This precursor to the analysis of the radiation exiting the ocean layer is called "atmospheric correction". As such, atmospheric correction consists of estimating the atmospheric radiance (L_{atm}), as well as the total transmittance of the air column (T_{atm} and T_{gas}), and then in subtracting those components from the satellite radiance $L_{satellite}$ to obtain L_W on the basis of equation [4.1].

The atmospheric correction of ocean color satellite images is a very difficult task in practice for several reasons. On the one hand, as the ocean areas are often dark targets, the component L_W only represents a maximum of 10% of the radiation $L_{satellite}$. Consequently, the atmospheric radiance L_{atm} must be estimated very accurately, typically with an error margin of less than 1%, in order to manage to estimate L_W with an error margin that will, however, be of 10% in this case. On the other hand, the estimation of L_{atm} becomes more complex because of the high variability of the radiation scattered by the aerosols present in the air column. In fact, contrary to the radiance scattered by air molecules, $L_{Rayleigh}$, which is rather stable in time and space, the concentration and nature of the aerosols widely vary according to the given regions and the season, which makes obtaining an accurate estimate of the radiance linked to aerosols $L_{aerosol}$ difficult.

The different steps involved in obtaining accurate atmospheric measurements of ocean color satellite data are as follows:

– cloud masking;

– removal of the component of the sun reflection on the sea (called *sunglint*);

– estimation of the radiance linked to molecules ($L_{Rayleigh}$) and aerosols ($L_{aerosol}$);

– estimation of the transmittances T_{gas} and T_{atm};

– estimation of the water-leaving radiance L_w.

4.3.1. *Cloud masking*

Cloud masking consists of ensuring that any clouds present on the satellite image are eliminated before the analysis of all the radiation components of equation [4.1]. Clouds are so opaque that they obscure the observation of the ocean surface. Image pixels corresponding to clouds must thus be excluded from the processing procedure. Cloud masking is a relatively simple occurrence. It is based on the fact that the radiation exiting the open ocean is very weak, even non-existent, when the scene is observed in the near infrared range, for example at a 870 nm wavelength. Conversely, clouds are reflective and induce a very strong radiation at this wavelength. Consequently, a simple method to detect clouds consists of establishing the threshold value of the satellite radiance $L_{satellite}$ for a near infrared wavelength. If the radiation measured at the top of the atmosphere is higher than that threshold value, the observed pixel is one composed of clouds. In this case, the pixel is removed by following the atmospheric correction process. An illustration of the cloud filtration procedure for a capture by the satellite sensor SeaWiFS (NASA) is indicated in Figure 4.5. However, we should note that the procedure described here is limited to when clouds are not very thick, for example, in case of cirrus or when the water is rich in suspended mineral matters like in coastal waters, for which the ascending radiation is no longer necessarily insignificant in the near infrared.

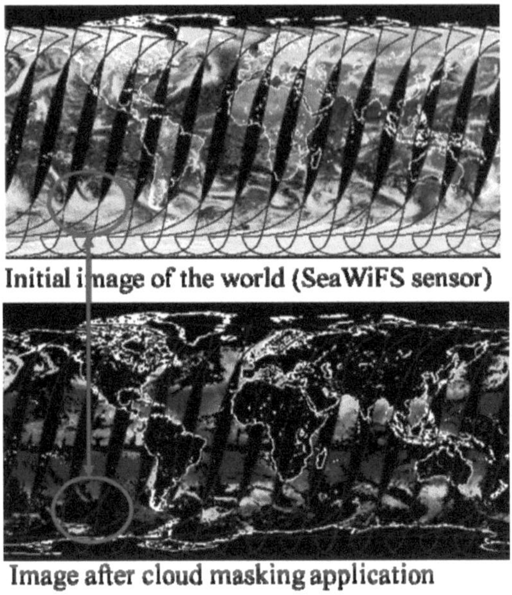

Figure 4.5. *Illustration of the step masking clouds on a SeaWiFS sensor satellite image (NASA) captured around the world. At the top, the false-color composite image (red-green-blue) shows the reflective clouds in white (red circle). At the bottom, the cloud pixels appear in black after the application of the cloud filter (in other words, the pixel is eliminated when the radiance $L_{satellite}$ at 870 nm is superior to a threshold value). For a color version of this figure, see www.iste.co.uk/baghdadi/5.zip*

4.3.2. *Eliminating sun reflection from the sea (L_G)*

The radiation linked to the reflection of the sun on the sea surface is highly dependent on the roughness of that surface, which directly depends on the wind speed. So, in case of a "flat" sea (wind speed = 0), the solar radiation falling directly (i.e. not scattered by air molecules and aerosols) will be reflected by the sea surface in the same direction as the solar zenith angle. This is called the specular reflection. However, when the wind speed is not zero, the direct solar radiation reflection will occur in directions around the solar zenith angle. In this case, which is the most common, a wave slope distribution model, aimed at providing the probability of the direct solar radiation reflection in a given observation direction according to the wind speed, should be used. The most commonly used model in the literature is the one developed by Cox and Munk [COX 54], which is based on the observation of the wave geometry from a series of aerial photographs of the sea surface. Though far from recent, the model was found to be very reliable in describing wave slope probability and to provide the necessary tools for modeling the radiation L_G linked to *sunglint*. In order to assess the radiance L_G, simulations were carried out by means of an "atmosphere–ocean" coupled radiative transfer model taking into account the wave slope distribution of Cox and Munk [COX 54]. Furthermore, simulations allow us to accurately esitimate the observation geometries affected by *sunglint* radiation. Thus, on the basis of the geometrical conditions in which the ocean color satellite image was acquired, theoretical simulations provide both L_G and the pixel geometry affected by the *sunglint*. There are very few radiative transfer models available to the scientific community which allow us to consider an atmosphere–ocean coupled system with a rough surface to calculate the radiation L_G. An example of this radiative transfer model is the "OSOAA" model (Ocean Atmosphere Succeeding Orders) which was developed by the Laboratoire d'Océanographie de Villefranche (LOV) at the University Pierre and Marie Curie, and the company CS-SI (Toulouse) based on funding provided by the French space agency, CNES (Centre National d'Études Spatiales) [CHA 15].

Most of the time, the satellite image pixels that will be geometrically affected by the sun reflection on the sea will be eliminated from the image process procedure, since they do not give any information on L_w. Since a large part of an image is affected by *sunglint*, methods have been developed in order to correct the radiance L_G, thus allowing the use of the pixels affected by *sunglint*. Thus, the "POLYMER" method [STE 08a] was found to be as efficient to correct the radiance $L_{satellite}$ as the radiance L_G. An example of the application of the POLYMER method is given in Figure 4.6, in which one can see that all the black pixels contaminated by *sunglint* (Figure 4.6(a)) are corrected in order to provide a consistent value of the phytoplanktonic biomass (Figure 4.6(b)).

4.3.3. *Estimation of the radiance linked to molecules ($L_{Rayleigh}$) and aerosols ($L_{aerosol}$)*

4.3.3.1. *Estimation of the molecule contribution ($L_{Rayleigh}$)*

The contribution of air molecules to the radiation received at the top of the atmosphere is the most significant among all the other components of equation [4.1], in the visible short wavelength range ($\lambda < 600$ nm). The light scattering by oxygen and nitrogen molecules is directly linked to solar radiation wavelength according to a λ^{-4} variation law. Air molecules are considered to be electric dipoles, the physical laws of which allow us to accurately quantify the quantity of scattered light according to wavelength. Thus, the optical thickness of air molecules (optical thickness called Rayleigh and written as $\tau_{Rayleigh}$) is determined with an accuracy of less than 0.1% as follows:

$$\tau_{Rayleigh} = 0.0089\lambda^{-4}P/Po \qquad [4.3]$$

where λ is the visible wavelength, P is the atmospheric pressure at a given altitude z and Po is the atmospheric pressure at sea level.

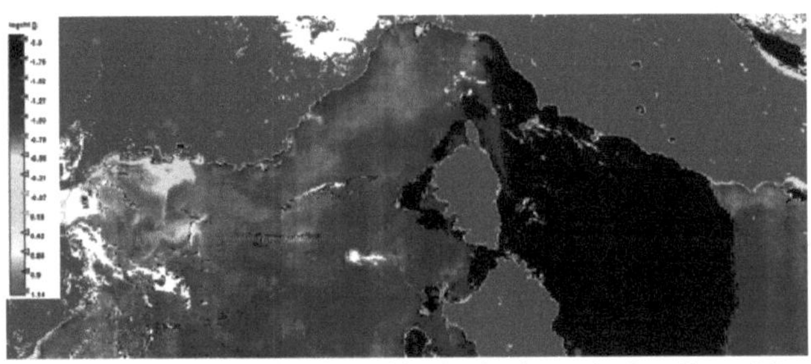

(a) Before the "POLYMER" application : black pixels = pixels affected by "sunglint"

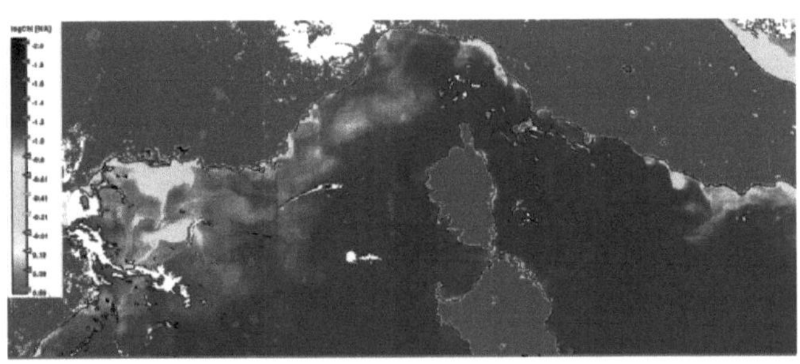

(b) After the "POLYMER" application

Figure 4.6. *Example of the POLYMER method used to correct the pixels affected by sunglint of an ocean color satellite image (chlorophyll concentration here) captured by the MERIS sensor (ESA) on July 25, 2003 above the Mediterranean sea (Corse, France): a) before the POLYMER application and b) after the POLYMER application [STE 08b]. For a color version of this figure, see www.iste.co.uk/baghdadi/5.zip*

4.3.3.2. *Estimatation of the aerosols contribution ($L_{aerosol}$)*

The determination of the aerosol radiance $L_{aerosol}$ is far more complex than for the other components, because of the high spatiotemporal variability of the aerosols present in the atmosphere. That difficulty is the reason for the constant development of $L_{aerosol}$ estimation methods by the scientific community, specific to the instrumental characteristics of each satellite sensor (for example, wavelength values or radiometric performances). Here, we present the general principle on which is based the great majority of the methods to estimate the component $L_{aerosol}$ in the prevailing case when observations take place over open ocean waters (i.e. outside the coastal areas). In the open ocean, the absorption of the marine radiation by the sea water molecules is so high in near infrared wavelengths (typically $\lambda > 700$ nm) that it is considered that the water-leaving radiance L_w is nil, or in other words, that the water is "black". It should be highlighted that such approximation is not true for coastal waters rich in mineral matters (which represent less than 10% of the world ocean), which can significantly scatter light in the infrared. In this chapter, we only consider open ocean waters. In the infrared, on the basis of a "black" water (where $L_w = 0$) and knowing that the scattering linked to air molecules is known (and very small), the radiation reaching the satellite is mostly caused by the scattering of aerosols. Consequently, the observation of open oceans at high wavelengths provides direct information on the component $L_{aerosol}$.

The estimation procedure of $L_{aerosol}$ is based on the following principle: two near infrared wavelengths for which $L_w = 0$, for example 750 and 870 nm for the SeaWiFS sensor (NASA), are first considered. In those conditions, the radiation measured by the satellite, $L_{satellite}$, directly provides the contribution $L_{aerosol}$. Second, the radiance ratio $L_{aerosol}$ between those two wavelengths, $L_{aerosol}(750 \text{ nm})/L_{aerosol}(870 \text{ nm})$ is calculated and then compared to theoretical simulations of radiative transfer that were performed for several aerosol models documented in the literature (urban, marine and continental aerosols). For example, the aerosol models of Shettle and Fenn [SHE 79], or more recently of Ahmad *et al.* [AHM 10], are representative of the optical properties (absorption, scattering and abundance) of a large number of aerosols, by taking into account the relative air humidity. The comparisons between the $L_{aerosol}(750 \text{ nm})/L_{aerosol}(870 \text{ nm})$ ratio measured by satellite and the one calculated for different aerosol models allow us to select the model which finds the best agreement with the measurements. Subsequently, aerosol optical properties, such as their optical thickness, their spectral dependency and their capacity to absorb radiation (called single scattering albedo) can be deducted. Once the aerosol model (or atmosphere model) is defined by this step, the radiative transfer simulations are used to calculate the radiance $L_{aerosol}$ at wavelengths other than the near infrared ($\lambda < 700$ nm) for which $L_w \neq 0$.

4.3.4. *Estimation of T_{atm} and T_{gas} transmittances*

The atmospheric transmittance T_{atm} [4.2] is easily calculated with the Beer–Lambert exponential law, once the atmospheric model has been determined, since the optical thicknesses $\tau_{Rayleigh}$ and $\tau_{aerosol}$ (whose sum provides the total optical thickness) are now known.

The transmittance T_{gas} is also determined by applying the Beer–Lambert exponential law, but this time using the optical thickness of the absorbing gases, in particular the one of the atmospheric ozone. The ozone optical thickness is typically provided by the data acquired by another satellite sensor specialized in ozone detection, for example the TOMS sensor (NASA).

4.3.5. *Estimation of the water-leaving radiance L_w*

Numerous atmospheric correction algorithms of ocean color observation satellite images have been developed since the launch of the first satellite, CZCS. Even if the great majority of those algorithms relies on the "black" water method in the near infrared described above (i.e. the radiation exiting the water is nil), each algorithm has its own features according to the technical specifications of the sensor to which it will have to apply. One of the pioneer atmospheric correction algorithms is the one developed by Gordon and Wang [GOR 94] for the SeaWiFS sensor [GOR 94]. However, given the difficulty to obtain radiance L_w with the required accuracy (typically 5%), in order to then determine the marine constituents, other methods were proposed. For example, the use of wavelengths located in the blue range [HU 12] or even in the mid infrared (*shortwave*) [WAN 07] was considered to improve the determination of the aerosol model. More recently, numerous numerical and statistical optimization techniques, for example neural networks, have been used to make the atmospheric correction methods operational, given the increasing progress of the microprocessor performances [JAM 05, AHN 07, DOE 07, KUC 09, STE 11, HAR 11, BRA 12]. Those methods allowed us to significantly improve the performance and accuracy of the satellite derivatives. The evaluation of different atmospheric correction algorithms was conducted by many authors according to the ocean color sensor studied. For example, Goyens *et al.* [GOY 13] compared the radiance L_w obtained after the atmospheric correction of the MODIS sensor (NASA) with *in situ* measurements (Figure 4.7). Their results showed that relative gaps of the range of 15% were observed between the radiance derived from the satellite data and the *in situ* measurements at 550 nm.

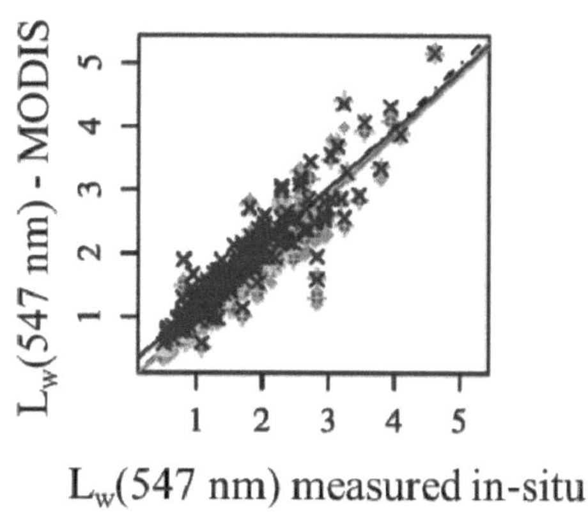

Figure 4.7. *Comparison of the radiance L_W at 547 nm estimated after the atmospheric correction of the images captured by the MODIS satellite sensor with in situ measurements [GOY 13]. The points in different colors on the figure correspond to different atmospheric correction methods [GOY 13] . For a color version of this figure, see www.iste.co.uk/baghdadi/5.zip*

Since all the atmospheric contributions to the satellite signals are known, it is possible to estimate the contribution of the ocean radiation L_W by means of [4.1].

Figure 4.8 shows the impact of the atmospheric correction on the radiance measured at the top of the atmosphere by a satellite sensor (MODIS). The corrected atmosphere image shows many more patterns and structures of the ocean radiation than the original image captured from space.

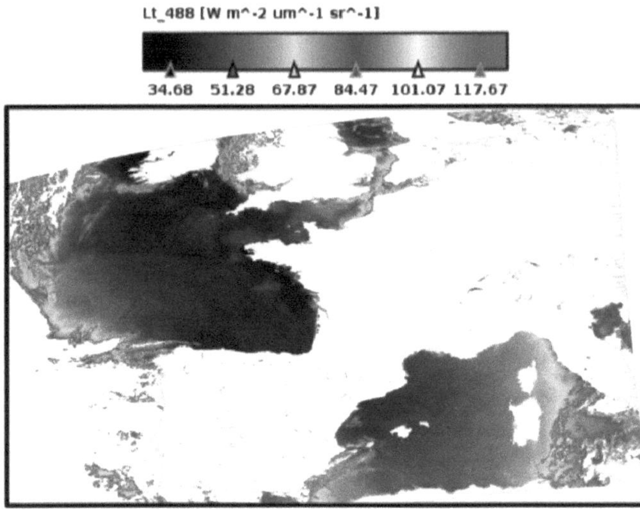

a)

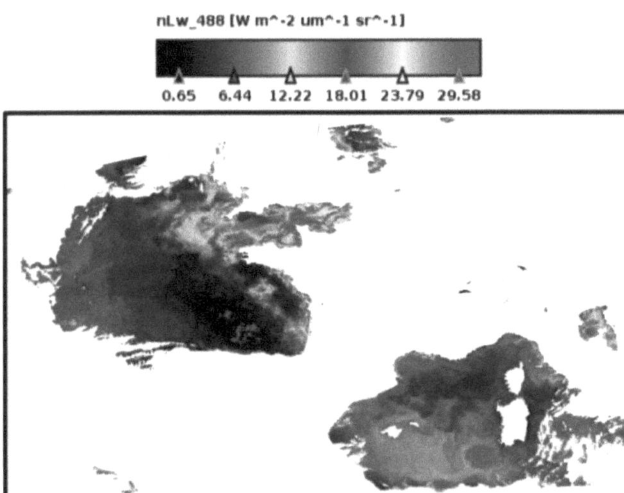

b)

Figure 4.8. *Illustration of the atmospheric correction impact on an ocean color satellite image (MODIS) captured above France: a) radiance $L_{satellite}$ measured at 488 nm (in W m^{-2} µm^{-1} sr^{-1}) before the atmospheric correction and b) radiance L_w at 488 nm obtained after the correction of the atmosphere effects. For a color version of this figure, see www.iste.co.uk/baghdadi/5.zip*

Now that the atmospheric correction step is completed and that the water-leaving radiance L_w is determined, we must focus on the bio-optical properties of the marine constituents, in order to link L_w with the suspended matter concentrations.

4.4. Bio-optical properties of seawater

The water leaving radiance is directly dependent on the processes of absorption and scattering by water molecules and hydrosols. Thus, L_w can be expressed as a function of the hydrosol optical properties as follows:

$$L_w = f(E_{\text{solar}}(\lambda), b(\lambda), a(\lambda)) \qquad [4.4]$$

where E_{solar} is the incident solar illumination (in W m^{-2}), and $b(\lambda)$ and $a(\lambda)$ are, respectively, the scattering and absorption coefficients of the water column.

Note that the absorption $a(\lambda)$ and scattering $b(\lambda)$ coefficients are mathematical and physical expressions of the way marine constituents interact with light. Thus, those coefficients, whose unity is m^{-1}, facilitate the quantification of the mechanisms described in Figure 4.1. In order to manage to link L_w to the marine constituent concentrations, one must previously know the spectral optical behavior ($a(\lambda)$ and $b(\lambda)$) of each of the constituents.

4.4.1. *Optical properties of water molecules*

Figure 4.9 shows the spectral variation of the water molecules spectral absorption coefficient, a_w, as measured by numerous authors.

It is important to note that the pure seawater absorption coefficient reaches very high values (up to several m^{-1}) for $\lambda > 600$ nm. As such, long wavelengths (red and near infrared) will be highly absorbed by water molecules and will not be able to deeply spread in the water column. As order of magnitude, if the seawater did not contain water molecules (no hydrosols), 90% of the sunlight entering the ocean would be absorbed at a depth of 500 m at 412 nm (blue) and only at a depth of 0.75 m at 760 nm (near infrared). The very small penetration of the solar radiation in the near infrared (few centimeters) caused by water molecules absorption is the main justification for the theory of the black waters in this spectral range on which relies the general principle of the atmospheric corrections (section 4.3).

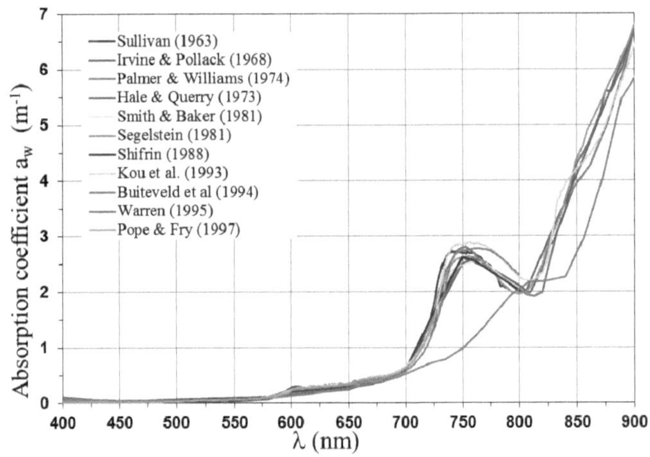

Figure 4.9. *Spectral variation of the pure sea water absorption coefficient (water molecules without hydrosols) a_W measured by different authors. For a color version of this figure, see www.iste.co.uk/baghdadi/5.zip*

Figure 4.10 illustrates the spectral behavior of the scattering coefficient by water molecules, b_W. It shows that those molecules highly scatter short wavelength radiation in the blue ($\lambda < 500$ nm), whereas the scattering decreases and tends to 0 at long wavelengths. This spectral variation of the scattering, which follows a type λ^{-4} law, coupled with high absorption properties in the red (Figure 4.9), explains why the dominant color in hydrosol-poor waters, like the ones encountered in most of the open ocean, is blue.

4.4.2. *Optical properties of phytoplankton*

The phytoplankton is the marine constituent for which the concentration estimate is crucial to improve our knowledge of the carbon cycle on a global scale and, subsequently, the ocean role in climate change. The interaction of the phytoplankton with light occurs through the photosynthetic pigments contained in phytoplanktonic cells. Those pigments are in particular responsible for the light absorption by the phytoplankton, which uses it to feed and reproduce itself through the photosynthesis mechanism. The production of organic matter (including carbon) by phytoplankton is called primary production.

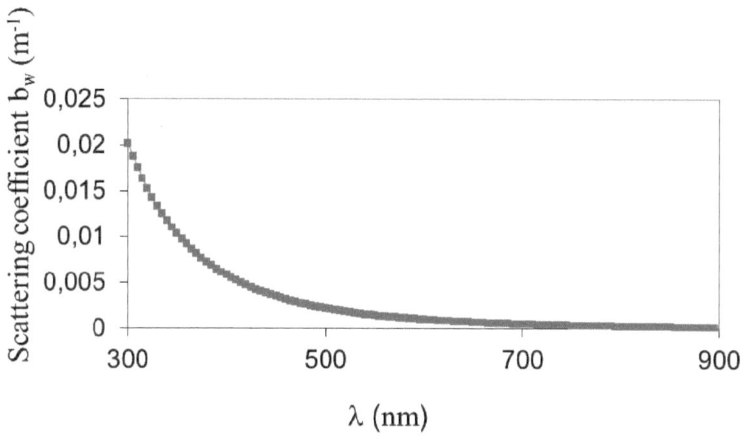

Figure 4.10. *Spectral variation of the pure sea water scattering coefficient b_w [MOR 74]*

As we pointed out in the introduction, the diversity of the phytoplankton species is very significant in the marine environment. However, all the species contain a pigment which plays a pivotal role in the ocean color, chlorophyll *a*. That pigment is in fact the only one that is common to all phytoplanktonic organisms. In addition, it interacts with the light radiation that reaches it by absorbing it. Consequently, it is used in ocean color as the main indicator of phytoplanktonic biomass. In other words, the value we are trying to estimate from space, when we wish to quantify the phytoplankton concentration, will be the chlorophyll *a* concentration, written as Chl-a (in mg m^{-3}). It is important to highlight that all the other pigments existing in the phytoplanktonic cell, for example chlorophyll *b*, chlorophyll *c* or carotenoids, have an impact on the light radiation. As those pigments are not necessarily in every species, they are sometimes used as markers for the presence of a specific phytoplankton species in the water mass.

The absorption spectrum of the phytoplankton, written as $a_{Chl\text{-}a}$, indicates the phytoplankton ability to absorb radiation for a given concentration in chlorophyll *a*. It is very variable according to the phytoplankton types (Figure 4.11). However, it presents two significant properties that are common to all these types. They represent a major absorption peak of the radiation at a wavelength of 440 nm (blue) and a secondary absorption peak around 675 nm (red).

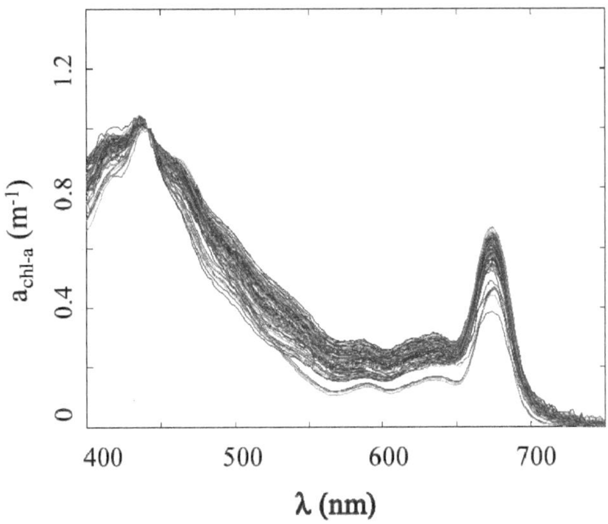

Figure 4.11. *Absorption spectra of the phytoplankton for different species (extract from the reference database of the MERIS sensor, European Space Agency). The absorption coefficient $a_{chl\text{-}a}$ was normalized by the value at 440 nm (i.e. $a_{chl\text{-}a}$ (440 nm) = 1) in order to reduce the impact linked to the variability of Chl-a. For a color version of this figure, see www.iste.co.uk/baghdadi/5.zip*

The major peak of absorption by chlorophyll, which can be seen in the blue, is the main reason for the ocean color change, when the water is concentrated in phytoplankton. The blue color, for which water molecules are responsible, is replaced by greenish colors, since red colors are also highly absorbed. The absorption spectrum $a_{Chl\text{-}a}$ displays a high variability according to the species in the spectral region located between the two peaks. One of the reasons for this variability is what we call the *package effect* [BRI 81]. The *package effect* is an effect that reflects the fact that chlorophyll pigments are located within the chloroplasts in the phytoplanktonic cell. Those chloroplasts create packages that lead to an uneven distribution of the pigments. That is why the absorption peak at 440 nm generated by chlorophyll *a* often has a smaller magnitude than in cases where the pigments are evenly distributed in the cell (in other words, not contained in the chloroplasts) for the same concentration in chlorophyll *a*. The *package effect* also indicates the fact that an increase in absorption is not linearly proportional to the concentration in chlorophyll *a*. The *package effect* depends on several factors including the physiological conditions of the cell and the intensity of the light radiation. Consequently, two absorption spectrums of the phytoplankton can be different for the same species according to the environmental conditions in which the species is. It is the reason why the determination of the concentrations in chlorophyll *a* through the absorption spectrum can be tarnished by strong uncertainties that will also be present when those concentrations are estimated from space.

The scattering coefficient of the phytoplankton, $b_{chl\text{-}a}$, follows a spectral variation law typically under the form λ^{-1}. In practice, the scattering coefficient of the phytoplankton is not often measured in seawater. It is the scattering coefficient of all the hydrosols (phytoplankton, mineral matters and detritus) that is measured and deconvolution mathematical techniques are used to deduce the phytoplankton contribution to the overall scattering.

4.4.3. *Optical properties of colored dissolved organic matter*

Colored dissolved organic matter, or CDOM, are complex molecules mainly composed of humic and fulvic acid. They are are directly linked to the degradation of living organic matter. In open ocean waters, those matters are in first order dependent on the phytoplankton concentration. However, in coastal waters, CDOM come from both the phytoplankton degradation and the degradation of land-based plant matters brought by river discharges. The optical properties of CDOM are limited to absorption properties, because they are supposedly non-scattering. In reality, they are a source of light scattering, but it is difficult to isolate the CDOM contribution to the scattering compared to the one of the other hydrosols present in water. The CDOM absorption coefficient, a_{CDOM}, varies according to an exponential decay law linked to wavelength (Figure 4.12):

$$a_{CDOM}(\lambda) = a_{CDOM}(\lambda_0)\, e^{-S_{CDOM}(\lambda-\lambda_0)} \qquad [4.5]$$

where λ_0 is a reference wavelength, of typically 440 nm, and S_{CDOM} is the spectral slope in which the variation range often seen in oceans lies between

0.01 and 0.03 nm^{-1}. The average value of S_{CDOM} observed in open ocean waters is 0.014 nm^{-1}. The S_{CDOM} variable can provide information on the CDOM origin (for example, phytoplankton origin, river discharges and bacterial origin) and on the photo-oxidation state of the dissolved matter.

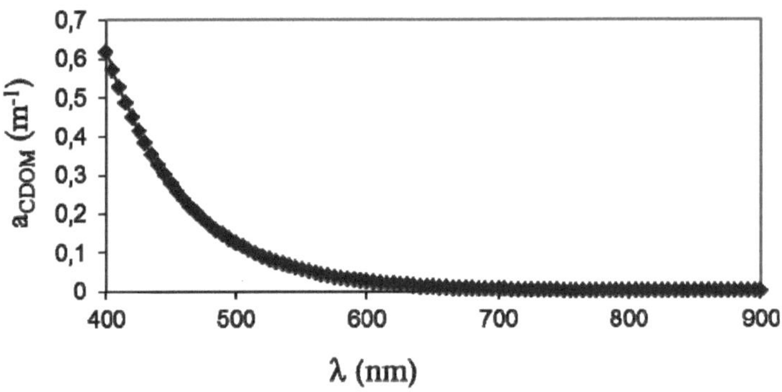

Figure 4.12. *Typical absorption spectrum of the colored dissolved organic matter (CDOM) [ROE 89]*

The light absorption by CDOM is very high in the blue and quickly subsides to become nearly nil when $\lambda > 600$ nm. This is the reason why those colored dissolved matters are also called "yellow substances", or *gelbstoff*, given the brownish color they give to sea waters, when they are present in high concentrations. It should be noted that CDOM absorption at short wavelengths covers the one of chlorophyll *a*, in particular at 440 nm, which makes it difficult to separate the contributions specific to CDOM and chlorophyll *a* from the total absorption.

4.4.4. *Optical properties of the detrital organic matter*

Detrital organic matters are matters derived from phytoplankton degradation, but of a large enough size, typically greater than 0.2 µm, in order to be considered as non-dissolved particulate matters. Those detrital matters have nearly the same optical properties as the CDOM dissolved matters, i.e. that their absorption coefficient, a_{det}, follows a decay exponential law linked to wavelength [4.4]. However, the slope S_{det} has a value slightly lower than the one for the CDOM. The average value observed in the oceans is typically of 0.011 nm^{-1}. The scattering properties of the detritus are also difficult to differentiate from the ones of the other hydrosols, since the *in situ* measurement is carried out with instruments that do not often allow us to do it. However, note that the scattering spectrum of the hydrosols varies according a type $\lambda^{-\alpha}$ potential decay law.

4.4.5. *Optical properties of mineral matters*

Mineral matters are mainly encountered in coastal regions, because of river discharge and coastline erosion. However, they can be present in open oceans in the form of desert dust deposits transported far from their emitting source through the atmospheric circulation. The absorption properties of mineral matters are still not well known and are a controversial subject within the scientific community, because they are so hard to measure. For a long time, they have been considered as non-absorbing, but recent studies tend to show that they can in fact be absorbing [PEN 13]. The scattering coefficient by the mineral matters, $b_{mineral}$, generally follows a type $\lambda^{-\alpha}$ potential decay law [BAB 03].

4.4.6. *Additivity of optical properties*

The optical properties of marine constituents are additive. In fact, the total absorption and scattering coefficient of a water mass is the sum of the contributions of each marine constituent. Thus, we can write:

$$a_{total}(\lambda) = a_w(\lambda) + a_{chl-a}(\lambda) + a_{\text{det}}(\lambda) + a_{CDOM}(\lambda) \quad [4.6]$$

$$b(\lambda) = b_w(\lambda) + b_{chl-a}(\lambda) + b_{\text{mineral}}(\lambda) + b_{\text{det}}(\lambda) \quad [4.7]$$

As an example, Figure 4.13 shows the absorption spectrum of the different marine constituents and the total spectrum. It shows that beyond 670 nm, the absorption by pure seawater molecules largely prevails compared to the one of the other constituents. The additivity property of the absorption and scattering coefficients is an advantage, because it allows the development of deconvolution method to identify, by means of mathematical techniques, the contributions of each marine constituent to the absorption and to the scattering from a measurement of the total coefficients.

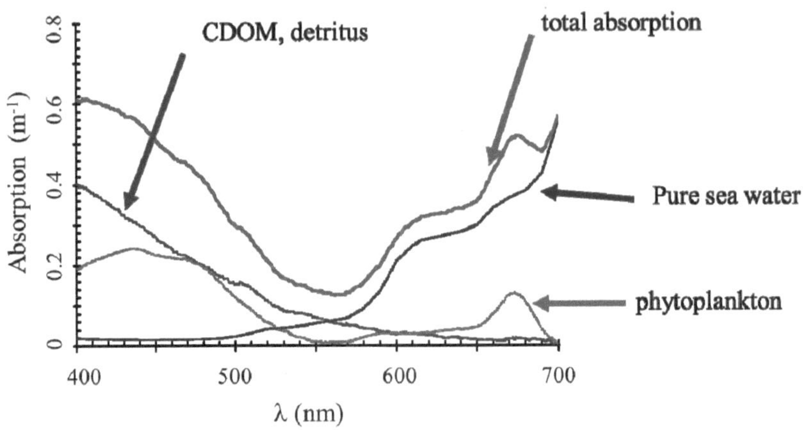

Figure 4.13. *Absorption spectra of the different marine constituents of the total absorption coefficient*

4.4.7. *Definition of the radiometric values used in remote sensing*

Radiometric values are physical values that depend on both the optical properties of hydrosols and ambient light radiation. They can be measured with a satellite sensor. The radiance, which we have already previously mentioned, is the basic radiometric value. It expresses the quantity of light that is received (or emitted) per surface unit for a cone focused on the observation direction (called solid angle). Its measurement unit is Watt m^{-2} sr^{-1}. From the radiance, several other radiometric values can be deduced.

Illumination (or irradiance) is the complete radiance on a set of directions. Its unit is Watt m^{-2}. For example, the descending plane irradiance, E_d, is the sum of the descending radiance in all the directions on a flat surface. Figure 4.14 schematically represents the definition of the descending and ascending plane irradiance (E_u), which are values often used in ocean color, in particular at the air–water boundary level.

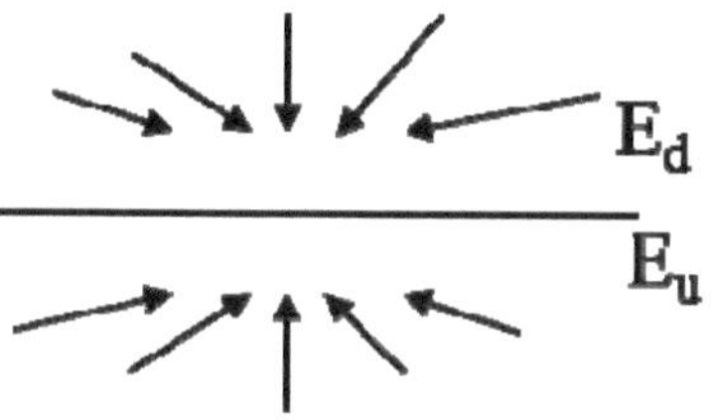

Figure 4.14. *Schematic representation of the descending plane irradiance E_d and the ascending plane irradiance E_u on a flat surface*

From the descending and ascending irradiances on the sea surface (or in the water column), the scattered marine reflectance R is defined by the E_u/E_d ratio:

$$R(\lambda) = \frac{E_u(\lambda)}{E_d(\lambda)} \qquad [4.8]$$

The point of dividing the ascending irradiance E_u by the descending irradiance E_d is to separate the radiation E_u from the solar illumination conditions and the atmospheric conditions in order to compare the water-leaving radiation of ocean waters. The marine reflectance R is a crucial value in the ocean color field, because it can directly link the illumination water-leaving radiation with the bio-optical properties of hydrosols. It can be easily measured on site, which allows us to establish *in situ* relationships between marine constituents and ocean radiation, as we will later see (section 4.5).

Given that a satellite aims in a given direction, it only measures one radiance and cannot measure an ascending irradiance. The *remote sensing* reflectance, Rrs, is then defined as the ratio of the water-leaving radiance L_w per irradiance unit descending on the surface (sr^{-1} unit):

$$R_{rs}(\lambda) = \frac{L_w(\lambda)}{E_d(\lambda)} \qquad [4.9]$$

The remote sensing reflectance, which is also a directional reflectance, is the one used to validate the quality of the atmospheric corrections of a satellite sensor by comparing *in situ* measurements of L_W with the ones estimated from the satellite, free of atmospheric conditions and solar radiation. The descending irradiance E_d is calculated by means of a radiative transfer model from the atmosphere optical properties known (for example, the optical thickness of aerosols), which were obtained during the atmospheric corrections step.

The reflectance R_{rs} is directly linked to the nature and abundance of the marine constituents in the water column and their respective optical properties, as illustrated in Figure 4.15. In particular, R_{rs} depends in first order on the ratio of the backscattering coefficient of hydrosols and the absorption coefficient of those hydrosols. Thus, when the quantity of backscattered light increases, the reflectance R_{rs} increases; when the hydrosols are highly absorbing, the reflectance R_{rs} decreases.

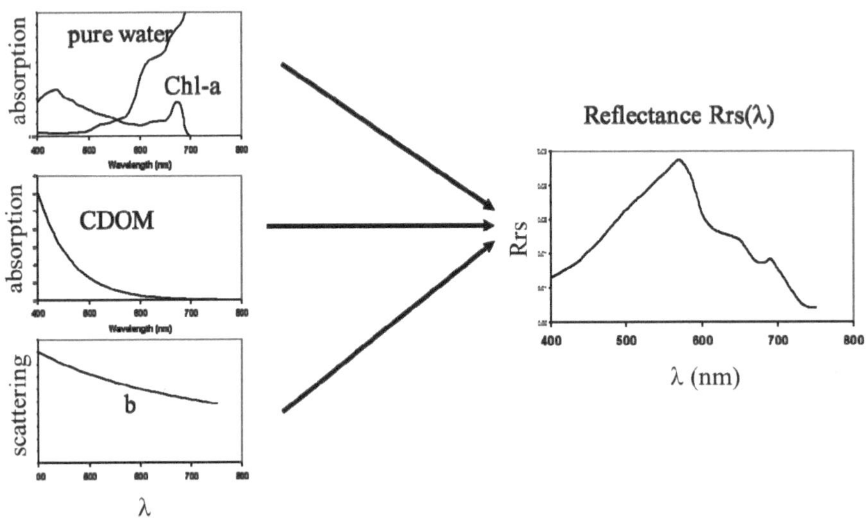

Figure 4.15. *Relationship between the optical properties of the marine constituents and the remote sensing reflectance R_{rs}*

4.5. Determination principle of hydrosol concentrations by satellite

From a radiometric value (radiance or reflectance R_{rs}) measured by the satellite sensor and corrected from the atmosphere effects, it is possible to estimate bio-optical properties, and even the concentration of marine constituents, given the strong link between optical properties and reflectance (Figure 4.15). We call the "inverse problem" of the ocean color the step consisting of converting the reflectance R_{rs} in terms of bio-optical values (absorption, scattering coefficients and concentrations), which are a major relevance for ocean ecosystems. The inverse problem in the case of open oceans consists of determining the suspended matter (Chl-a) present in the water column from the R_{rs} measured by satellite (Figure 4.16).

The methods that are used to determine the concentrations in hydrosols are different if the satellite images are captured over the open ocean or over coastal waters. In this chapter, we are only discussing the case of open oceans, i.e. the case where waters are mainly dominated by phytoplankton and the matters associated with its degradation (CDOM and detritus). The inverse problem of coastal waters is far more complex because of the presence of mineral matters and land-based dissolved matters with optical properties that no longer co-vary with the ones of the phytoplankton.

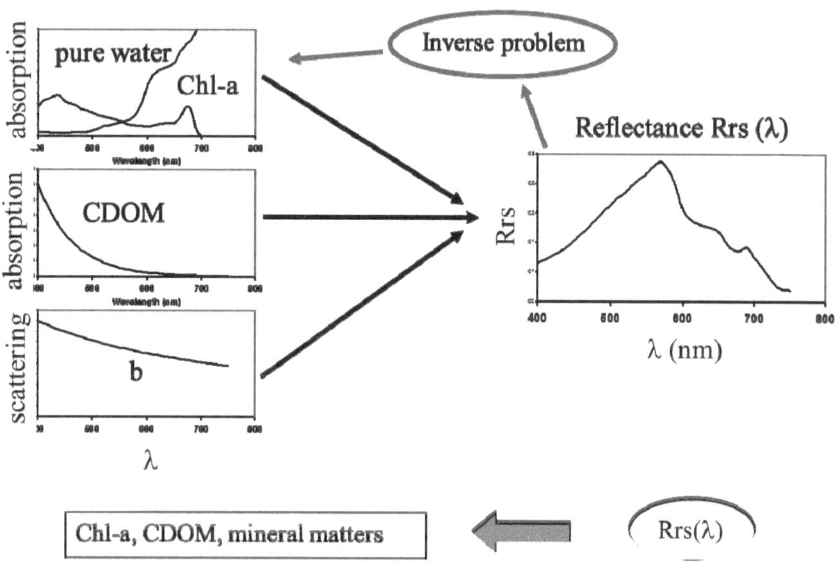

Figure 4.16. *Illustration of the inverse problem of the ocean color*

4.5.1. *Spectral variation of the reflectance according to chlorophyll a*

The marine reflectance necessarily depends on the concentration in chlorophyll *a* given the spectral variations of this pigment absorption

(Figure 4.11). This spectral dependence on chlorophyll *a* and on reflectance represents the basic principle of the multispectral inversion techniques to determine the concentration in chlorophyll *a*. Figure 4.17 shows the spectral variation of the scattered reflectance R for several concentrations in chlorophyll *a*.

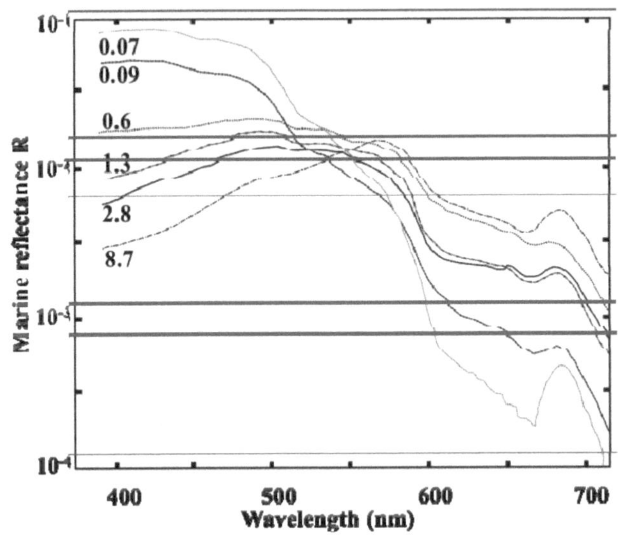

Figure 4.17. *Variation of the marine reflectance R for different concentrations in chlorophyll a (Chl-a) [BER 80].The concentration values in chlorophyll a (from 0.07 to 8.7 mg m^{-3}) are indicated on the left of each curve*

The reflectance spectra demonstrate several aspects that should be highlighted:

– a significant decay of R according to the concentration in Chl-a in the blue wavelengths (λ<500 nm);

– a relative insensitivity of R with Chl-a in wavelengths close to 550 nm (neutral point);

– a substantial growth of the reflectance in long wavelengths (λ> 600 nm) when Chl-a increases.

The spectral variation of R largely reflects the reverse of the spectral variation of the phytoplankton absorption (a_{chl-a}) (Figure 4.11), since the radiation that is not absorbed contributes to the reflectance. However, it should be noted that the magnitude of R (in other words, absolute values) is stronger than a factor 10 in the blue compared to the red.

4.5.2. *Estimation of the concentration in Chl-a*

On the basis of the variation properties of the reflectance spectrum with Chl-a (Figure 4.17), the existence of a relative insensitivity of the reflectance around 550 nm, which is similar to a "neutral point", is used as a "pivot" to calculate the spectral slope of R between 440 (wavelength or R variations with Chl-a are the most significant) and 550 nm, R_{440}/R_{550}. The value of the R spectral slope is highly dependent on the values of Chl-a (Figure 4.18).

In particular, the spectral blue/green (R_{440}/R_{550}) ratio increases when the concentration in Chl-a decreases. Thus, the slope variations can be used to estimate Chl-a. Chl-a can be expressed as a function of the R_{440}/R_{550} reflectance ratio.

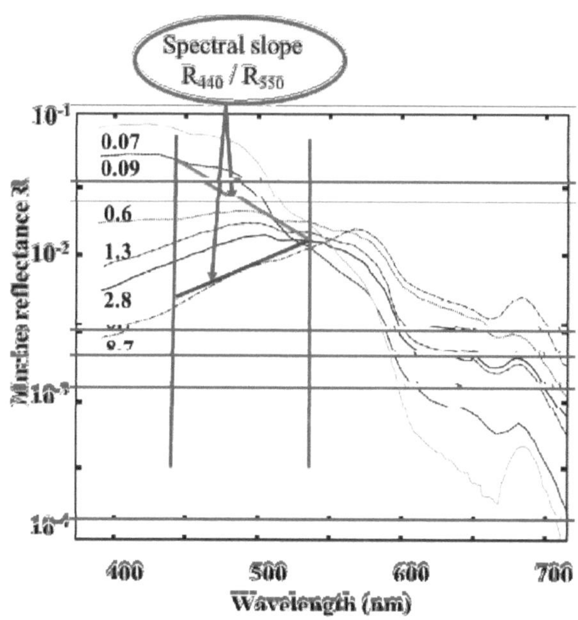

Figure 4.18. *Variation of the spectral slope of the marine reflectance R calculated between 440 and 550 nm, R_{440}/R_{550}, according to the concentration in Chl-a (from Figure 4.17). The blue and dark green straight lines indicate different values of the R_{440}/R_{550} slope for two given concentrations in Chl-a (0.09 and 8.7 mg m^{-3}) . For a color version of this figure, see www.iste.co.uk/baghdadi/5.zip*

In practice, the relationship between R_{440}/R_{550} and Chl-a is documented by numerous *in situ* measurements in order to obtain a strong empirical statistical relationship. The empirical relationship observed in the field is of a logarithmic type. This relationship serves as a standard. The value of the R_{440}/R_{550} ratio measured by a satellite is transferred on the standard curve in order to provide the estimated value of the concentration in Chl-a. From an operational point of view, the parameters of the standard relationship are determined to establish the equation that links the Chl-a to the R_{440}/R_{550} ratio. That equation is then used to convert satellite data in Chl-a. For example, the empirical relationship, which was developed for the SeaWiFS satellite sensor, often called "OC4", is a fourth-order polynomial of the form [ORE 98, ORE 00]:

$$\text{Chl-a} = 10^{\left(a_0 + a_1 * \rho + a_2 * \rho^2 + a_3 * \rho^3\right)} + a_4 \qquad [4.10]$$

with:

$$\rho = \log_{10}\left[\max\left(\frac{Rrs_{443}}{Rrs_{555}}, \frac{Rrs_{490}}{Rrs_{555}}, \frac{Rrs_{510}}{Rrs_{555}}\right)\right]$$

and $a_0 = 0.47$; $a_1 = -3.85$; $a_2 = 4.53$; $a_3 = -2.44$; $a_4 = -0.0414$

The parameter ρ uses the maximal value among three reflectance ratios in order to ensure a value not close to zero. In fact, when waters are rich in phytoplankton, the reflectance at 443 nm becomes very weak, because of the high radiation absorption by chlorophyll *a*. In this scenario, the reflectance ratio Rrs_{490}/Rrs_{555} is preferred. If the value of that last ratio is still too close to zero, then the Rrs_{510}/Rrs_{555} ratio is used to allow the algorithm to provide a value for chlorophyll *a*. An illustration of the methodology used is given in Figure 4.19.

Other polynomial relationships of the same type as the ones of equation [4.10] were used by other satellite sensors (MODIS, MERIS, etc.), by adapting to the measurement wavelengths of those sensors. However, it is important to highlight that the uncertainties of those empirical relationships in order to determine the Chl-a are great. They can reach ±50% [CAM 06]. So, it is necessary to keep such error bars in mind, when the concentrations in Chl-a derived from a satellite are especially used as inputs for primary production models aiming at quantifying the amount of carbon produced by the phytoplankton on a global scale. The main advantage of the empirical algorithms to determine the Chl-a comes from the fact that those relationships are based on field truths (*in situ* measurements). This type of method is efficient for open ocean waters, where the hydrosol that has a main optical impact on radiation is the phytoplankton. However, the main limit of the empirical relationships is the lack of performance in coastal waters for which the phytoplankton is not the main hydrosol. For those waters, the inversion methods used are called "semi-analytical" (and no longer empirical). The semi-analytical methods consist of using the radiative transfer theory to support *in situ* measurements to establish a strong mathematical relationship between the reflectance and the optical properties (absorption and scattering) of the hydrosols. The concentrations of the hydrosols are deduced in a second step from the optical properties. The advantage of the semi-analytical methods (for example, the method developed by Nechad *et al.* [NEC 10]) is to be able to perform reflectance inversions for all the types of marine conditions (open and coastal ocean). A limit to those methods is the necessity to have numerous *in situ* measurements, including in coastal waters, in order to validate the optical properties and the estimated concentrations in hydrosols.

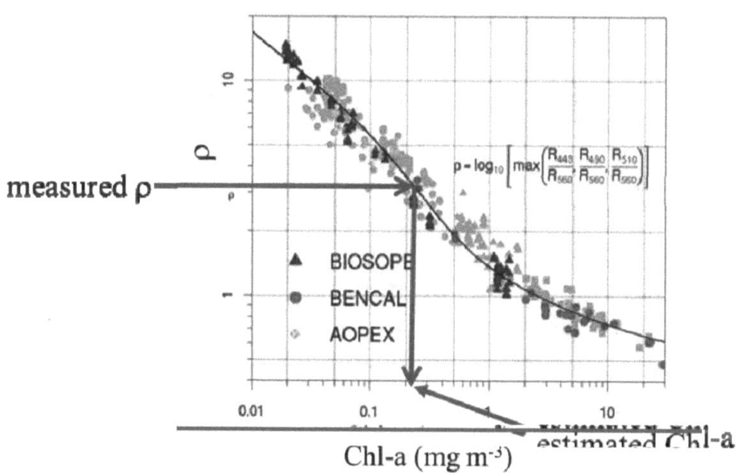

Figure 4.19. *Empirical relationship between the ρ reflectance ratio [4.10] and the concentration in Chl-a obtained from numerous in situ measurements (BIOSOPE, BENCAL, AOPEX etc.) [MOR 07]. For a color version of this figure, see www.iste.co.uk/baghdadi/5.zip*

4.6. Examples of ocean color satellite sensors

As previously mentioned, the first ocean color sensor is the CZCS sensor, which was put into orbit in 1978 by NASA on board of the Nimbus 7 satellite. It provided data until 1986. That sensor's initial objective was to map coastal areas in chlorophyll *a*. It was finally mainly used to characterize the world ocean waters. From 1997, a second generation of sensors was developed with technical performances highly superior to the ones of CZCS. The pioneer sensor of that second generation was the SeaWiFS sensor (NASA) (1997–2012), which embarked on the OrbView2 platform to provide the first long time series of the concentration in chlorophyll *a* on a global scale. The main specifications of SeaWiFS were as follows: global cover in 1 day, sun-synchronous orbit, altitude of 705 km, inclination of 98°, revolution period of 99 min, space resolution of 1 km, swath of 2,800 km and eight spectral bands (from 410 to 860 nm). The SeaWiFS launch represented the development of a free data distribution system for the whole scientific community, with a dedicated operational software. Consequently, the data acquired by that sensor were used for numerous applications, which made the sensor world famous. In the wake of SeaWiFS, several other sensors were put into orbit in order to allow the creation of time series for the ocean bio-optical parameters. For example, the MODIS sensor (NASA) can cover the world surface in 1 day due to a swath of 2,300 km. Its spatial resolution is 1 km and it has 36 spectral bands, including in the thermal infrared. The MERIS sensor (ESA) (2002–2012), which was followed by its twin sensor OLCI (ESA) (2016–2020), has a smaller swath (1,150 km) than SeaWiFS or MODIS, which implies a time revisit over a site of between 2 and 3 days instead of 1 day. However, MERIS was the first ocean color sensor to provide a spatial resolution less than 1 km (300 m) for numerous wavelengths (15 in total) with good radiometric performances. Figure 4.20 lists the main past and current ocean color sensors. From 2020, the PACE mission (NASA) will provide a new sensor for the ocean color observation.

4.7. Some applications of ocean color remote sensing

4.7.1. *Detection of phytoplanktonic proliferations*

Phytoplanktonic proliferations, called *blooms*, are frequent in the ocean given the various environmental conditions (for example, meteorology and ocean circulation) affecting certain regions. The ocean color satellite sensors allow us to observe on a large scale those phytoplanktonic proliferations, which have a significant impact on the production of carbon in the water mass.

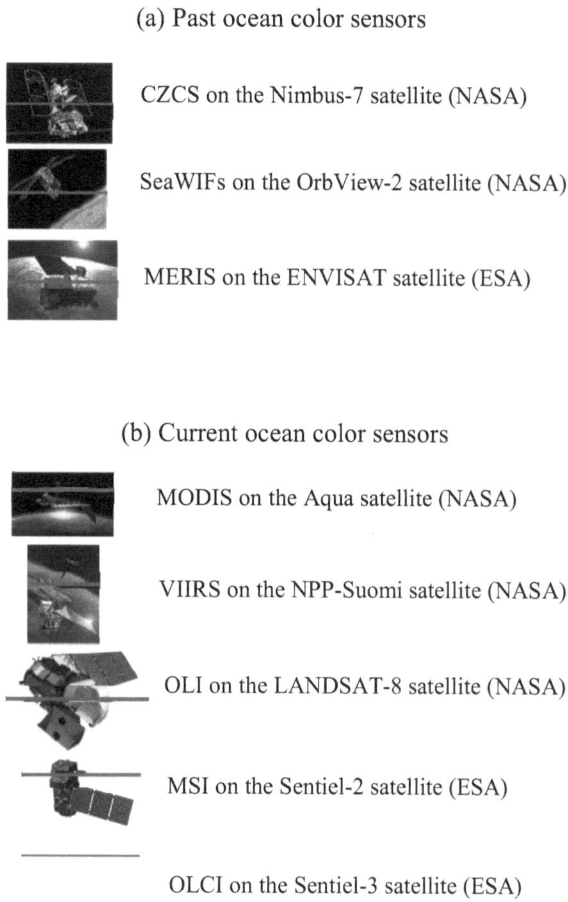

Figure 4.20. *Examples of ocean color sensors: a) past sensors and b) current sensors (2015)*

Thus, phytoplanktonic proliferations of diatoms were observed in the Austral ocean near the Antarctic continent (Weddell sea, Figure 4.21). In this example, the formation of polynyas, which are open ocean areas surrounded by sea ice due to the effect of violent winds, has an impact on the mixing conditions of surface waters by facilitating the development and proliferation of some phytoplanktonic species, like the ones belonging to the diatom types. However, the phytoplankton type and concentration depend on the size and the formation period of the polynyas. The environmental variations of polynyas and their concentration in phytoplankton ultimately affect the quantity of food available to the penguins.

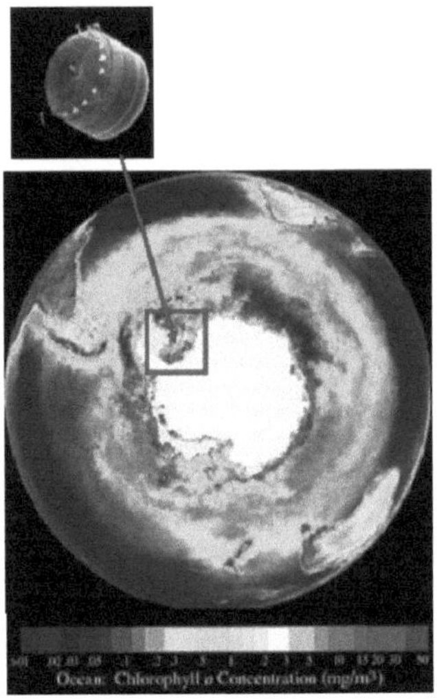

Figure 4.21. *Phytoplanktonic bloom observed in the Austral ocean in the Weddell sea (Antarctica) by the SeaWiFS sensor (NASA) by means of the "concentration in Chl-a" parameter (in mg m^{-3}) (website: http://oceancolor.gsfc.nasa.gov. For a color version of this figure, see www.iste.co.uk/baghdadi/5.zip)*

Due to the satellite observation, annual spring proliferations such as the ones in the North Atlantic can be observed (Figure 4.22).

4.7.2. *Estimation of the phytoplankton functional types by satellite*

As the diversity of the phytoplanktonic organisms is very high in the ocean, a proposal was made to define the different types, called *phytoplankton functional types*, on the basis of the species functionality (export of organic carbon toward the deep ocean and local recycling) or on the basis of other characteristics, such as the size of the phytoplankton (pico-, nano- or microphytoplankton). Those functional types represent indicators of the marine ecosystem operation, in particular regarding the efficiency of ocean carbon sequestration. Consequently, the consideration of those functional types in biochemical models will allow us to improve the predictive abilities of those models. The functional types can be estimated from satellite data on the ocean color, given that the composition changes of the phytoplankton can lead to changes in their absorption and scattering properties, and thus in the reflectance spectrum exiting water.

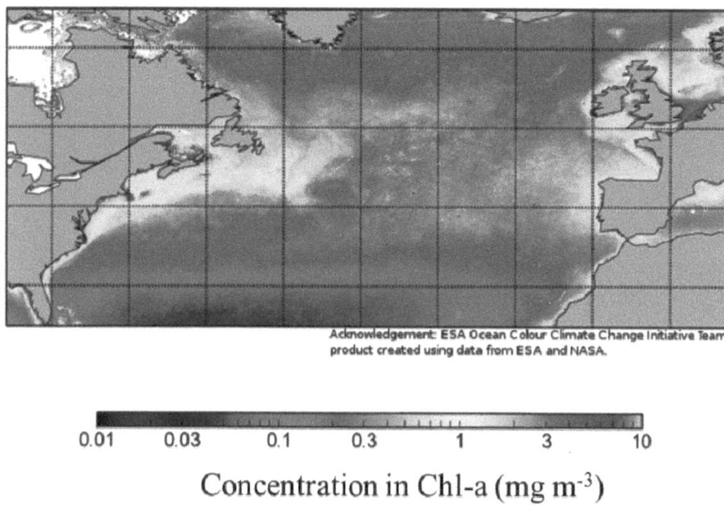

Figure 4.22. *Spring proliferation annually observed in the North Atlantic ocean due to the MERIS sensor (ESA) (website: http://www.esa-oceancolour-cci.org). For a color version of this figure, see www.iste.co.uk/baghdadi/5.zip*

Numerous methods were developed to estimate the phytoplankton functional types by satellite [SAT 14]. Due to the use of the time series of the satellite images created by the SeaWiFS sensor, in parallel with *in situ* measurements, the world mapping of the different phytoplankton functional types was possible. For example, Alvain *et al.* [ALV 08] established, by means of SeaWiFS, a global ocean mapping composed of five functional types: phaeocystis, diatoms, Synechococcus, Prochlorococcus and nanoeucaryotes (Figure 4.23).

Figure 4.23 shows that the nanoeucaryotes species is the main species on a global scale, in particular in middle and high latitudes. Its contribution to the global biomass varies between 45% in winter and 75% in summer [ALV 08]. Prochlorococcus and Synechococcus are the most significant types in tropical waters. Synechococcus represents nearly 15% of the global biomass and is mainly seen in highly oligotrophic waters (in other words, poor in phytoplankton), whereas Prochlorococcus is encountered in the most productive regions, even though it only represents 10% of the total biomass [ALV 08].

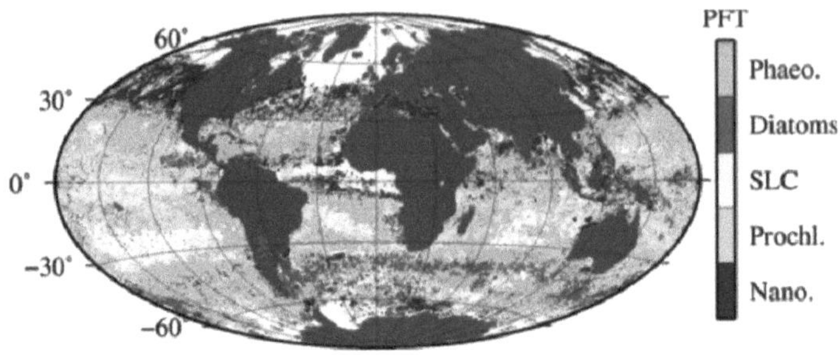

Figure 4.23. *Estimate of the phytoplankton functional types (written PFT) by satellite (SeaWiFS sensor) [ALV 08]: phaeocystis, diatoms, Synechococcus (written as SLC on the figure), Prochlorococcus and nanoeucaryotes. For a color version of this figure, see www.iste.co.uk/baghdadi/5.zip*

4.7.3. *Estimation of oceanic primary production*

In the presence of nutrients and light, the phytoplankton converts carbon dioxide (CO_2) and water molecules in organic matters. This is called the primary production process. Oceanic primary production is defined as the total mass of organic matters created by photosynthesis in carbon grams (gC) $m^{-2}day^{-1}$. The chemical equation of the primary production is as follows:

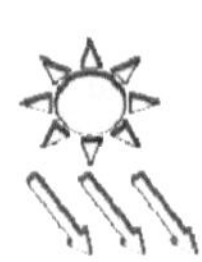

$$\underbrace{CO_2 + NO_3 + PO_4 + H_2O}_{\text{mineral}} \longrightarrow \underbrace{CH_2O\ NP}_{\text{living}} + O_2 \quad [4.11]$$

where NO_3 and PO_4 are nutrient elements (respectively, nitrates and phosphates).

The determination of the primary production from space is possible due to the knowledge of several parameters, in particular the concentration in chlorophyll *a*. The geophysical products derived from the satellite data (Chl-a, light radiation and surface temperature) are associated with a primary production model that calculates the quantity of carbon produced by the phytoplankton. There are different primary production models. The main inputs of those models are:

– concentration in chlorophyll *a*;

– radiation available for photosynthesis (written as PAR), which is estimated through a specific algorithm detailed in Frouin *et al.* [FRO 89];

– sea surface temperature;

– absorption by dissolved matters, which is determined with specific algorithms [SIE 05].

For example, the model of Smyth *et al.* [SMY 05], which is inspired from the model of Antoine and Morel [ANT 96], was specifically designed to calculate primary production from ocean color satellite data. Figure 4.24 illustrates the general outline of the primary production estimation by satellite using the model of Smyth *et al.* [SMY 05].

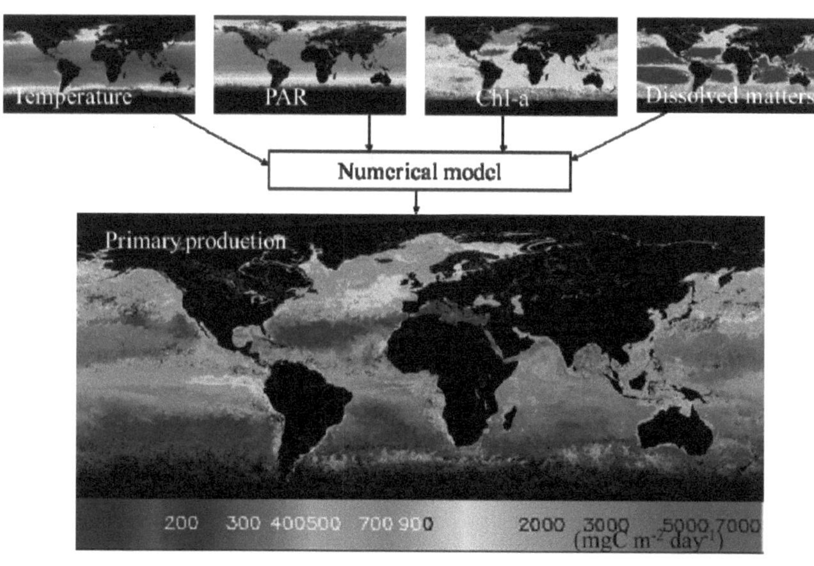

Figure 4.24. *Steps for the estimation of the oceanic primary production by satellite. The satellite images shown as an example here are derived from the MODIS sensor (NASA) . For a color version of this figure, see www.iste.co.uk/baghdadi/5.zip*

On the basis of the different primary production models and satellite data, the global average annual primary production varies between 45 and 60 gigatons of carbon [CAR 06].

4.8. Prospects

The prospects in the field of ocean color remote sensing are interesting in the short and long term. Indeed, the next few years will see the launch of several sensors that will ensure the time tracking of the ocean role in the carbon cycle. The Sentinel-3 operational missions of the European Space Agency (ESA) will supply two sensors (OLCI-1 and OLCI-2) embarked on two platforms a few months apart, Sentinel 3A and 3B (2016 and 2017), for a duration of approximately 10 years in total, in order to ensure the continuity of the observations acquired by the MERIS sensor (2002–2012). For this reason, the instrumental and technical specifications of OLCI are highly similar to the ones of the MERIS sensor, in order to avoid a maximum of intercomparison problems between the two time series (2002–2012 and 2016–2025). A significant and substantial difference will be that the OLCI images will be routinely supplied with a high spatial resolution (300 m) for all the world pixels, which is not the case for MERIS, which routinely supplied data with a spatial resolution of 1.2 km and 300 m in certain conditions only. However, the availability of full resolution data for OLCI will mean that the quantity of data acquired by that sensor will be very high (several Terabytes every day), which will create some problems for the users in order to archive and store it during scientific exploitations. Processing softwares for the OLCI sensor images will normally be put online by the ESA and will be openly accessible to all users. In line with the Sentinel-3 missions, NASA is preparing the launch (planned for 2020) of an ocean color sensor through the PACE mission. The sensor should be equipped with a polarimeter, which will improve in particular the detection of aerosols over the oceans, and in the end, will increase the quality of the geophysical products derived from the sensor (for example, Chl-a). Studies carried out by the scientific community and the French Space Agency (CNES) for the development of an ocean color satellite sensor (called GEOCAPI), which can capture images with a high spatial resolution (300 m) from a geostationary orbit in order to significantly increase the time repeatability of the observations (images every 30 min), have been ongoing for several years. This type of sensor could allow us to study more specifically the daily variations of the phytoplankton lifecycle (night cycle). It would also allow for further studies on coastal ecosystems that are subject to very variable river discharges during the day. The geographical area covered by GEOCAPI would be Europe, Africa and the East of the American continent. This sensor, which is currently a priority for the French ocean color community, would be available in 2020–2025.

With the launch of a new generation of satellite sensors, the development of methods associated with these in order to guarantee their exploitation and the provision of quality data, will continue in the years to come. This will be done in particular to improve the algorithms in order to significantly reduce the uncertainties associated with the derived geographical products. Indeed, those uncertainties are currently still too significant in order to be able to insert the parameters estimated by satellite in the data assimilation models capable of predicting their spatial and seasonal variations. So, the use of a concentration in Chl-a value estimated by satellite that is too vague could lead to significant errors concerning the primary production arising from it. Thus, the carbon footprint estimated on a global scale and the role played by the ocean in climate change still suffer from those uncertainties that will have to be reduced in the future. Finally, the comprehensive study of littoral and coastal ecosystems must continue, given the significant socioeconomic role that they play with, for example, the density increase in the human population living near the coast on a global scale. The very high spatial resolution data acquired by the satellite missions will have to be used in order to improve the observations close to the coast. For example, the Sentinel 2 (ESA) or Landsat 8 (NASA) missions have a resolution in the range of approximately 10 m; the data of the Pleiades mission of the CNES offer a metric resolution. Significant applications are concerned by coastal environment remote sensing: estimating bathymetry (bottom depth), mapping and identification of the benthic species (bottom-dwelling), detection of harmful or toxic algae (for example, the red tides caused by the phytoplanktonic species Karenia Brevis), studying marine pollution with the monitoring of oil spills, etc. In order to obtain good quality geophysical products derived from remote sensing, it is necessary to continue the methodological work for the inversion of the radiation exiting those complex coastal waters from an optical point of view. The semi-analytical methods inspired from the radiative transfer modeling or the ones inspired from artificial intelligence (neural network and genetic algorithm) could contribute to the improvement of the inversion performances and, consequently, to the reduction of the uncertainties associated with the bio-optical properties of hydrosols in those environments.

4.9. Key points

– The interaction of oceanic radiation with hydrosols influences the color of the ocean.

– The main parameter that we are estimating in the open ocean through the observation of the ocean color is the concentration in the chlorophyll *a* pigment, which is the only pigment present in all the phytoplankton species and is thus representative of the total phytoplanktonic biomass.

– The step prior to ocean observation by means of a satellite is the correction of the atmospheric effects on the images. The general principle consists of using the radiation measured at the top of the atmosphere in the red and in the near infrared in order to determine the optical properties of the aerosols, since the radiation exiting water is nil at those wavelengths. Then, the atmospheric radiation is quantified through the extrapolation of the radiation obtained at long wavelengths toward short wavelengths with the help of the radiative transfer theory.

– The bio-optical properties of the hydrosols (absorption and scattering) condition the marine reflectance (the ascending irradiance/descending irradiance ratio on the sea surface). Water molecules scatter a lot of light in the blue and highly absorb it in the red. Chlorophyll *a* highly absorbs in the blue and absorbs less in the red. Dissolved and detrital matters highly absorb at short wavelengths.

– The general principle to determine chlorophyll *a* by satellite in the open oceans relies on empirical relationships established from field measurements of the marine reflectance and the concentration in chlorophyll *a*. Those relationships are of a logarithmic type. The estimate uncertainty concerning the concentration in chlorophyll *a* is of the range of $\pm 50\%$.

– The main ocean color sensors that allowed us (and still allow us) to map the world in chlorophyll *a* are: CZCS (1978–1986, NASA), SeaWiFS (1997–2012, NASA), MERIS (2002–2012, ESA), MODIS-Aqua (since 2002, NASA), OLCI (since 2016, ESA). Their typical spectral range of ocean color observation is 400–860 nm.

– The sensors dedicated to ocean color allow multiple applications: detection of phytoplanktonic proliferations on a global or regional scale, detection of phytoplankton functional types and estimation of the average carbon production annually created by the phytoplankton on a global scale (approximately 50 GtC/year).

4.10. List of acronyms

CDOM	color dissolved organic matter
CZCS	coastal zone color scanner
ESA	European Space Agency
IOCCG	International Ocean Color Coordinating Group
NASA	National Aeronautics and Space Administration
MERIS	medium resolution imaging spectrometer
MODIS	moderate resolution imaging spectroradiometer
MSI	multi-spectral imaging instrument
OLCI	ocean and land color instrument
OLI	operational land imager
PACE	pre-aerosol, clouds and ocean ecosystem
PAR	photo-synthetically available radiation
PFT	phytoplankton functional types
SeaWiFS	sea-viewing wide field of view sensor
TOMS	total ozone mapping experiment spectrometer
VIIRS	visible infrared imaging radiometer suite

4.11. Bibliography

[AHM 10] AHMAD Z., FRANZ B.A., MCCLAIN C.R. *et al.*, "New aerosol models for the retrieval of aerosol optical thickness and normalized water-leaving radiances from the SeaWiFS and MODIS sensors over coastal regions and open oceans", *Applied Optics*, vol. 49, pp. 5545–5560, 2010.

[AHN 07] AHN Y-H, SHANMUGAM P., "Derivation and analysis of the fluorescence algorithms to estimate phytoplankton pigment concentrations in optically complex coastal waters", *Journal of Optical A: Pure and Applied Optics*, vol. 9, pp. 352–362, 2007.

[ALV 08] ALVAIN S., MOULIN C., DANDONNEAU H. *et al.*, "Seasonal Distribution and Succession of Dominant Phytoplankton Groups in the Global Ocean: A satellite view", *Global Biogeochemical Cycles*, vol. 22, no. 3, pp. GB3001, 2008.

[ANT 96] ANTOINE D., MOREL A., "Oceanic primary production: I. Adaptation of a spectral light-photosynthesis model in view of application to satellite chlorophyll observations", *Global Biogeochemical Cycles*, vol. 10, pp. 43–55, 1996.

[BAB 03] BABIN M., MOREL A., FOURNIER-SICRE V. *et al.*, "Light scattering properties of marine particles in coastal and open ocean waters as related to the particle mass concentration", *Limnology and Oceanography,* vol. 48, no. 2, pp. 843–859, 2003.

[BER 80] BERGMANN M., PETERS R.H., "A Simple" Reflectance Method for the Measurement of Particulate Pigment in Lake Water and its Application to Phosphorus–Chlorophyll–Seston Relationships", *Canadian Journal of Fisheries and Aquatic Sciences*, vol. 37, no. 1, pp. 111–114, 1980.

[BRA 12] BRAJARD J., SANTER R., CREPON M., THIRIA S., "Atmospheric correction of MERIS data for case-2 waters using a neuro-variational inversion", *Remote Sensing of Environment*, vol. 126, pp. 51–61, 2012.

[BRI 81] BRICAUD A., MOREL A., PRIEUR L., "Absorption by dissolved organic matter of the sea (yellow substance) in the UV and visible domains", *Limnology and Oceanography*, vol. 26, pp.43–53, 1981.

[BRI 16] BRIOTTET X., "Radiometry in the optical domain", in BAGHDADI N., ZRIBI M. (eds), *Optical Remote Sensing of Land Surfaces*, ISTE Press Ltd, London and Elsevier Ltd, Oxford, 2016.

[CAM 06] CAMPBELL J.W., O'REILLY J.E., "Metrics for Quantifying the Uncertainty in a Chlorophyll Algorithm: Explicit equations and examples using the OC4. v4 algorithm and NOMAD data", *Ocean Color Bio-optical Algorithm Mini Workshop*, vol. 4, pp. 1–15, 2006.

[CAR 06] CARR M.-E., FRIEDRICHS M.A.M., SCHMELTZ M. *et al.*, "A comparison of global estimates of marine primary production from ocean color", *Deep-Sea Research II*, vol. 53, pp. 741–770, 2006.

[CHA 15] CHAMI M., LAFRANCE B., FOUGNIE B. *et al.*, "OSOAA: a vector radiative transfer model of coupled atmosphere-ocean system for a rough sea surface. Application to the estimates of the directional variations of the water leaving reflectance to better process multi-angular satellite sensors data over the ocean", *Optics express*, vol. 23, no. 21, pp. 27829–27852, 2015.

[COX 54] COX C., MUNK W., "Measurements of the roughness of the sea surface from photographs of the sun's glitter", *Journal of Optical Society of America*, vol. 44, pp. 838–850, 1954.

[DOE 07] DOERFFER R., SCHILLER H., "The MERIS case 2 water algorithm", *International Journal Remote Sensing*, vol. 28, pp. 517–535, 2007.

[FRO 89] FROUIN R., LINGNER D., GAUTIER C. *et al.*, "A simple analytical formula to compute clear sky total and photosynthetically available solar irradiance at the ocean surface", *Journal of Geophysical Research*, vol. 94, no. C7, pp. 9731–9742, 1989.

[GOR 94] GORDON H.R., WANG M., "Retrieval of water-leaving radiance and aerosol optical thickness over the oceans with SeaWiFS: A preliminary algorithm", *Applied Optics*, vol. 33, pp. 443–452, 1994.

[GOY 13] GOYENS C., JAMET C., SCHROEDER T., "Evaluation of four atmospheric correction algorithms for MODIS-Aqua images over contrasted coastal waters", *Remote Sensing of Environment*, vol. 131, pp. 63–75, 2013.

[HAR 11] HARMEL T., M. CHAMI, "Influence of polarimetric satellite data measured in the visible region on aerosol detection and on the performance of atmospheric correction procedure over open ocean waters", *Optics Express*, vol. 19, pp. 20960–20983, 2011.

[HOV 80] HOVIS W.A, CLARK D.K., ANDERSON F. *et al*, "Nimbus-7 Coastal Zone Color Scanner: System Description and Initial Imagery", *Science*, vol. 210, no. 4465, pp. 60–63, 1980.

[HU 12] HU C., LEE Z., FRANZ B.A., "Chlorophyll-a algorithms for oligotrophic oceans: A novel approach based on three-band reflectance difference", *Journal of Geophysical Research*, vol. 117, no. C1, pp. C01011, 2012.

[JAM 05] JAMET C., THIRIA S., MOULIN C., CRÉPON M., "Use of a neurovariational inversion for retrieving oceanic and atmospheric constituents from ocean color imagery: A feasibility study", *Journal of Atmospheric and Oceanic Technology*, vol. 22, pp. 460–475, 2005.

[KUC 09] KUCHINKE C.P., GORDON H.R., HARDING L.W. *et al.*, "Spectral optimization for constituent retrieval in Case 2 waters II: Validation study in the Chesapeake Bay", *Remote Sensing of Environment*, vol. 113, pp. 610–621, 2009.

[LON 95] LONGHURST A., SATHYENDRANATH S., PLATT T. *et al.*, "An estimate of global primary production in the ocean from satellite radiometer data", *Journal of Plankton Research,* vol. 17, no. 6, pp. 1245–1271, 1995.

[MOR 74] MOREL A., *Optical properties of pure water and pure seawater*, In JERLOV, STEEMAN NIELSEN E. (eds), Optical aspects of oceanography.Academic, pp. 1–24, 1974.

[MOR 07] MOREL A., CLAUSTRE H., ANTOINE D. *et al.*, "Natural variability of bio-optical properties in Case 1 waters: attenuation and reflectance within the visible and near-UV spectral domains, as observed in South Pacific and Mediterranean waters", *Biogeosciences*, vol. 4, pp. 913–925, 2007.

[NEC 10] NECHAD B., RUDDICK AND K.G., PARKY., "Calibration and validation of a generic multisensor algorithm for mapping of total suspended matter in turbid waters", *Remote Sensing of the Environment*, vol. 114, pp. 854–866, 2010.

[ORE 98] O'REILLY J.E., MARITORENA S., MITCHELL B.G. *et al.*, "Ocean color chlorophyll algorithms for SeaWiFS", *Journal of Geophysical Research,* vol. 103, no. C11, pp. 24937–24953, 1998.

[ORE 00] O'REILLY J.E., MARITORENA S., O'BRIEN M.C. *et al.*, 2000. "SEAWIFS POSTLAUNCH CALIBRATION AND VALIDATION ANALYSES, PART 3". NASA TECH. MEMO, 206892 (11), PP.3–8.SeaWiFS postlaunch calibration and validation analyses: Part 3, vol. 11, pp. 9–23, 2000.

[PEN 13] PENG F., EFFLER S.W., "Spectral absorption properties of mineral particles in western Lake Erie: Insights from individual particle analysis", *Limnology and Oceanography,* vol. 58, no. 5, pp. 1775–1789, 2013.

[ROE 89] ROESLER C.S., PERRY M.J., CARDER K.L., "Modeling in situ phytoplankton absorption from total absorption spectra in productive inland marine waters", *Limnology and Oceanography*, vol. 34, no. 8, pp. 1510–1523, 1989.

[SAT 14] SATHYENDRANATH S. (ed.), Phytoplankton Functional Types from Space. Reports of the International Ocean-Colour Coordinating Group (IOCCG), no. 15, Dartmouth, Canada, 2014.

[SHE 79] SHETTLE E.P., FENN R.W., "Models for the Aerosols of the Lower Atmosphere and the Effects of Humidity Variations on Their Optical Properties", Report No, AFGL-TR-79-0214, U.S. Air Force Geophysics Laboratory, Hanscom Air Force Base, Mass, 1979.

[SIE 05] SIEGEL D., MARITORENA S., NELSON N.B. *et al.*, "Colored dissolved organic matter and its influence on the satellite-based characterization of the ocean biosphere", *Geophysical Research Letters*, vol. 32, no. 20, pp. L20605, 2005.

[SMY 05] SMYTH T.J., TILSTONE G.H., GROOMS. B., "Integration of radiative transfer into satellite models of ocean primary production", *Journal of Geophysical Research*, vol. 110, no.C10, pp. C10014, 2005.

[STE 08a] STEINMETZ F., "Etude de la correction de la diffusion atmosphérique et de la réflexion du rayonnement solaire réfléchi par la surface agitée de la mer pour l'observation de la couleur de l'océan depuis l'espace", PhD Thesis, Lille University of Science and Technology, 2008.

[STE 08b] STEINMETZ F., DESCHAMPSPP.Y., RAMON D., Atmospheric correction in presence of sun glint, Application to MERIS -POLYMER", Oral presentation in MERIS (A) ATSR workshop, Frascati, Italy, 2008.

[STE 11] STEINMETZ F., DESCHAMPSPP.Y, RAMON D., "Atmospheric correction in presence of sun glint: application to MERIS", *Optics Express*, vol. 19, pp. 9783–9800, 2011.

[WAN 07] WANG M., W. SHI, "The NIR-SWIR combined atmospheric correction approach for MODIS ocean color data processing", *Optics Express*, vol. 15, pp. 15722–15733, 2007.

5

LiDAR Measurements and Applications in Coastal and Continental Waters

5.1. Introduction: history and typology of LiDARs applied to aquatic environments

The first LiDAR systems applied to aquatic environments were developed for underwater imaging by the US navy, This pioneer LiDAR system was developed by Radio Corporation of America (RCA) for conducting moreefficient military operations and improving the safety of divers. This primitive LiDAR was characterized by a blue-green laser with a laser pulse a pulse repetition smaller than 0.01 Hz [OTT 65, GIL 66]. In the late 60s, LiDAR technologies underwent major progress due to the development of scanners and advanced photodetectors (*Avalanche Photo Diode* (APD) or *Photomultiplier Tubes* (PMT) type) [EVA 78, AUL 96]. It was during this period that detection of submerged targets based on LiDAR systems was first attempted. This capability was originally evaluated in the United States, Canada, Australia and Sweden [HIC 68, SIZ 75, CLE 78, FEI 92] and was later exapanded to other countries such as China and the Soviet Union [LIU 90, STE 97].

In the 80s and the 90s, a major application of dual wavelength (i.e., 532 and 1,064 nm) LiDAR systems was the detection of underwater mines and submarines. During the same period, the pulse repetition rate of LiDARs increased from 200 Hz to 70 kHz, and telescopes were improved by using

---

Chapter written by Jean-Stéphane BAILLY, Martin MONTES-HUGO, Yves PASTOL and Nicolas BAGHDADI.

shutters. Since then, LiDAR applications in aquatic environments followed two directions: the bathymetry of coastal waters and the study of oceanographic processes. Indeed, a major effort was made in hydrography for deriving better nautical charts by producing high-resolution maps of water-depth [GUE 78, HOG 80]. In the United States, the bathymteric lidar system AOL (airborne oceanographic ldiar) was jointly developed by National Aeronautics and Space Administration (NASA), National Oceanic and Atmospheric Administration (NOAA) and the US navy. In Canada, the first bathymetric LiDAR system was built by Optech during the 1980s, and managed afterward by the Canada Center for Remote Sensing (CCRS). A commercial version of this system, the LARSEN 500, produced for the first time a bathymetric map of the Vancouver Island. A generation of instruments that were able to measure water depth was then born: These LiDAR systems are commonly known as airborne LiDAR bathymeter (ALB), hydrographic LiDAR or bathymetric LiDAR. They are also called "hydrographical LiDAR" or "bathymetric LiDAR". In general, ALBs measurements can be enhanced by coupling passive optical measurements [LYZ 85]. Notice that expansion of ALBs coincided with the development of several lidar systems in Germany for characterizing sea surface roughness and mapping fluorescence substances in the first top meters of the water column [FRA 82] and on the characterization of the effects of the seas roughness on the LiDAR signal [LUC 87]. During the 1990s, helicopters were used for the first time for obtaining LiDAR measurement. Also during this decade, the first bathymetric charts of rivers as derived from LiDAR measurements were produced. Since 2010, a new generation of LiDAR systems called 'topo-bathymeytric LiDAR' has been adopted by various research groups and companies for performing depth sounding of shallow and deep littoral environments. Unlike ALBs, topo-bathymetric LiDARs have a stronger laser at 532 nm that allow a deeper penetration into the water column [COT 08, MAN 11].

Since 1990, the number and diversity of studies related to airborne LiDAR surveys on fluvial and marine ecosystems has witnessed a tremendous boom. Most common applications are summarized below:

– Ecological investigations including the mapping of benthic habitats, turbidity, vegetation and characterization of marine food chains, [HOG 83, CHU 01, CHU 04a, BRO 04, COL 10, CHU 14a];

– Physical oceanography studies related to characterization of thin layers, internal waves, wind and bubbles [HWA 00, CHU 01, CHU 04b, CHU 09, RUD 14];

–Monitoring of shipping routes and optimization of dredging operations [IRI 99];

– Integrated Coastal Zone Management (ICZM) to assess coastal submergence risks [PEE 11];

The use of space borne LiDAR (e.g., satellite, space shuttle and international space station) for studying aquatic environments is a more recent research arena. The feasibility of spaceborne LiDAR systems for hydrographic applications was studied for the first time in 1997. These theoretical experiments were initiated by German scientists who worked with fluorescence LiDAR systems [REU 97]. During the same timespan, the LiDAR system BALKAN-1 was set up as part of the scientific payload of the russian space station MIR [BAL 97]. This mission was mainly focused on measuring ocean surface parameters.

The Geoscience laser altimeter system (GLAS), a spaceborne LiDAR developed by NASA, was in operation between 2002 and 2009 [ZWA02]. This sensor was part of the ICEsat mission and was specifically dedicated for monitoring the elevation of ice caps. However, GLAS applications also include cloud dynamics, sea waves and coastal topography [DUO 09]. More recently, the spaceborne LiDAR CALIOP (*Cloud-Aerosol LiDAR with Orthogonal Polarization*), measurements have been used in several disciplines for studying water turbidity and different optical particulate components [BAR 11, CHU 13a, LU 14].

In addition to airborne and spaceborne platforms, there are LiDAR systems that have been evaluated from ships, land facilities [KAR 06], drones [TUE 07] and gliders [ROY 13]. A major expansion of these LiDAR systems is expected in upcoming years due mainly to the reduced cost and miniaturization of LiDARs' components.

5.2. Equations and parameters of LiDAR systems applied to aquatic environments

Here, details about LiDAR equations are only provided for the most common LiDAR system in aquatic environments. The equation of a wave profile (i.e., power received as a function time t) as derived by an ALB is consistent with the LiDAR expression presented in another chapter of this book [CHA 16]. However and unlike other parameterizations, the bathymetric wave has a unique propagation throughout two optical medium (air and water) and reflection against two boundaries (water surface and bottom). The total power of a bathymetric LiDAR waveform (P) can be divided into five contributions (Figure 5.1):

$$P(t) = P_S(t) + P_{CE}(t) + P_F(t) + P_{CA}(t) + P_N(t) \qquad [5.1]$$

where P_S, P_{CE} and P_{CA} and P_N terms correspond to water surface, water column, bottom, atmosphere, and sensor noise contributions respectively. Each waveform component is the result of convolving the transmitted waveform (for example, a Gaussian shaped pulse) and the environmental response of the interface or optical medium. Here, only LiDAR beam interactions with air-water and water-bottom boundaries are described [ABD 12]. Additional propagation effects due to air and detector noise components are further explained in Chapter 5, Volume 1 of this series.

5.2.1. *Water surface return*

A simplification of the impulse response of the water surface [GUE 85, TUL 99, ABD 12, MON 14] is shown below:

$$P_s = \frac{P_e T_{atm}^2 A_R \eta_e \eta_R L_S}{\pi \left(\dfrac{H}{\cos\theta} \right)^2} \qquad [5.2]$$

where P_e is the power transmitted by the laser in Watt = *E0/T0*. *E0* (J) is the energy transmitted by the LiDAR, *T0* (s) is the pulse length or the width at mid-height of the Gaussian pulse (i.e., FWHM: *Full Width at Height Maximum*), $T^2{}_{atm}$ is the atmospheric transmittance (squared because of the round trip of the laser pulse), A_R is the sensor surface (m^2), η_e and η_R are the optical efficiencies of the transmitter and the receiver, respectively, L_S is the reflectance at the air/water interface (surface albedo), θ (rad) is the incident angle of the laser beam on the water surface and H is the sensor altitude relative to the water surface.

The key parameter of equation [5.2] is the factor L_s. Several formulations were proposed to estimate this parameter based on "sea state" information (i.e., water surface characteristics).

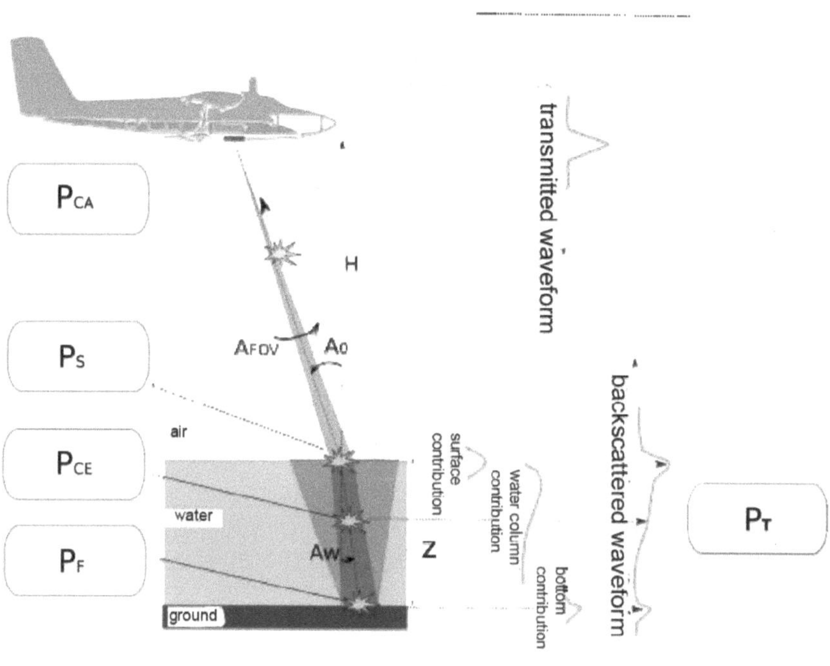

Figure 5.1. *Propagation of a bathymetric LiDAR waveform over an aquatic environment [BAI 12]*

Guenther [GUE 85] proposed to estimate L_S based on an empirical parameterization derived from wind speed and direction. Alternatively, Lesaignoux *et al.* [LES 07] and Abdallah *et al.* [ABD 12] suggested computing L_s according to the *Bidirectional Reflectance Distribution Function* (BRDF) of Cook and Torrance [COO 82]:

$$L_S = \frac{k_d}{\pi} + \frac{k_s DOF_r}{\pi \cos^2\theta} \qquad [5.3]$$

In this model, k_s is the specular reflection coefficient and k_d is the scattered reflection coefficient of the facets ($k_s + k_d = 1$). D is the microfacet distribution function that models the surface roughness like geometrical facets and O is the geometrical attenuation factor of the BRDF linked to the self-shading of the facets. F_r is the Fresnel reflection on each microfacet. According to Beckmann [BEC 87], D is defined as follows:

$$D = \frac{1}{r^2 \cos^4\theta} e^{-(\tan\theta / r)^2} \qquad [5.4]$$

where r is the surface roughness defined as the standard deviation of the microfacet [COO 82]. A full description of those parameters can be found in Lesaignoux *et al.* [LES 07] and Abdallah *et al.* [ABD 12].

5.2.2. *Water column return*

The water column impulse response $P_{CE}(z)$ at a depth z can be written as follows:

$$P_{CE}(z) = \frac{P_e T_{atm}^2 A_R \eta_e \eta_R F (1 - L_s)^2 \beta(\pi, z) \exp\left(-\frac{2k_d z}{\cos\theta_w}\right)}{\left(\frac{n_w H}{\cos\theta} + \frac{z}{\cos\theta_w}\right)^2} \qquad [5.5]$$

where θ_w is the LiDAR incident angle inside the water, F is a loss factor due to the *Field of View* (FOV) of the telescope, n_w is the water refraction index, $\beta(\pi,z)$ is the volume scattering function calculated at the angle π and k_d is the vertical diffuse attenuation of downwelling irradiance. An approximation of the coefficient k_d was provided by Guenther [GUE 85] in order to show the link between k_d and the inherent optical properties (i.e., those that are independent of the light fieldgeometry) of water or IOPs:

$$k_d = c(\lambda)[0.19(1-\omega_0)]^{\omega_0/2} \qquad [5.6]$$

where $\omega_0 = b(\lambda)/c(\lambda)$ is the simple scattering albedo, $b(\lambda)$ is the total (water + particulates) scattering coefficient, $c(\lambda) = a(\lambda) + b(\lambda)$ is the beam attenuation coefficient in m^{-1} and $a(\lambda)$ is the total absorption coefficient. Note that IOPs can be computed as the sum of optical contributions due to chromophoric dissolved organic matter, phytoplankton and non-algal particulates (e.g., sediment, organic detritus).

5.2.3. *Water Bottom return*

By assuming a flat morphology, the impulse response of the bottom (P_F) can be computed as:

$$P_F = \frac{P_e T_{atm}^2 A_R \eta_e \eta_R F(1-L_s)^2 R_b exp\left(-\frac{2k_d Z}{cos\theta_w}\right)}{\pi\left(\frac{n_w H}{cos\theta}+\frac{Z}{cos\theta_w}\right)^2} \qquad [5.7]$$

where Z is the water depth, R_b is the reflectivity of the bottom (or bottom reflectance). Bouhdaoui *et al.* [BOU 14] suggested a mathematical expression expression for simulating assymetrical bottom contributions within the LiDAR footprint.

Figure 5.2 shows a theoretical LiDAR waveform and its power components as calculated using the LiDAR signal simulator WaLID. In this example, the instrumental parameters correspond to those characteristics of the HawckEye II sensor.

5.3. LiDAR acquisitions systems

5.3.1. *Airborne LiDAR Bathymeter (ALB) systems*

All the ALBs include diode-pumped lasers of Nd:Yag type and a Q-switch (optical shutter) that transmits at an initial wavelength of 1,064 nm (infrared or IR) and whose frequency is doubled to a wavelength of 532 nm (green), which provides a better capacity to penetrate the water column. In order to increase the penetration of the laser beam into the water, high-power ALB systems were first developed. Like all the LiDAR systems, the ALB systems are composed by a transmitting (diode lasers) and receiving (telescope) component. The characteristics of those systems are shown in Table 5.1.

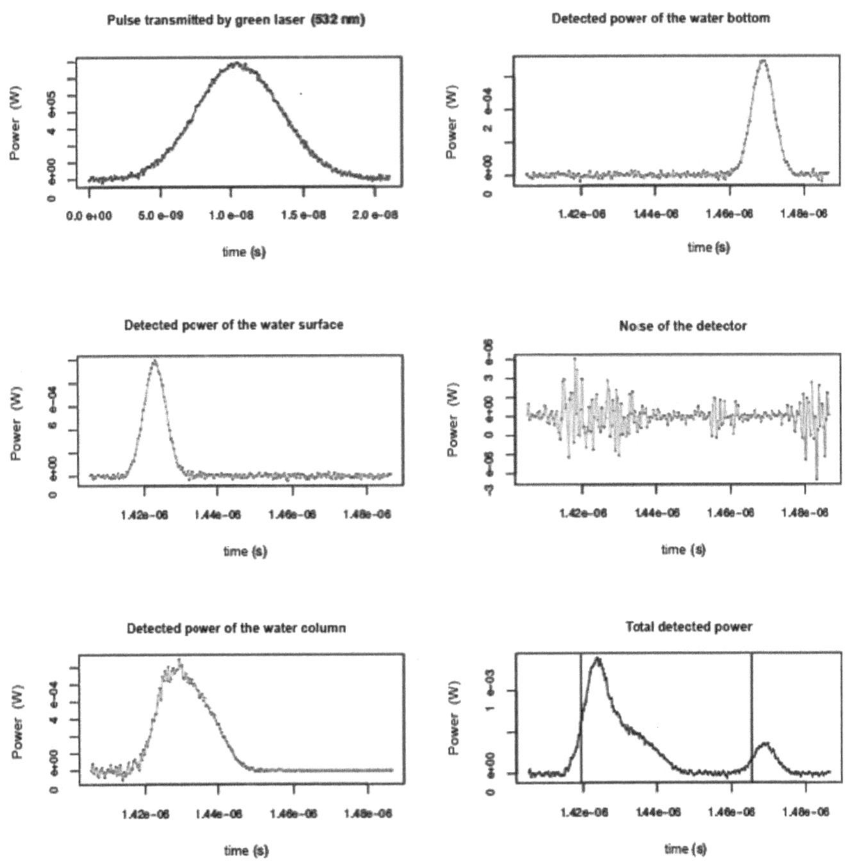

Figure 5.2. *Simulation of LiDAR backscattering measurements. Transmitted pulse (top left), bottom return (top right), water surface return (middle left), detector noise (middle –right) and water return (bottom left) and total power (bottom-right) LES 07]. All runs were made with a bottom depth of 5 m and Rb = 0.17*

5.3.1.1. *High-power ALBs*

Numerous ALBs transmit both in infrared (IR) and in the green spectral range, with a variable laser shot rate between 1 and 20 kHz (Table 5.1) and characterized by relatively short laser pulses (from 2 to 6 ns). The double wavelength configuration offers three options for detecting the water surface position: the IR backscattering, the "green" (i.e., wavelength = 532 nm) backscattering and the Raman backscattering (i.e., wavelength = 647 nm for a "green" laser).

Unlike systems with relatively low-power (see section 5.3.1.2), the green laser beam of high-power ALBs may penetrate deeper layers (up to 50 m in oceanic waters) and illuminate the bottom if measurements are performed over relatively shallow and clear environments.

The accurate position of the water surface is essential for mapping bathymetry as speed differences in air and water, and variations on beam direction due to air–water refraction can be corrected. To optimize bottom detection, two types of highly sensitive detectors are used: Avalanche photo diodes (APDs) and *Photo Multiplier Tubes* (PMT). In general and unlike low-power ALBs, high-power ALBs can measure the entire signal returning from the target (full waveform) at very high resolutions (i.e., ~ 1 Ghz).

The swath of ALBs measurements is determined by a mirror that is also responsible for modifying the shot rate. In this regard, an important advancement has been made by shifting from an rotating prism technology to a rotating prism technology. These later systems mainly differ in terms of scanning patetrns (e.g. arcs, circles, ellipses). For water applications, it is important to have an inclination between the laser beam and the water surface in order to minimize the water reflection and subsequent saturation of the LiDAR detector. Another LiDAR configuration improvement is the FOV division into smaller imaging units. This modification has been implemented in HawkEye II systems and allows an increase in the density of measurements at the ground level. In the case of a typical high-power ALB configuration (i.e., FOV = 20 mrad, carrier height = 300 m), the illuminated spot at the water surface is around 2 m. That footprint is enough for power transmission without compromising eye safety standards, standards at the laser emission wavelengths. In addition, potential eye damage risk at laser wavelength 532 and 1,064 nm is minimized during flights by using an automatic laser transmission shutdown system.

Classically, the position of the aerial LiDAR platform (e.g., plane, helicopter and drone) is controlled by using a change global positioning system GPS" by "Global Navigation Satellite System GNSS. Likewise, the roll, pitch and yaw variations are modulated by the inertial measurement unit. The path of the light wave is always recalculated *a posteriori* and the final position of the bathymetric points is computed based on position and attitude parameters of the transmitter, position of water surface and travel time of light wave throughout air and water paths.

The main manufacturers of high-power ALBs are: Teledyne-Optech (Canada), Hexagone-AHAB (Sweden) and FugroLads (Australia). Unlike the other companies, FugroLads not only develops the instruments but also operates them. Teledyne-Optech is the manufacturer of SHOALS (Scanning Hydrographic Operational Airborne LiDAR Survey) systems, CZMIL, and CZMIL nova LiDAR systems. Unlike its predecessors, CZMIL systems are characterized by a multiple field of view and circular scanning patterns [TUE 10]. Finally, the Swedish AHAB, after the HawkEye IIa and IIb, has constructed HawkEye III. All the characteristics of those systems are shown in Table 5.1.

High-power ALBs are instruments weighing a few hundred kilograms. The transmitted pulse energy is several milijoules, which results in high cooling requirements (given the sensitivity of these sensors to the ambient temperature). Also, these systems have a substantial electricity (i.e., several hundred amps). Lastly, the platforms that can accommodate high-power ALBs are those with large dimensions such as planes with two-engine propellers. These planes must have two doors and be able to fly at relatively low speeds (i.e., 140 knots or approximately 250 km h-1?). Also, it is technically possible to use helicopters as platforms for high -power ALBs but their operating cost is often limiting.

High-Power ALB	CZMIL (Canada)	LADS mk3 (Australia)	Hawk Eye III (Sweden)	
Laser	532 nm and 1,064 nm	532 nm	532 nm and 1,064 nm	
Footprint	Circular	Circular arc	Elliptic	
Inclination	20° front back	Front setting 0° to 8°	14° front back 20° right left	
Swath (400 m altitude)	290 m	300 m	280 m	
Energy per pulse	3 mJ	7 mJ	3 mJ	
Bathy rate	10 kHz	1.5 kHz	10 kHz (Deep) 35 kHz (Shallow)	
Topo rate	70 kHz	----	500 kHz	
Weight	390 kg	232 kg	170 kg	
Characteristic	Hyperspectral camera + new lighter system (CZMIL Nova)	Shifting to a 2.5 kHz rate soon	Camera RCD30 80MP	
Low-Power ALB	**EAARL (USA)**	**Aquarius (Canada)**	**RIEGL VQ820G (Austria)**	**Chiroptera (Sweden)**
Laser	532 nm	532 nm	532 nm	532 nm and 1,064 nm
Footprint	Zigzag	Elliptic arc	Elliptic arc section	Elliptic
Inclination	-	7° front	21°	14° front back 20° right left
Swath (400 m altitude)	250 m	370 m	400 m	280 m
Energy per pulse	70 µJ	0.1 mJ	0.02 mJ	0.1 mJ
Bathy rate	10 kHz	33–50–70 kHz		35 kHz (shallow)
Topo rate			520 kHz (200 usable measurements per second)	500 kHz
Weight	113 kg	53 kg	26 kg	80 kg
Characteristic	Sampling at 1 GHz		VQ880G with full elliptic footprint	Camera RCD30 80MP

Table 5.1. *Main characteristics of the bathymetric and oceanographic LiDAR systems* systems. *The new RIEGL VQ880G is now running: 532 nm, Rotating Polygon Mirror, inclination +-20°, Swath 400 m for 400 m altitude, 550 kHz à 400 m, 60 kg*

5.3.1.2. *Low-power ALBs*

The Experimental Advanced Airborne Research LiDAR (EAARL) was a pioneer system designed by the National Aeronautics and Space Administration and the United States Geological Survey [BRO 04]. The EAARL was built in 2011 as part of a new generation of low- power bathymetric lasers [PAR 12]. Those systems have a green laser (wavelength = 532 nm) and a topo-bathymetric operational mode. They are known to be light () and have a low-power consumption (Table 5.1). In general, existing low-power ALBs are not profiling systems (i.e., capable of measuring full waveforms) since they can only analyze specific echos or LiDAR returns Volume 1, Chapter 5. As part of the evolution of these systems, the laser shot rate has been increased from 30 to 550 KHz.

Apart from those characteristics, low-power ALBs the same operating principles as the high-power ALBs. Main manufacturers of low-power ALBs are Teledyne-Optech (Aquarius and Triton sensors), Hexagone-AHAB (Chiroptera sensor) and RIEGL (VQ820G and VQ880G sensors). Miniaturization is the current technological challenge of the ALB systems. This size reduction will be an advantage for LiDAR surveys based on unmanned aerial vehicles due to the smaller weight of future ALBs.

5.3.2. *Oceanographic LiDAR systems*

The first type of oceanographic LiDARs are the fluorosensors or sensors that can detect time-resolved changes of fluorescence In general, LiDAR fluorosensors have three wavelengths (532, 355, and 266 nm) after doubling, tripling and quadrupling the original Nd:Yag laser beam at 1,064 nm. These systems showed up during the same period that ALBs came to light. In general, the excitation wavelengths of LiDAR fluoro-sensors are chosen within the UV-blue spectral range in order to maximize the light absorption by fluorescence components. Typically, these systems have pulse repetition rates between 1 and 20 kHz, and can be used for obtaining profiles. The first backscattering lidar system was the Airborne Oceanographic Lidar (AOL) [BUF 83]. The second type of oceanographic lidars are those measuring backscattering. Unlike ALBs, LiDAR fluorosensors do not have scanners, thus measurements are performed along the flight-path. The initial configuration included a laser transmitter with two excitation wavelengths (355 and 532 nm), and two receivers for the fluorescence and Raman scattering wavelengths. During the 1980s, The AOL system was mainly used for characterizing optical properties of coastal waters and studying mixing processes of the upper ocean [HOG 81]. In the 1990s, the NOAA engineers built the *Fish LiDAR Oceanographic Experimental* or FLOE. This system was originally conceived to minimize the accidental capture of dolphin during tuna fishing, and was later applied to studies related to thin layers and internal waves [CHU 14a]. Unlike AOL, the FLOE system only works with one single wavelength centered at 532 nm, and a receiver having two linear polarization modes. [CHU 01]. Due to its pulse length (10–15 ns), power (100-150 mJ) and shot rate (20 Hz), FLOE surveys have a vertical resolution of 10 cm, a horizontal resolution of 5 m and a maximum penetration depth of 20 and 50 m in coastal and oceanic waters, respectively. Finally, an interesting oceanographic LiDAR system that has not widely used is the Airborne Topographic Mapper or ATM [HWA 00]. This system is equipped (532 nm) with a shot rate of 5 kHz. At a height of 700 m, the laser footprint of ATM is 1 m.

5.3.3. *Spaceborne LiDAR systems in oceanography*

Although never included as part of oceanographic missions, spaceborne LiDARs were found to be useful sensors for studying relationships between sea surface reflectance and wind speed [LAN 05], bathymetry [BAR 11] and detection of suspended particulates in oceanic [BEH 13, CHU 13a, LU 14] and coastal waters [SHI 08]. In most cases, space-based measurements are done at two wavelengths, 532 and 1,064 nm. The Cloud-Aerosol Lidar with Orthogonal Polarization (CALIOP) has a shot rate of 20 Hz, a pulse energy of 110 mJ, and a footprint of 100 m. In aquatic environments, CALIOP waveforms may only provide an integrated look of vertical distribution of optical components. This limitation is related to the coarse vertical resolution (i.e., 200 ns or 22.4 m inside the water) linked to the relatively low sampling rate of the digitizer (i.e., 10 MHz) of the analog signal and the subsequent averaging of measurements. Also, these studies revealed that spaceborne LiDAR measurements are prone to signal saturation and sunlight contamination.

5.4. Optical variables derived from LiDAR waveforms

This section is organized into two main parts describing the most common bottom (i.e., depth) and water column (e.g., inherent optical properties) parameters that can be extracted from LiDAR waveforms. The processing methods presented here are consistent with those already described by Chazette *et al.* [CHA 16]. The first processing step of LiDAR waveforms is the logarithmic conversion of voltage to intensity by using specific gain and offset settings for each corresponding radiometric channel (e.g., co- and cross-polarized) [MON 10].

5.4.1. *Bathymetry*

Water depths can be estimated from vertical LiDAR profiles by computing the time difference between the backscattering peaks associated with the water surface (t_s) or first strong return, and the bottom or last return (t_b). This later time position can also coincide with the mid-height of the leading edge of the bottom Gaussian maximum as shown in Figure 5.3:

$$Z = \frac{c_w(t_b - t_s)}{2} \cos \theta_w \quad [5.8]$$

where c_w is the speed of light in the water (2.25 10^8 m s^{-1}) and θ_w is the pointing angle of the refracted laser beam inside the water and with respect to the vertical component. These angular changes are important for correcting measurements derived from LiDARs with circular scanners due to the variable incidence angle of energy pulses. Thus, the main difficulty for measuring Z is the identification of surface and bottom peaks [ALL 10]. In general, there are four approaches for estimating bathymetry based on a vertical LiDAR profiles [WAN 15, ALL 10].

The first approach is based on peak detection. This calculation is not affected by peak amplitude and can be done based on thresholds, gravity centers [ZWA 02], position of local maximum [WAG 07], adaptive filters [BRE 08] and correlations functions between transmitted and received waveforms [RON 08].

The second approach deals mainly with the estimation of surface and bottom position based on mathematical functions. The most popular method is the use of a mixture of Gaussian-type probability density functions (2-6 functions). However, the evaluation of alternative functions (e.g., log-normal, Weibull and triangular, quadrilateral) has been attempted [ABD 12].

The third approach is less common and does not require an empirical model. In this case, the variable dimensions of a LiDAR profile are reduced by using techniques such as wavelet decomposition [PEE 07, ALL 10]. This method is particularly used in very shallow environments where surface and bottom peaks are hard to discriminate based on deconvolution of LiDAR waveforms. Here, the goal is to retrieve the target response by eliminating the characteristics of the transmitted pulse. Examples of deconvolution technqiues are the Wiener filter [JUT 06], the Richardson–Lucy decomposition [WAN 15] and the expectation-maximization method [PAR 11b].

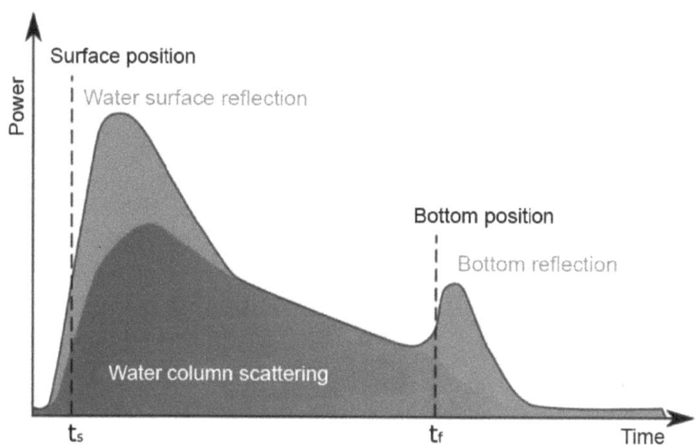

Figure 5.3. *Timing of surface and bottom peaks in a typical LiDAR waveform [ALL 10]*

5.4.2. *Water and bottom optical properties*

By analogy with passive optical sensors (see Chapter 4 of Volume 5), LiDAR measurements can be used for estimating optical properties of water (e.g., vertical diffuse attenuation coefficient of downwelling irradiance k_d) as long as surface effects and bottom morphology have a minor influence [MCL 96]. Likewise, passive systems provide an integrated view of the vertical variation of properties within the first optical depth (i.e., od = geometric path-length × beam attenuation coefficient). Conversely, LiDAR profiles allow a vertical description of optical parameters and associated biogeo-optical components within 3 optical depths.

For LiDARs measuring "elastic" backscattering (Chazette *et al.*, Chapter 5, Volume 1), three main properties can be derived from waveforms ($S_e(z)$: elastic backscattering): the LiDAR attenuation coefficient (α) [CHU 98, KOP 08], the volumetric backscattering coefficient at 180° $\beta(\pi,z)$ and the LiDAR ratio B [CHU 14a], also known as LiDAR attenuation-backscattering ratio:

$$S_e(z) = A\,\beta(\pi,z)\,e^{-2\alpha z} \quad [5.9]$$

$$B = \frac{\alpha}{\beta(\pi,z)} \quad [5.10]$$

where S_e is the elastic backscattering power arriving to the detector after been attenuated twice by the water components. A is a constant related to LiDAR system parameters (see equation 1 in [CHU 14a]) and z is the path-length inside the water. Notice that z is the water depth if the path-length is not corrected by angular variations with respect to the vertical.

The magnitude of α can be calculated from the slope of S_e as as a function of range and polarization state. One example is shown in Figure 5.4 for measurements obtained with the FLOE system. In this case, α is calculated at a depth between 3 and 13 m in order to avoid near-surface backscattering contributions derived from bubbles and noisy LiDAR values at depths higher than 13 m. As pointed out by other studies [LEE 13], magnitude of alpha values computed in Figure 5.4 varied with respect to the polarization state of the backscattering signal. Also in this example, it was found that cross-polarized alpha estimates had a magnitude that was closer to k_d as computed at 532 nm.

Although equation [5.9] assumes single-scattering collisions, α is a parameter that is also influenced by multiple scattering effects [BIS 96]. α varies between c and k_d depending on water turbidity and LiDAR system parameters. Gordon [GOR 89] noted that α tends to approximate c when the product between c and the receiver footprint (i.e., the product between c and the receiver footprint). In this circumstances, c tends to approximate K_d as the scattering regime changes with (Gordon's parameter > 2). However, when turbidity increases, multiple scattering becomes a major contribution. Thus, a new scattering regime as described by Gordon's parameter, began to rule. α became tends to resemble that corresponding to K_d calculations.

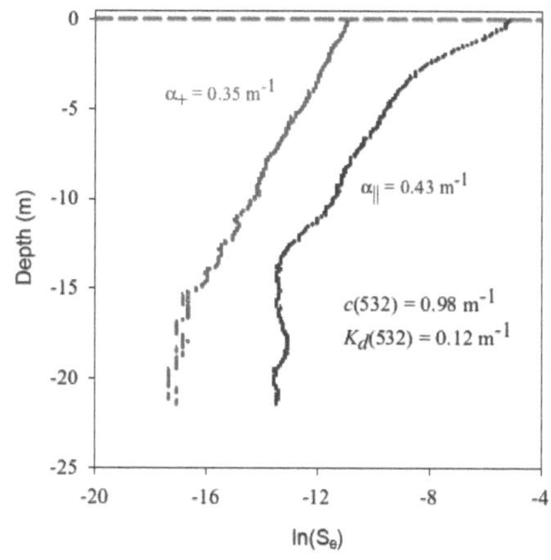

Figure 5.4. *Examples of FLOE profiles derived for cross- (+ red) and copolarized (|| blue) waveforms. Data were collected in East Sound, Unites States. Average values of K_d and c are computed within 3–13 m depth. The pink line indicates the sea surface*

Assuming α tends to approximate K_d, the total absorption coefficient value (a) can be estimated by linking $\beta(\pi,z)$ with α [GOR 89]:

$$K_d(\lambda_i) = 1.04\left[a(\lambda_i) + b_b(\lambda_i)\right] \quad [5.11]$$

$$a(\lambda_i) = \frac{K_d(\lambda_i)}{1.04} - \frac{\beta_p(\pi,z)}{\rho_p(\pi)} b_b^{eff} - b_{bw}(\lambda_i) \quad [5.12]$$

In equations [5.11] and [5.12], λ_i is the wavelength of the laser source (532 nm), b_{bw} is the water backscattering coefficient in m^{-1}, β_p is the volume scattering function of particulates ($m^{-1}\ sr^{-1}$) [PET 77], $\rho_p(\pi)$ is the phase scattering function (sr^{-1}) of particulates and b_b^{eff} is the backscattering efficiency of particulates (=b_{bp}/b_p where b_{bp} and b_p are, respectively, the backscattering and scattering coefficients of particulates.

The approximations in equations [5.11] and [5.12] are valid only if the following conditions are met: K_d is calculated near the surface (i.e., from 0 m to half of the euphotic depth at which the downwelling irradiance is reduced to 10% of its initial surface value):

– the variation of downwelling irradiance with depth follows a Lambert–Beer law;

– the calculation of light distribution functions is performed assuming a clear sky and a solar zenith angle equal to 0°;

– light attenuation is dominated by light absorption (i.e., *b*/*c* is relatively small, *b* is the total scattering coefficient).

Equation [5.11] has a maximum root mean square error of 10% and can be applied in case I waters (i.e., where inherent optical properties are mainly influenced by phytoplankton and phytoplankton-derived constituents).

In case I waters, Churnside *et al.* [CHU 14 b] developed an empirical model for estimating chlorophyll concentration based on *B* values. The uncertainty of this inversion is less than 10% with respect to calculations based on passive optical measurements [LEE 96].

For an inelastic lidar waveform (Si) (Raman or fluorescence scattering), the equation of a mono-static system is summarized as follows:

$$S_i(\lambda_{em},z)=A'N_i(z)\sigma_i(\lambda_{ex},\lambda_{em},z)\left(\frac{Y}{\alpha_w(\lambda_{ex},z)+\alpha_w(\lambda_{em},z)}\right) \quad [5.13]$$

with:

$$Y=e^{-(\alpha_a(\lambda_{ex})+\alpha_a(\lambda_{em}))R_w} \quad [5.14]$$

where *A*' is a system constant which includes the transmitted power of the laser, the surface area of detector, the optical efficiencies for the source and the detector, the influence of the water surface roughness on the energy arriving to the detector (overlap or field of view loss function) [CHU 13b] and the signal transmission through the air-water interface. R_w is the distance between the active light source and the water surface, and N_i is the molecule number density (e.g., pure water, oil, chlorophyll) inelastic scattering cross-section σ_i calculated at specific excitation (i.e., subscript ex) and emission (i.e., subscript em) wavelengths.

The sum of the extinction coefficients at λ_{em} and λ_{ex} is indicated for water (α_w) and air (α_a) paths. For a LiDAR fluorosensor, equation [5.13] is often normalized by the Raman signal in order to eliminate the fluorescence variability linked to changes in water temperature and turbidity [REU 93, FIO 06]. The detection of elastic (backscattering) and inelastic (Raman, fluorescence) optical signals allows the retrieval of water optical properties (e.g., *c*, *bb*), biogeochemical components (e.g., chlorophyll a, trytophan, chromophoric dissolved organic matter, suspended particulate matter) and physiological parameters of phytoplankton (e.g., fluorescence yield) [BAR 98, FIO 06, HOG 03, REU 93]. The use of LiDAR backscattering and fluorescence measurements for classifying bottom types is still premature. However, this is probably feasible since it has been demonstrated based on underwater laser line scanners [MAZ 03]. The characterization of bottom shape and morphology has been already attempted based on multiple FOVs [TUE 10]. Interestingly, this technique has been also evaluated for estimating size distributions of suspended particulates [VES 06]. Finally, different polarizations of LiDAR backscattering can be analyzed for estimating different biophysical parameters of suspended particulates (such as sphericity) [CHU 08].

Additional water column variables that can be studied with LiDAR measurements are bubbles, fish distribution, internal waves and thin layers [CHU 14a]. Multispectral bathymetric LiDARs can also be used to differentiate seagrass beds from sediments based on their reflectivities at 1,064 and 532 nm [COT 08, COL 10]. It should be noted that unlike Cottin [COT 08], Collin *et al.* [COL 10] do not use the gain function, so the waveform is analyzed in terms of voltage and not intensity. Finally, spaceborne LiDAR backscattering systems, initially developed for the atmosphere (e.g., CALIOP), are now used for mapping pigmented and non-pigmented particulates in oceanic and coastal waters [BAR 11, LU 14].

5.5. Case studies of airborne LiDAR applications in hydrography and oceanography

5.5.1. *Examples in coastal waters*

The data acquired by airborne bathymetric LiDARs are mainly used for applications in littoral Environments including spatial planning, management and hydrographic mapping. ALBs have been adopted by several countries for performing quick and continuous land-sea topo-bathymetric data. ALB measurements allow a new global view of geomorphological features and shore relief in coastal areas. Indeed, the survey of shallow environments (i.e., less than 3 m depths) is not efficient and in some cases impossible based on conventional acoustic methods (e.g., multibeam, lateral probes).

Also and unlike ALBs, acoustic surveys may be risky in terms of sensor deployment and technical work. The temporal variability of bottom depth may sometimes be very large (e.g., sandy shores) and acoustic sounding is not a suitable technique for deriving an almost instantaneous view of the study area. Thus, ALBs measurements are indispensable in these conditions.

For example, the framework of the Litto3D® project aims to form a continuous land-sea topographic reference under the form of an accurate and continuous digital elevation model, jointly created by the French Naval Hydrographic and Oceanographic Service (SHOM) and the French National Geographic Institute (*Institut national de l'information géographique et forestière*– IGN) [PAS 07]. The SHOM uses bathymetric lasers to process the marine part (Figure 5.5). The data acquired in this project framework are used as reference data for applications that require elevation data. Other countries such as the United States are currently undertaking similar work within the framework of the coastal Mapping project (http://shoals.sam.usace.army.mil/Mapping.aspx). Likewise and since 1990, the Royal Australian Navy (RAN) has been using bathymetric laser systems for mapping their coastline.

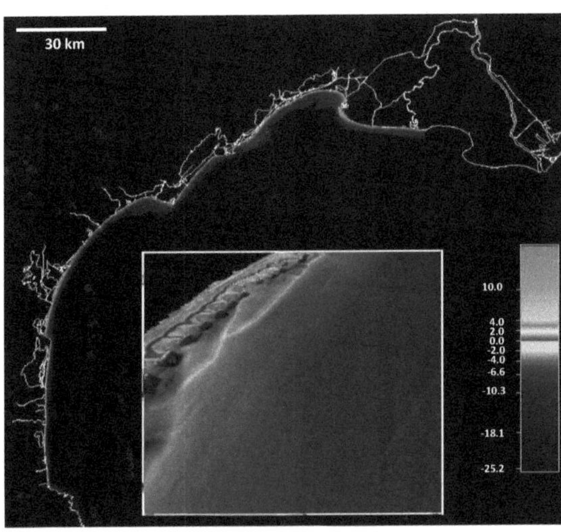

Figure 5.5. *Litto3D® Languedoc-Roussillon 2011 survey, South of France: 220 km of coastline and bathymetric survey (in blue) by Lads MK3 for depths between 0 and 10 m. (Source: SHOM. For a color version of this figure, see www.iste.co.uk/baghdadi/5.zip)*

However, high costs may result from using bathymetric airborne LiDARs. In order to optimize those costs, it is important to process extensive areas. The cost is divided into three main parts: mobilization, acquisition and processing. The mobilization cost includes the installation of the LiDAR system setup, staff transportation expenses, and plane relocation. That cost is fixed and independent of the surface to be sampled. In order to optimize the acquisition, it is important to plan the longest flight lines possible (typically more than 20 km) and avoid u-terns as much as possible. It is considered that a u-turn consumes on average 5 min, whereas 5 min are necessary to fly on a line of 20 km. Finally, the processing cost is variable and depends on the final processing level desired. The cost of such works is very dependent on the area complexity (turbidity, shoreline, topography relief, etc). In any case, the bathymetric is the only one technique that is capable of deriving bottom depth in areas that are inaccessible by engine boats due to the presence of banks and other geomorphological obstacles. (Figure 5.6). The use of bathymetric lasers and low-power lasers allows us to map those areas in safe conditions. Thus, the airborne bathymetric LiDAR are important tools for surveying shallow depths (less than 10 m), in areas difficult to access and when an extensive spatial domain must be quickly surveyed, for example, the sandy areas of the Languedoc-Roussillon (Figure 5.4).

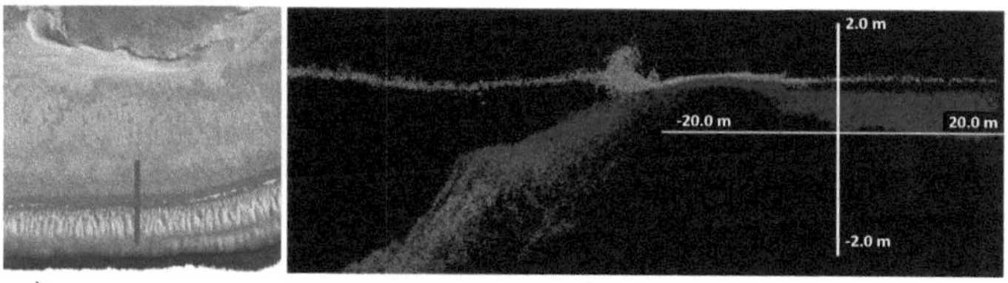

a) b)

Figure 5.6. *2D view a) and transect (red line) of a coral reef b). Lagoon with depths less than 1.5 m and swash area (system: Riegl VQ820G). The water surface and the swash (Surveying Wide-Area Shorelines) are represented by white points (right). (Source: SHOM. For a color version of this figure, see www.iste.co.uk/baghdadi/5.zip)*

5.5.1.1. *Detection capabilities and spatial details*

Bathymetric lasers are generally operated at altitudes between 350 and 650 m. High-power ALBs can detect objects measuring as little as 2 m × 2 m. Smaller objects can be detected, but detection is not guaranteed [PAS 11]. Due to their smaller footprint and higher point density, low-power bathymetric lasers can detect objects as small as 1 m × 1 m. As a example of

SHOALS 1000's system capabilities, SHOM conducted target detection tests in 2005. Two artificial targets (cubes with sides of 1 m) were placed over the sea bottom. The target located at 6 m depth was detected but target at deeper levels could not being imaged. The water turbidity and the relatively low energy of laser returns were the main reasons behind the observed lack of detection in relatively deep waters. Along the flight-track, a third object was detected. Visual recognition by divers identified an granite boulder of 1.2 × 1.6 × 2 m that was situated at 3 m depth. Regardless of the LiDAR system used, there are areas that are often associated with missing data (non-detection). Missing data are commonly attributed to water turbidity and the presence of white water (e.g., ship wakes, wave splash). Bottom detection by high-power bathymetric lasers is usually performed within depths that are 2.5 to 3 times greater than those corresponding to Secchi depth. In general for clear waters, low-power bathymetric lasers reach maximum depths of 10 m. The presence of white water at sea surface prevents any detection by masking the targeted object (Figure 5.7).

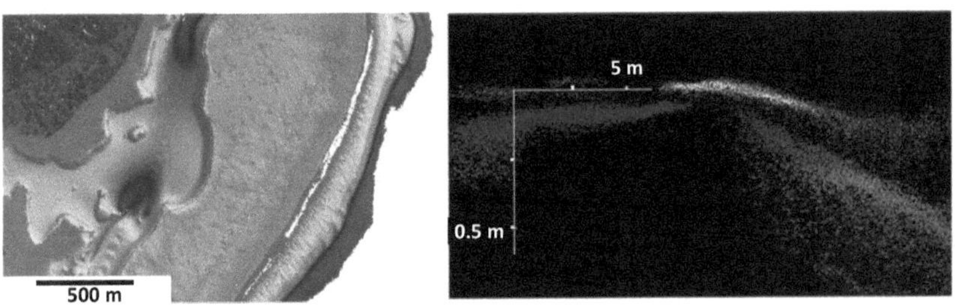

a) b)

Figure 5.7. *Missing data on a 2D view of the Moorea reef and lagoon (white area) due to the swash on a coral reef a) The white points on a transect across the reef b) represent the water surface and the swash (Lads MK3 and Riegl VQ820G systems). (Source: SHOM. For a color version of this figure, see www.iste.co.uk/baghdadi/5.zip)*

5.5.1.2. *Examples in hydrography*

5.5.1.2.1. Nautical charts

The data acquired first allow us to improve the hydrographic knowledge of a coastal area, which is often not well known. They improve the management of that area, which is subject to multiple pressures (bathymetric knowledge, creation of hydrodynamic circulation model, etc.). Figure 5.8 shows an example of an improved hydrographic knowledge.

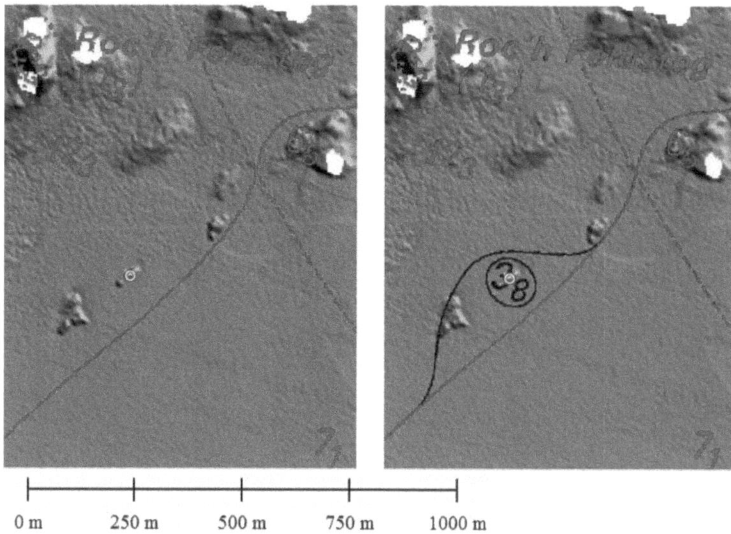

Figure 5.8. *Detection and reporting on a marine map (SHOM 7024) of a new dangerous rock (Plestin-les-Grèves, Côtes-d'Armor, France) at a depth of more than 10 m during the bathymetric laser surveying of the Finistère (lat : 48.700 and long: –3.657) by the Lads MK3 system. Map comparison before the laser surveying (magenta, on the left) and after laser surveying (curve and new probe in black, on the right). (Source: SHOM) . For a color version of this figure, see www.iste.co.uk/ baghdadi/5.zip*

5.5.1.2.2. Ponds and lagoon monitoring

Low-power ALBs with short laser pulse durations are adapted to work above immersed areas with shallow depths (ponds, lagoons, etc.). The presence of algae or sea grass beds, the presence of suspended matters and bloom risks can affect the bottom detection (Figure 5.9).

5.5.1.2.3. Mapping of ports and management of facilities

Work in port areas can often be tricky. The bathymetric laser is often defective in those areas because of the presence of boats, turbidity, anchorage chains, ropes and pontoons (see white areas in Figure 5.10). It is often better to use other techniques (e.g., acoustic sensors).

Figure 5.9. *Topographic and bathymetric survey of the Canet pond in Roussillon, France (lat : 42.667, and long : 3.025) by the HawkEye III system. The topo-bathymetric LiDARs allow us to process extensive areas with shallow depths. (Source: SHOM)*

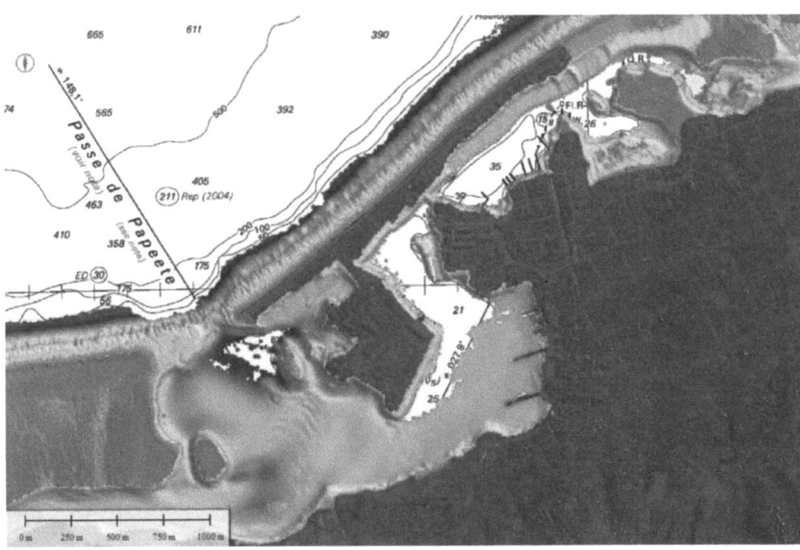

Figure 5.10. *Papeete port (Tahiti, lat. : -17.534 and long. : -149.575) by LiDAR survey using the Lads MK3 and Riegl VQ820G systems. The laser reaches depths of more than 40 m outside the lagoon and loses the bottom of around 20 m in the port. (Source: SHOM)*

However, the acquisition of altimetric data simultaneously in emerged and immersed areas facilitates the management of port facilities. Due to this information, the infrastructures can be globally managed. This knowledge is a precious help for monitoring, and during interventions. In addition, a topographic capacity can be added to the capacity of the bathymetric laser. This simultaneous acquisition of topography and bathymetry gives a global view of the infrastructures, such as a dike.

5.5.1.2.4. Coastal geomorphology

The possibility to measure the bathymetry and topography of a coastline at the same time provides useful knowledge in order to protect those areas against the risks that threaten them (erosion, tsunami, flooding and submersion). Due to the use of a topo-bathymetric laser combined with a bathymetric laser, we can closely describe the topography (several points per square meter), a high density in a shallow area (1 point per square meter) and in deep area, a lower density (0.2 point per square meter). The results obtained allow us to globally manage it on a regional scale. The short timeframe of the surveying and the lower cost, compared to other techniques, mean those operations can be repeated regularly and can monitor the coastal sediment transfer processes (erosion and accretion) (Figure 5.11).

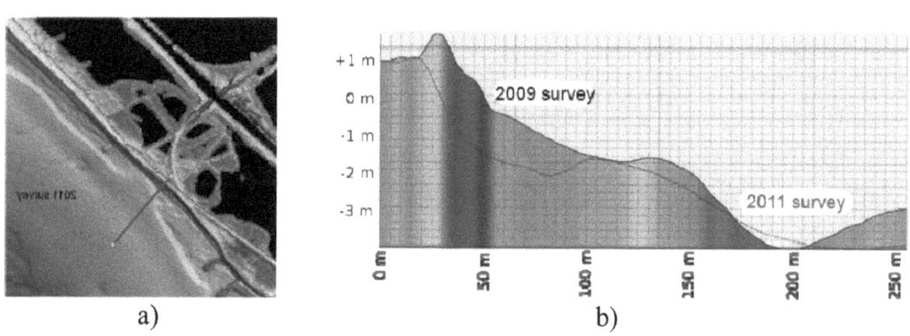

Figure 5.11. *Evolution of the topo-bathymetric profile between 2009 and 2011 in the Aresquiers area in Frontignan (Hérault – France; lat. 43.461, and long : 3.821). Data acquired by the LADS MKII and then MKIII systems. (Source: SHOM)*

5.5.2. *Examples in coastal oceanography*

In this section, two applications using FLOE measurements are discussed: detection of subsurface layers (SSLs) and characterization of the optical properties of water. For each application, an introduction is made followed by a literature review of experimental and theoretical results.

5.5.2.1. *Detection of subsurface layers (SSLs)*

The optical detection of biogeo-optical layers in the water column (SSLs or subsurface layers) is an important topic in studies related to ecology (e.g., harmful algae blooms) and defense (e.g., mine detection). SSLs are associated with drastic vertical changes of biochemical and physical parameters, and may be present in oceanic and littoral waters. Natural SSLs are created by physical (e.g., stratification and shear linked to internal waves) and biological (e.g., diurnal vertical migration and phytoplankton growth) processes [MCM 03, SUL 10].

SSLs can last several days and mainly develop along the horizontal component due to water stratification and diffusion [CHU 09]. SSLs are mainly investigated based on airborne LiDAR profiles due to the limitations of other remote sensing methods (e.g., acoustic signals cannot cross the air–water interface) [MON 11]. Despite these advantages, LiDAR detection of SSLs can be challenging in two cases:

– when the vertical location of the SSL is deeper than the LiDAR penetration depth;

– when the optical thickness of the SSL is large enough to fully attenuate the LiDAR signal (e.g., a false bottom associated with an underwater turbidity plume).

A specific type of SSLs are the thin layer or layers with a thickness between 0.3 and 3 m. Thin layers are present at water depths between 5 and 50 m, and are commonly associated with the accumulation of plankton cells and areas where the horizontal shear is high and the vertical mixing is reduced [DUR 12]. LiDAR surveys performed in Alaska, Washington and Oregon shelves have shown that thin layers vary in terms of thickness, vertical position and maximum LiDAR backscattering intensities [MON 10, MON 11]. Those characteristics were estimated by using complementary passive optical measurements. The third- and fourth-order moments (skewness and kurtosis) of the blue ratio (quotient between the remote sensing reflectance at 443 and 490 nm) were correlated with water depth and thickness of SSLs. At 250-m horizontal resolution, deep SSLs corresponded with negative values of skewness as derived using blue ratios. These variations seem to be related with an unequal distribution of optical components and changes of light scattering and absorption properties between the layer and the surrounding medium. These empirical relationships were confirmed with in-water optical profiles derived from an undulating towed vehicle deployed in different coastal regions (Black Sea, and Monterey Bay). Since thin layers were associated with the pycnocline depth, airborne LiDAR backscattering measurements may also be useful for tracking vertical changes in density gradients and studying water column mixing processes [MON 11]. Because of the sensitivity of the total optical depth to the density change depth, which marks the SSM vertical location, the airborne LiDAR is a tool that potentially allows us to follow the drastic changes of the density gradients in the water superior layers [MON 11].

5.5.2.2. *Water optical properties*

Oceanographic LiDAR profiles carry information regarding the optical system, the signal noise (environment and system) and the optical characteristics of the aquatic environment. Likewise, these waveforms can be deconvoluted in order to estimate optical properties that depend (e.g., a) or do not depend (e.g., K_d) on the illumination and detection geometry of the optical system. By analogy, bathymetric LiDAR waveforms may sometimes provide additional information regarding bottom properties (e.g., reflectivity). This inversion is ill-posed since it is impossible to simultaneously estimate water optical properties and bottom characteristics due to the dependency of their impulse responses (see equations [5.5] and [5.7] in section 5.2). A solution to this limitation is the use of ancillary variables derived from passive remote sensing and in-water optical measurements. In order to derive biogeochemical variables, it is important to known a priori some optical parameters (e.g., cross sections and fluorescence yields) that allow conversion of optical properties to biogeochemical products. In general, uncertainties derived from LiDAR fluorescence inversions are larger (e.g., 25% error for chlorophyll concentration as derived from LiDAR fluorosensor built by ENEA[1] [BAR 98]) with respect to those obtained from LiDAR backscattering (i.e., ~12 and 10% errors in K_d and chlorophyll concentration, respectively, as derived from FLOE) [CHU 98, LEE 13]. Those additional uncertainties are partially related to the double estimation of optical parameters from inelastic signals (e.g., fluorescence): a first error linked to the extinction coefficient at the excitation wavelength (i.e., α_{ex}), and the second error associated with the

1 Italian National Agency for New Technologies, Energy and Sustainable Economic Development.

extinction coefficient at the emission wavelength (i.e., α_{ex}) (see equations [5.13] and [5.14]).

5.5.3. *Examples in continental waters*

LiDAR contributions over continental waters (i.e., those including lakes, streams and rivers) are summarized below. Numerous satellite, airborne and land LiDAR data were used in fluvial and lacustrine environments [NOT 09, FEU 08], but apart from the systematic surveying of the American great lakes (http://shoals.sam.usace.army.mil/silverlight.aspx), few studies are related to immersed surfaces or water properties [BAI 10]. For those continental aquatic environments, the objective of the LiDAR measurements was either the measurement of the water level dynamics, or the measurement of the immersed surface topography, in particular of minor river and stream beds for hydraulic modeling or hydrobiology (for example, fish habitat modeling), or the estimation of bed macroroughness, a parameter which is also involved in hydraulic and hydro-biological modeling. As the type and scale of the LiDAR application on aquatic environments is linked to the used platform (satellite, airborne or land), the presentation of the various LiDAR feedbacks on continental waters is made below for each platform.

5.5.3.1. *Spaceborne measurements in continental waters*

The advent of satellite LiDARs (e.g., GLAS sensor on ICESat platform) has represented a major progress for studying water surface elevation and with respect to SAR systems [ABD 11]. In fact, SAR altimeters (e.g., Topex-Poseidon) have a coarser horizontal resolution with respect to spaceborne LiDARs, thus SAR measurements may only be useful relatively large rivers. Due to its small divergence angle (e.g., 5 mrad for ICESat), spaceborne LiDAR measurements offer a horizontal resolution of 70 m that greatly improve the mapping of water surface elevation of relatively small environments.

For larger bodies of water (e.g., great lakes in US), GLAS measurements are averaged in order to improve the accuracy of water surface elevation estimates [BRA 01, SWE 09, ABD 11, ZHA 11b, ZHU 14]. Water surface elevation estimates may be associated with large errors if the satellite overpass occurs across very different surfaces in terms of reflectivity (e.g., transition from flat water to coastal banks). This ‘hysteresis’ phenomenon linked with signal saturation at the LiDAR receiver may last a couple of kilometers and is difficult to correct even after using algorithms [LEM 11] (Figure 5.12).

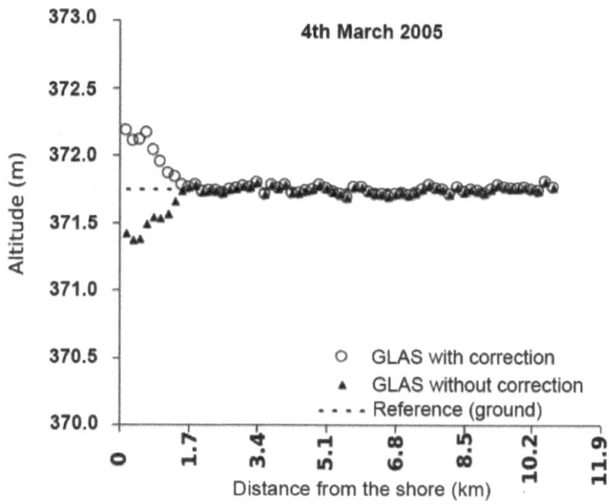

Figure 5.12. *Water surface elevation changes on Lake Leman, France. Datapoints correspond to GLAS/ICESat measurements obtained during March 4, 2005 [BAG 11]*

Therefore and despite the high spatial resolution of GLAS-derived water surface elevation estimates [BAG 11], their RMSE was relatively large on rivers. These results correspond to measurements made on rivers of metropolitan France, with a riverbed width of 60 m, and for the period of 2003-2009. The error of spaceborne LiDAR elevations was computed with respect to ground data derived from the national hydrological database, and was likely related to the hysteresis effects described above. Thus, these findings point out that spaceborne LiDAR sensors with adjustable gain such as GLAS are not recommended for monitoring small and not very wide bodies of water.

5.5.3.2. *Airborne LiDAR examples in continental waters*

Airborne LiDAR studies on continental waters are few and have been done based on hydrographic sensors (i.e., those currently used in coastal waters and characterized by a green laser source). This scarcity of LiDAR measurements on continental environments is related to the difficulty to obtain meaningful signatures in very turbid waters, and planning low altitude missions over. [HIL 08, BAI 12]. Likewise and similar to coastal areas, airborne LiDAR is a better mapping alternative with respect to acoustic methods due to the working limitations of non-navigable waterways.

To date, most airborne LiDAR surveys in rivers have been focused on characterizing the topography of minor and major riverbeds. These studies mainly encompass measurements over waters with steep slopes, high speed and not very turbid (e.g., pebble braided rivers and torrents). Experiments in US rivers have been accomplished by using SHOALS 1000T [MIL 05], AQUARIUS, GEMINI [PAN 15] or EAARL [KIN 07, MCK 08, MCK 14] systems, or in Europe with the HawkEye IIa [BAI 10] and RIEGL VQ820G [AUF 15] systems. In general, in all these contributions, the main measuring limitation are the topographic constraints.

Bailly *et al.* [BAI 10] mapped the topography of the Gardon River, France based on aerial surveys performed at 250 m above the ground. The flights were done along a cross section of 1.5 km, with no river bank vegetation, and using a HawkEye II sensor. Three parallel flight lines were required to cover the river section in 30 min. The density of the bathymetric measurements obtained was 1.1 measurement per square meter. Due to the quality of in situ groundtruth measurements, the obtained accuracy of topographic estimates for the immersed riverbed was 32 cm (with an average slight overestimate of 1.7 cm), and the maximum depth was 4.45 m (i.e., 6.8 times the Secchi depth). The highest measurement uncertainty remains for very shallow depths (Figure 5.13). For this specific study, airborne LiDAR bathymetry closely reproduced minor bed forms (Figure 5.13). However this approach was insuitable for bottom depths shallower than 32 cm, which may represent 28% of the average bottom depth in that section of the river. This empirical results confirmed other theoretical studies that have shown major limitations of LiDAR bathymetry at very shallow depths [LES 07].

Kinzel *et al.* [KIN 07] and McKean *et al.* [MCK 14] applied the experimental green LiDAR EEARL for estimating the topography of North-American braided rivers. In this case, LiDAR measurements were assimilated into hydraulic models and compared with in situ measurements as derived from a tacheometer. McKean *et al.* reported an excellent correlation between LiDAR simulations and field measurements when the difference among estimates never exceeded 1 cm. These results were corroborated by Dobler *et al.* [DOB 14] who analyzed 20 km of a torrent situated in the Italian tyrol region.

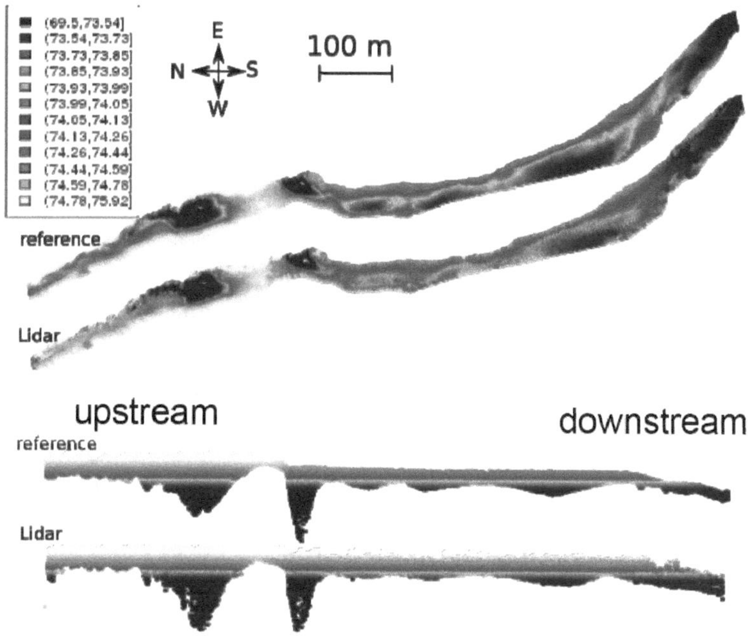

Figure 5.13. *Riverbed topography obtained from LiDAR and field measurements (in situ) on 1.5 km of the Gardon River, France; top: aerial view – bottom: lateral cross-section viewview from lengthwise profile [BAI 10]*

5.5.3.3. *Terrestrial LiDAR examples in continental waters*

Unlike airborne studies, the use of land-based LiDAR measurements has been extensive in the last decade for investigating fluvial [NOT 09, LAG 13]. In general, these studies are based on land topographic LiDARs that operate in the infrared (i.e., 1,064 nm) and may be applicable to aquatic environments. Lam *et al.* is the only study that evaluated a green LiDAR system to characterize the geometry and macroroughness of a river bed section of 5 m. The calculations were made by considering a point density of 5,600 points/m² for the depth of the immersed bed. This point density allowed them to clearly identify coarse components of the bed substrate with a vertical resolution of up to 3 cm.

5.6. Prospectives of spaceborne LiDAR mapping of aquatic environments

Space-based LiDAR systems are a unique remote sensing tool for mapping and monitoring environmental parameters at the regional and global scale. However, one important limitation of these systems at the present is the relatively low sampling density over large spatial domains.

Within the framework of the future ICESAt-2 mission (launch planned in 2017), the ATLAS LiDAR system (*Advanced Topographic Laser Altimeter System*) can potentially be used in coastal oceanography due to its ability to simultaneously measure six parcels of water with a footprint size of 10 m each. This high density of spatial measurements is achieved by the scanning of six laser beams operating at 532 nm and at different viewing angles. The beams are spaced every 3.3 km and are oriented in perpendicular way with respect to the satellite path. Unlike CALIOP, ATLAS measurements can help to better understand the spatial patterns (e.g., gradients) of suspended particulates in littoral waters (i.e., sediments, phytoplankton) and during day or night conditions. However, given the low energy of the transmitted laser pulses (25 to 100 mJ), the penetration depth of ATLAS waveforms in relatively turbid waters is expected to be small (i.e., first meter), thus ATLAS measurements should be spatially integrated in order to increase the signal to noise ratio. In addition to water optical properties, ATLAS is expected to provide a better mapping of wind speed with respect to other spaceborne sensors such as GLAS due to its relatively high spatial resolution and sampling rate.

Looking further into the future, the new generation of spaceborne hyperspectral LiDARs (HSLs) will be a revolutionary tool for characterizing suspended particulates in littoral and oceanic waters. Similar to ATLAS, HSLs will allow the simultaneous measurement of other environmental variables (e.g., water temperature). Unlike traditional "elastic" backscattering LiDARs, HSLs will be also able to process inelastic scattering contributions. Lastly, HLSs, HLSs can separate molecular from particulate backscattering contributions based on Brillouin peaks [SHI 83, LIU 12]. The Brillouin signal is first used to estimate α. Then, the particulate backscattering coefficient is independently computed at each water depth by applying the α estimates previously computed at 532 nm.

Although there are not specific spaceborne LiDAR sensors designed for aquatic environments, the following would be highly recommended for a future profiling system devoted to the study of water composition:

– a high energy by pulse (more than 100 mJ);

– a laser transmitting in the infrared spectrum (1,064 nm) for the detection of the air–water interface;

– a waveform sampling frequency of 0.5–1 GHz and a pulse duration of 5–7 ns corresponding to a minimum vertical resolution Of approximately 5 to 7 cm.

– a wide dynamic range for the LiDAR signal (e.g., > 60 dB);

– detectors with two linear polarizations (parallel and perpendicular) at a wavelength of 532 nm;

– at least three receivers per wavelength (i.e., green and IR channels) with different and small FOVs (i.e., 100-160 mJ).

This last criterion is important to minimize atmospheric scattering and allows optical inversions related to second-order attributes of suspended particulates (e.g., size spectra) [HUT 94].

For water surface elevation monitoring (limnimetry), the optimal LiDAR wavelength is located in the infrared with a very low water penetration. Regarding bathymetric applications in coastal and continental waters, a future spaceborne LiDAR must transmit a laser pulse with a wavelength capable of penetrating the water column and to reach the bottom. The most appropriate wavelength should be centered in the green spectral region due to the minimum attenuation (maximum transmission) of energy within this optical window. This task can be successfully achieved by using an Nd-Yg laser (neodymium-doped yttrium aluminum garnet). Due to the strong depolarization characteristics of the bottom [GIL 13], the use of LiDAR systems with a cross-polarization would allow a better estimation of bottom depths in very shallow environments [MIT 10].

It will take many years of technological progress before spaceborne LiDAR measurements with a high spatial and temporal resolution became operational. In that regard, ATLAS will be useful for mapping bathymetry in littoral areas deeper than 1 m. Also and unlike GLAS, ATLAS will allow a higher spatial resolution that will be translated in a more through monitoring of continental waters (e.g., smaller streams). These improvements along with the spatial integration of measurements will result in more accurate water surface elevation estimates in lakes and rivers [ABD 11], with respect to those derived from GLAS on ICESAT-1 measurements. Finally, the smaller FOV of the telescope on ATLAS-ICESAT-2 will will contribute further to improve the accuracy of bathymetric retrievals by reducing the signal contamination associated with solar radiation.

The performance of spaceborne LiDARs for estimating bottom depth of coastal and continental environments was explored by Abdallah *et al.* [ABD 13] These results were obtained at two laser transmitting wavelengths (355 and 532 nm), different water optical properties, and recommended eye safety threshold levels. This study reveals that it is possible to estimate bathymetry when the concentration of sediments (i.e., <0.5 g m^{-3}) phytoplankton (< 1 mg m^{-3}), and chromophoric dissolved organic matter (<0.5 m^{-1}) is relatively low. The performance of two spaceborne LiDAR configurations (i.e., at 355 and 532 nm, respectively) was evaluated by using synthetic LiDAR waveforms, different water optical types, and laser energy levels that comply with eye safety regulations.

5.7 Key points

The LiDAR systems used in oceanic, littoral and continental waters are mainly airborne profilers that are equipped with pulsed lasers transmitting at a centered wavelength of 532 nm. These active optical systems are commonly configured to measure elastic scattering (e.g., volume backscattering), are mainly dedicated for bathymetric applications, and may have an additional receiver for detecting inelastic optical signatures (e.g., Raman and fluorescence). In general, the maximum range of these LiDAR systems is 3-4 times longer than that obtained by passive optical imagers. Unlike passive optical surveys, LiDAR measurements allow depth-resolved distributions of water optical properties and mapping of bottom depth in shallow areas that are hard to study by other by other means.

5.8. Bibliography

[ABA 14] ABADY L., BAILLY J.-S., BAGHDADI N. *et al.*, "Assessment of quadrilateral fitting of the water column contribution in LiDAR waveforms on bathymetry estimates", *IEEE Geoscience and Remote Sensing Letters*, vol. 11, no. 4, pp. 813–817, 2014.

[ABD 10] ABDALATI W., ZWALLY H.J. BINDSCHADLER R. *et al.*, "The ICESat-2 laser altimetry mission", *Proceedings of the IEEE*, vol. 98, pp. 735–751, 2010.

[ABD 11] ABDALLAH H., BAILLY J.S., BAGHDADI N., "Improving the assessment of ICESat water altimetry accuracy accounting autocorrelation", *ISPRS Journal of Photogrammetry and Remote Sensing*, vol. 66, pp. 833–844, 2011.

[ABD 12] ABDALLAH H., BAGHDADI N., BAILLY J.S. *et al.*, "Wa-LiD: a new LiDAR simulator for waters", *IEEE Geoscience and Remote Sensing Letters*, vol. 9, no. 4, pp. 744–748, 2012.

[ABD 13] ABDALLAH H., BAILLY J.S., BAGHDADI N. *et al.*, "Potential of space-borne LiDAR sensors for global bathymetry in coastal and inland waters", *IEEE Journal of Selected Topics in Applied Earth Observations and Remote Sensing*, vol. 6. no. 1, pp. 202–216, 2013.

[ALL 10] ALLOUIS T., BAILLY J., PASTOL Y. *et al.*, "Comparison of LiDAR waveform processing methods for very shallow water bathymetry using Raman, near-infrared and green signals", *Earth Surface Processes and Landforms*, vol. 35, no. 6, pp. 640–650, 2010.

[AUF 15] AUFLEGER M., BARAN R., STEINBACHER F. *et al.*, "High resolution bathymetric LiDAR data as base for hydraulic-modeling of a mountain stream by bathymetric LiDAR data", *HIC 2014-11th* International Conference on Hydroinformatics, New-York City, USA, 2015.

[AUL 96] AULL B.F., "Geiger-mode avalanche photodiode arrays for imaging laser radar", *Solid State Research Report*, Lincoln Laboratory, pp. 39–41, 1996.

[BAG 11] BAGHDADI N., LEMARQUAND N., ABDALLAH H. *et al.*, "The relevance of GLAS-ICESat elevation data for the monitoring of river networks", *Remote Sensing*, vol. 3, pp. 708–720, 2011.

[BAI 10] BAILLY J.-S., LECOARER Y., LANGUILLE P. *et al.*, "Geostatistical estimation of bathymetric LiDAR errors on rivers", *Earth Surface Processes and Landforms*, vol. 35, pp. 1199–1210, 2010.

[BAI 12] BAILLY J.-S., KINZEL P.J., ALLOUIS T. *et al.* (eds), "Airborne LiDAR methods applied to riverine environments", in *Fluvial Remote Sensing for Science and Management*, John Wiley & Sons, Ltd, pp. 141—161, 2012.

[BAL 97] BALIN Y.S., TIKHOMIROV A.A., MELNIKOV V.E., "Some peculiarities in determining the range to the earth's surface with a space borne LiDAR "Balkan-1" from the space station "Mir"", *Proceedings of SPIE*, vol. 3218, pp. 131–138, 1997.

[BAR 98] BARBINI R., COLAO F., FANTONI R. *et al.*, "Design and application of a LiDAR fluorosensor system for remote monitoring of phytoplankton", *ICES Journal of Marine Science*, vol. 55, no. 4, pp. 793–802, 1998.

[BAR 11] BARTON J.S., JASINKI M.F., "Sensitivity of depth-integrated satellite LiDAR to subaqueous scattering", *Remote Sensing*, vol. 3, pp. 1492–1515, 2011.

[BEC 87] BECKMANN P., SPIZZOCHINO A., *The Scatter of Electromagnetic Waves from Rough Surfaces*, Artech House, Norwood, MA, 1987.

[BEH 13] BEHRENFELD M.J., HU Y., HOSTETLER C.A. *et al.*, "Space-based LiDAR measurements of global ocean carbon stocks", *Geophys. Res. Lett.*, vol. 40, pp. 4355–4360, 2013.

[BIS 96] BISSONNETTE L., "Multiple-scattering LiDAR equation", *Appl. Opt.*, vol. 35, pp. 6449–6465, 1996.

[BOU 14] BOUHDAOUI A., BAILLY J.-S., BAGHDADI N. *et al.*, "Modeling the water bottom geometry effect on peak time shifting in LiDAR bathymetric waveforms', *IEEE Geoscience and Remote Sensing Letters,* vol. 11, no. 7, pp. 1285–1289, 2014.

[BRA 01] BRAUN A., CHENG K.C., CSATHO B. *et al.*, "ICESat laser altimetry in the Great Lakes", *Proceedings of the 60th Annual Meeting of the Institute of Navigation*, pp. 409–416, 2001.

[BRE 08] BRETAR F., CHAUVE A., MALLET C. *et al.*, "Managing full waveform LiDAR data: a challenging task for the forthcoming years", *International Archives of Photogrammetry, Remote Sensing and Spatial Information Sciences,* Beijing, China, vol. 37, no. 1, pp. 415–420, July 2008.

[BRO 04] BROCK J.C., WRIGHT C.W., CLAYTON T.D. *et al.*, "Optical rugosity of coral reefs in Biscayne National Park, Florida", *Coral Reefs*, vol. 23, pp. 48–59, 2004.

[BUF 83] BUFTON J.L., HOGE F.E., SWIFT R.N., "Airborne measurements of laser backscatter from the ocean surface", *Appl. Opt.*, vol. 22, no. 17, pp. 2603–2618, 1983.

[CAD 02] CADALLI N., MUNSON D.C., SINGER A.C. "Bistatic receiver model for airborne returns incident on an imaging array from underwater objects", *Applied Optics*, vol. 41, pp. 3638–3649, 2002.

[CHA 16] CHAZETTE P., TOTEMS J., HESPEL L., *et al.*, "Principle and physics of the LiDAR measurement", in BAGHDADI N., ZRIBI M. (eds), *Optimal Remote Sensing of Land Surfaces*, ISTE Press Ltd, London and Elsevier Ltd, Oxford, 2016.

[CHU 98] CHURNSIDE J.H., TATARSKII V.V., WILSON J.J., "Oceanographic LiDAR attenuation coefficients and signal fluctuations measured from a ship in the Southern California Bight", *Applied Optics*, vol. 37, pp. 3105–3112, 1998.

[CHU 01] CHURNSIDE J.H., WILSON J.J., TATARSKII V.V., "Airborne LiDAR for fisheries applications", *Opt. Eng.*, vol. 40, pp. 406–414, 2001.

[CHU 04a] CHURNSIDE J.H., OSTROVSKY L.A., "LiDAR observation of a strongly nonlinear internal wave train in the Gulf of Alaska", *Int. J. Remote Sens.*, vol. 26, pp. 167–177, 2004.

[CHU 04b] CHURNSIDE J.H., WILSON J.J., "Airborne LiDAR imaging of salmon", *Applied Optics,* vol. 43, pp. 1416–1424, 2004.

[CHU 08] CHURNSIDE J.H., "Polarization effects on oceanographic LiDAR", *Opt. Express*, vol. 16, pp. 1196–1207, 2008.

[CHU 09] CHURNSIDE J., DONAGHAY P.L., "Thin scattering layers observed by airborne LiDAR", *ICES J. Mar. Sci.*, vol. 66, pp. 778–789, 2009.

[CHU 13a] CHURNSIDE J.H., MCCARTY B.J., LU X., "Subsurface ocean signals from an orbiting polarization LiDAR", *Remote Sensing*, vol. 5, pp. 3457–3475, 2013.

[CHU 13b] CHURNSIDE J.H., "Effect of surface roughness on LiDAR overlap function", *Proc. SPIE 8724,* Ocean Sensing and Monitoring V, 872409, 3 June 2013.

[CHU 14a] CHURNSIDE J.H., "Review of profiling oceanographic LiDAR", *Optical Engineering*, vol. 53, no. 5, 51405, pp. 1–13, 2014.

[CHU 14b] CHURNSIDE J.H., SULLIVAN J.M., TWARDOWSKI M.S., "LiDAR extinction-to-backscatter ratio of the ocean", *Opt. Express*, vol. 28, pp. 18698–706, 2014.

[CLE 78] CLEGG J.E., PENNY M.F., "Depth sounding from air by laser beam", *Journal of Navigation*, vol. 31, pp. 52–61, 1978.

[COL 10] COLLIN A., LONG B., ARCHAMBAULT P., "Salt-marsh characterization, zonation assessment and mapping through a dual-wavelength LiDAR", *Remote Sensing of Environment*, vol. 114, pp. 520–530, 2010.

[COO 82] COOK R.L., TORRANCE K.E., "A reflectance model for computer graphics", *ACM Transactions on Graphics*, vol. 1, pp. 7–24, 1982.

[COT 08] COTTIN A., Adaptation du SHOALS à la cartographie sédimentaire côtière peu profonde dans la Baie des Chaleurs, PhD Thesis, INRS, Quebec, Canada, 2008.

[DOB 14] DOBLER W., BARAN R., STEINBACHER F. *et al.*, "Using high resolution bathymetric LiDAR data for a Telemac2D simulation", *EGU General Assembly Conference Abstracts*, vol. 16, p. 1026, 2014.

[DUO 09] DUONG H., LINDERBERGH R., VOSSELMAN G., "ICESat Full-Waveform altimetry compared to airborne laser scanning altimetry over the Netherlands", *IEEE Transactions on Geoscience and Remote*, vol. 47, pp. 3365–3378, 2009.

[DUR 12] DURHAM W.M., STOCKER R., "Thin phytoplankton layers: characteristics, mechanisms, and consequences", *Annu. Rev. Mar. Sci.*, vol. 4, pp. 177–207, 2012.

[EVA 78] EVANS W.E., "Evaluation of a near-infrared photomultiplier", *NASA report*, contract NAS1-15114, SRI project 6919, p. 55, March 1978.

[FEU 08] FEURER D., BAILLY J., PUECH C. *et al.*, "Very high resolution mapping of river immersed topography by remote sensing", *Progress in Physical Geography*, vol. 32, no. 4, pp.1–17, 2008.

[FEI 92] FEIGELS V.I., "LiDARs for oceanological research: criteria for comparison, main limitations, perspectives", *Proc. SPIE XI*, vol. 1750, pp. 473–484, 1992.

[FIO 06] FIORANI L., "Environmental monitoring by laser radar", in LARKIN S.B. (ed.), *Lasers and Electro-Optics Research at the Cutting Edge*, pp. 123–175, Nova Science Publishers Inc., Hauppauge, 2006.

[FRA 82] FRANZ H., GEHLHAAR U., GÜNTHER K.P. *et al.*, "Airborne fluorescence LiDAR monitoring of tracer dye patches – a comparison with shipboard measurements", *Deep-Sea Research part A: Oceanographic Research Papers*, vol. 29, no. 7, pp. 893–901, 1982.

[GIL 66] GILBERT G.D., STONER T.R., JERNIGAN G.L., "Underwater experiments on the polarization, coherence, and scattering properties of a pulsed blue-green laser", *Proc. SPIE 0007, Underwater Photo Optics I*, vol. 8, p. 3, 1966.

[GIL 13] GILERSON A.A., STEPINSKI J. IBRAHIM A.I. *et al.*, "Benthic effects on the polarization of light in shallow waters", *Applied Optics*, vol. 52, pp. 8685–8705, 2013.

[GOR 89] GORDON H.R., "Can the Lambert-Beer law be applied to the diffuse attenuation coefficient of ocean water?", *Limnol. Oceanogr.* vol. 34, no. 8, pp 1389–1409, 1989.

[GUE 78] GUENTHER G.C., GOODMAN L.R., "Laser applications for near-shore nautical charting", *Proc. SPIE Ocean Optics V*, vol. 160, pp. 174–183, 1978.

[GUE 85] GUENTHER G.C., "Airborne laser hydrography: system design and performance factors", *NOAA professional paper series*, 1. U.S. Dept. of Commerce, National Oceanic and Atmospheric Administration, National Ocean Service, Charting and Geodetic Services, Rockville, Md, p. 385, 1985.

[HIC 68] HICKMAN G.D., HOGG J.E., "Application of an airborne pulsed laser for near-shore bathymetric measurements", *Remote Sensing of Environment*, vol. 1, pp. 47–58, 1968.

[HIL 08] HILLDALE R.C., RAFF D., "Assessing the ability of airborne LiDAR to map river bathymetry", *Earth Surface Processes and Landforms*, vol. 33, no. 5, pp. 773–783, 2008.

[HOG 80] HOGE F.E., SWIFT R.N., FREDERICK E.B., "Water depth measurement using an airborne pulsed neon laser system", *Applied Optics*, vol. 19, no. 6, pp. 871–883, 1980.

[HOG 81] HOGE F.E., SWIFT R.N., "Absolute tracer dye concentration using airborne laser-induced water Raman backscatter", *Appl. Opt.*, vol. 20, pp. 1191–1202, 1981.

[HOG 83] HOGE F.E., SWIFT R.N., "Airborne detection of oceanic turbidity cell structure using depth-resolved laser-induced water Raman backscatter", *Appl. Opt.*, vol. 1, pp. 3778–3786, 1983.

[HOG 03] HOGE F.E., "Validation of satellite-retrieved oceanic inherent optical properties: proposed two-color elastic backscatter LiDAR and retrieval theory", *Applied Optics*, vol. 2, pp. 7197–7201, 2003.

[HWA 00] HWANG P.A., WANG D.W., WALSH E.J. *et al.*, "Airborne measurements of the wave number spectra of ocean surface waves. Part I: spectral slope and dimensionless spectral coefficient", *J. Phys. Oceanogr.,* vol. 30, pp. 2754–2767, 2000.

[HUT 94] HUTT D.L., BISSONETTE L.R., DURAND L., "Multiple field of view LiDAR returns from atmospheric aerosols", *Applied Optics*, vol. 33, pp. 2338–2348, 1994.

[IRI 98] IRISH J.L., WHITE T.E., "Coastal engineering applications of high-resolution LiDAR bathymetry", *Coastal Engineering*, vol. 35, pp. 47–71, 1998.

[IRI 99] IRISH J.L., LILLYCROP W.J., "Scanning laser mapping of the coastal zone: the SHOALS system", *ISPRS Journal of Photogrammetry & Remote Sensing*, vol. 54, no. 2, pp. 123–129, 1999.

[JUT 06] JUTZI B., STILLA U., "Range determination with waveform recording laser systems using a Wiener filter", *ISPRS Journal of Photogrammetry and Remote Sensing*, vol. 61, no. 2, pp. 95–107, 2006.

[KAR 06] KARPICZ R., DEMENTJEV A., KUPRIONIS Z. *et al.*, "Oil spill fluorosensing LiDAR for inclined onshore or ship-board operation", *Appl. Opt.*, vol. 45, pp. 6620–6625, 2006.

[KIN 07] KINZEL P.J., WRIGHT C.W., NELSON J.M. *et al.*, "Evaluation of an experimental LiDAR for surveying a shallow, braided, sand-bedded river", *Journal of Hydraulic Engineering*, vol. 133, no. 7, pp. 838–842, 2007.

[KOP 08] KOPELEVICH Y.I., SURKOV A.G., "Mathematical modeling of the input signals of oeanological LiDARs", *J. Opt. Technol.*, vol. 75, pp. 321–326, 2008.

[LAG 13] LAGUE D., BRODU N., LEROUX J., "Accurate 3D comparison of complex topography with terrestrial laser scanner: application to the Rangitikei canyon (N-Z)", *ISPRS journal of Photogrammmetry and Remote Sensing*, vol. 82, pp. 10–26, 2013.

[LAM 15] LAM N., NATHANSON M., LUNDGREN N. *et al.*, "A cost-effective laser scanning method for mapping stream channel geometry and roughness", *JAWRA Journal of the American Water Resources Association*, vol. 51, pp. 1211–1220, 2015.

[LAN 05] LANCASTER R., SPINHIRNE J. *et al.*, "Laser pulse reflectance of the ocean surface from the GLAS Satellite LiDAR", *Geophysical Research Letters*, vol. 32, no. L22S10, 2005.

[LEE 96] LEE Z.P., CARDER K.L., PEACOCK T.G. *et al.*, "Method to derive ocean absorption coefficients from remote-sensing reflectance", *Appl. Opt.,* vol. 35, pp. 453–462, 1996.

[LEE 13] LEE J.H., CHURNSIDE J.H., MARCHBANKS R.D. *et al.*, "Oceanographic LiDAR profiles compared with estimates from in situ optical measurements", *Applied Optics*, vol. 52, pp. 786–794, 2013.

[LEM 11] LEMARQUAND N., BAGHDADI N., BAILLY J., "Étude de la précision du satellite LiDAR GLAS-ICESat pour l'altimétrie des eaux continentales", *Revue Française de Photogrammétrie et de Télédétection,* vol. 2, no. 194, pp. 45–52, 2011.

[LES 07] LESAIGNOUX A., BAILLY J.-S., ALLOUIS T. *et al.*, "Épaisseur d'eau minimale mesurable en rivière sur fronts d'ondes LiDAR simulés", *Revue Française de Photogrammétrie et de Télédétection*, vol. 186, pp. 48–53, 2007.

[LIU 90] LIU Z.S., "Estimate of maximum penetration depth of LiDAR in coastal water of the China sea", *Proc. SPIE Ocean Optics X*, vol. 1302, pp. 655–661, 1990.

[LIU 12] LIU D., HOSTETLER C., MILLER I. *et al.*, "System analysis of a tilted field-widened Michelson interferometer for high spectral resolution LiDAR", *Optics Express*, vol. 20, pp. 1406–1420, 2012.

[LU 14] LU X., HU Y., TREPTE C. *et al.*, "Ocean subsurface studies with the CALIPSO space borne LiDAR", *Journal of Geophysical Research*, vol. 119, pp. 4305–4317, 2014.

[LUC 87] LUCHININ A.G., "Certain properties of backscattered signal during laser sounding of the upper layer of the ocean through a rough surface", *IzvestiaAkademiinauk SSSR.Fizikaatmosferyiokeana*, vol. 23, pp. 976–983, 1987.

[LYZ 85] LYZENGA D.R., "Shallow-water bathymetry using combined LiDAR and passive multispectral scanner data", *Remote Sensing of Environment*, vol. 6, pp. 115–125, 1985.

[MAN 11] MANDLBURGER G., PFENNIGBAUER M., STEINBACHER F. *et al.*, "Airborne hydrographic LiDAR mapping-Potential of a new technique for capturing shallow water bodies", *19th International Congress on Modeling and Simulation*, Perth, Australia, 2011.

[MAZ 03] MAZEL C.H., STRAND M.P., LESSER M.P. *et al.*, "High-resolution determination of coral reef bottom cover from multispectral fluorescence laser line scan imagery", *Limnology and Oceanography*, vol. 48, pp. 522–534, 2003.

[MCK 08] MCKEAN J., ISAAK D., WRIGHT C., "Geomorphic controls on salmon nesting patterns described by a new, narrow-beam terrestrial-aquatic LiDAR", *Frontiers in Ecology and the Environment*, vol. 6, no. 3, pp. 125–130, 2008.

[MCK 14] MCKEAN J., TONINA D., BOHN C. *et al.*, "Effects of bathymetric LiDAR errors on flow properties predicted with a multi-dimensional hydraulic model", *Journal of Geophysical Research: Earth Surface*, vol. 119, no. 3, pp. 644–664, 2014.

[MCL 96] MCLEAN J.W., FREEMAN J.D., "Effects of ocean waves on airborne LiDAR imaging", *Applied Optics*, vol. 35, no. 18, pp. 3261–3269, 1996.

[MCM 03] MCMANUS M.A., ALLDREDGE A.L., BARNARD A.H. *et al.*, "Changes in characteristics, distribution and persistence of thin layers over a 48-hour period", *Marine Ecology Progress Series*, vol. 261, pp. 1–19, 2003.

[MEA 92] MEASURES R.M., *Laser Remote Sensing*, Krieger Publishing, Malabar, 1992.

[MIL 05] MILLAR D., *Using airborne LiDAR bathymetry to map shallow water river environments*, Coastal GeoTools, Myrtle beach, SC, 2005.

[MIT 10] MITCHELL S., THAYER J., HAYMAN M., "Polarization LiDAR for shallow water depth measurement", *Applied Optics*, vol. 49, no. 36, pp. 6995–7000, 2010.

[MON 10] MONTES-HUGO M., CHURNSIDE J., GOULD R. *et al.*, "Spatial coherence between remotely-sensed ocean color data and vertical distribution of LiDAR backscattering in coastal stratified waters", *Remote Sensing of Environment*, vol. 114, pp. 2854–2593, 2010.

[MON 11] MONTES-HUGO M.A., WEIDEMANN A., GOULD R. *et al.*, "Ocean color patterns help to predict depth of optical layers in stratified coastal waters", *Journal of Applied Remote Sensing*, vol. 5, p. 053548, 2011.

[MON 14] MONTES-HUGO M., BAILLY J.S., BAGHDADI N. *et al.*, "Modeling the effects of surface and bottom geometries on LiDAR bathymetric waveforms", *IEEE Geoscience and Remote Sensing Symposium,* pp. 2706–2708, 2014.

[NOT 09] NOTEBAERT B., VERSTRAETEN G., GOVERS G. *et al.*, "Qualitative and quantitative applications of LiDAR imagery in fluvial geomorphology", *Earth Surface Processes and Landforms*, vol. 34, no. 2, pp. 217–231, 2009.

[OTT 65] OTT L., "Underwater ranging measurements using blue-green laser", Technical report, Naval Air Development Center NADC-AE-6519, Warminster, PA, 1965.

[PAN 15] PAN Z., GLENNIE C., HARTZELL P. *et al.*, "Performance assessment of high resolution airborne full waveform LiDAR for shallow river bathymetry", *Remote Sensing*, vol. 7, no. 5, pp. 5133–5159, 2015.

[PAR 11a] PARK J.Y., Expansion of Countermine LiDAR UAV-based System (CLUBS), Optech International Inc., Kiln Ms, 2011.

[PAR 11b] PARRISH C.E., JEONG I., NOWAK R.D. *et al.*, "Empirical comparison of full-waveform LiDAR algorithms: range extraction and discrimination performance", *Photogrammetric Engineering and Remote Sensing*, vol. 77, no. 8, pp. 824–838, 2011.

[PAR 12] PARRISH C.E., WHITE S.A., ASLAKSEN M. *et al.*, "Topographic-bathymetric LIDAR evaluation for integrated ocean and coastal mapping", *European LiDAR Mapping Forum (ELMF)*, 4–5 December 2012, Salzburg, Austria, 2012.

[PAS 07] PASTOL Y., LE ROUX C., LOUVART L., "Litto3D – a seamless digital terrain model", *International Hydrographic Review*, vol. 8, no. 1, pp. 38–44, 2007.

[PAS 11] PASTOL Y., "Use of airborne LIDAR bathymetry for coastal hydrographic surveying: the French experience", *Journal of Coastal Research*, vol. 62, pp. 6–18, 2011.

[PEE 07] PE'ERI S., PHILPOT W., "Increasing the existence of very shallow-water LiDAR measurements using the red-channel waveforms", *IEEE Transactions on Geoscience and Remote Sensing*, vol. 45, no. 5, pp. 1217–1223, 2007.

[PEE 11] PE'ERI S., LONG B., "LIDAR technology applied in coastal studies and management", *Journal of Coastal Research*, vol. 62, pp. 1–5, 2011.

[PET 77] PETZOLD T.J., "Volume scattering functions for selected ocean waters", in TYLER J.E. (ed.), *Light in the Sea*, Dowden, Hutchinson & Ross, 1977.

[REU 93] REUTER R., DIEBEL D., HENGSTERMANN T., "Oceanographic laser remote sensing: measurements of hydrographic fronts in the German Bight and in the northern Adriatic Sea", *International Journal of Remote Sensing*, vol. 14, no. 5, pp. 823–848, 1993.

[REU 97] REUTER R., ZIELINSKI O., "Perspective of a hydrographic LiDAR in space: specifications and results of a simulation", *Proceedings of SPIE*, vol. 3104, pp. 145–151, 1997.

[RON 08] RONCAT A., WAGNER W., MELZER T. *et al.*, "Echo detection and localization in full-waveform airborne laser scanner data using the averaged square difference function estimator", *The Photogrammetric Journal of Finland*, vol. 21, pp. 62–75, 2008.

[ROY 13] ROY G., MATHIEU P., CAO X. *et al.*, "Development of an underwater fiber-optic LiDAR for the characterization of sea water and ice properties", *Proc. SPIE 8872,* vol. 14, p. 887201, 2013.

[RUD 14] RUDOLF A., WALTHER T., "Laboratory demonstration of a Brillouin LiDAR to remotely measure temperature profiles of the ocean", *Opt. Eng.*, vol. 53, p. 051407, 2014.

[SHI 83] SHIPLEY S.T., TRACY D.H., ELORANTA E.W., "High spectral resolution LiDAR to measure optical scattering properties of atmospheric aerosols. 1: Theory and instrumentation", *Appl. Opt.*, vol. 22, no. 23, pp. 3716–3724, 1983.

[SHI 08] SHI W., WANG M., "Three-dimensional observations from MODIS and CALIPSO for ocean responses to cyclone Nargis in the Gulf of Martaban", *Geophysical Research Letters*, vol. 35, no. 21, 21603, 2008.

[SIZ 75] SIZGORIC S., CARSWELL A.I., "Underwater probing with Laser radar. In Water Quality Parameters", ASMT, STP 573, *American Society for Testing Materials*, pp. 398–413, 1975.

[STE 97] STEINVALL O., KLEVEBRANT H., LEXANDAR J., *et al.*, "Laser depth sounding in the Baltic sea", *Applied Optics*, vol. 20, no. 19, pp. 3284–3286, 1997.

[SUL 10] SULLIVAN J.M., DONAGHAY P.L., RINES J.E.B., "Coastal thin layer dynamics: consequences to biology and optics", *Continental Shelf Research*, vol. 30, no. 1, pp. 50–65, 2010.

[SWE 09] SWENSON S., WAHR J., "Monitoring the water balance of Lake Victoria, East Africa, from space", *Journal of Hydrology*, vol. 370, no. 1, pp. 163–176, 2009.

[TUE 07] TUELL G., "Countermine LiDAR UAV-based Sytem (CLUBS)", Annual Report for Project N00014-05-C-0417, FY06, 2007.

[TUE 10] TUELL G., BARBOR K., WOZENCRAFT J., "Overview of the coastal zone mapping and imaging LiDAR (CZMIL): a new multisensor airborne mapping system for the U.S. Army Corps of Engineers", *Proc. SPIE 7695*, 76950R (12 May 2010), 2010.

[TUL 99] TULLDAHL H.M., STEINVALL K.O., "Analytical waveform generation from small objects in LiDAR bathymetry", *Applied Optics*, vol. 20, pp. 1021–1039, 1999.

[VAS 01] VASILKOV A.P., GOLDIN Y.A., GUREEV B.A., *et al.*, "Airborne polarized LiDAR detection of scattering in the ocean", *Applied Optics,* vol. 40, pp. 4353–4364, 2001.

[VES 06] VESELOVSKII I., KOLGOTIN A., GRIAZNOV V., *et al.*, "Inversion of multiwavelength Raman LiDAR data for retrieval of bimodal aerosol size distribution", *Applied Optics*, vol. 43, pp. 1180–1195, 2006.

[WAG 07] WAGNER W., RONCAT A., MELZER T., *et al.*, "Waveform analysis techniques in airborne laser scanning", *International Archives of Photogrammetry and Remote Sensing*, vol. 36, no. 3, pp. 413–418, 2007.

[WAN 15] WANG C., LI Q., LIU Y. *et al.*, "A comparison of waveform processing algorithms for single-wavelength LiDAR bathymetry", *ISPRS Journal of Photogrammetry and Remote Sensing*, vol. 101, pp. 22–35, 2015.

[WOZ 03] WOZENCRAFT J.M., LEE M., TUELL G., *et al.*, "Fusion of hyperspectral and bathymetric laser data in Kaneohe Bay, Hawaii", *Proceedings of SPIE*, 2003.

[ZHA 11a] ZHANG X., TWARDOSWSKI M., LEWIS M., "Retrieving composition and sizes of oceanic particle subpopulations from the volume scattering function", *Applied Optics*, vol. 50, pp. 1240–1259, 2011.

[ZHA 11b] ZHANG G., XIE H., KANG S., *et al.*, "Monitoring lake level changes on the Tibetan Plateau using ICESat altimetry data (2003–2009) ", *Remote Sensing of Environment*, vol. 115, no. 7, pp. 1733–1742, 2011.

[ZHU 14] ZHU W., JIA S., LV A., "Monitoring the fluctuation of Lake Qinghai using multi-source remote sensing data", *Remote Sensing*, vol. 6, no. 11, pp. 10457–10482, 2014.

[ZWA 02] ZWALLY H.J., SCHUTZ B., ABDALATI W. *et al.*, "ICESat's laser measurements of polar ice, atmosphere, ocean, and land", *Journal of Geodynamics*, vol. 34, pp. 405–445, 2002.

6

Contributions of Airborne Topographic LiDAR to the Study of Coastal Systems

6.1. Introduction

The use of airborne LiDAR data in coastal zones has mostly developed since the mid-1990s. Today, a large number of datasets are available that can be used to refine our knowledge of the terrain between marine and land environments, which was, for a long time, little-known.

The coast is a vulnerable zone where future development could be very sensitive to climate change. The knowledge of coastal elevation over vast areas and its sometimes significant changes as a result of storms has become crucial to better predict its future. LiDAR's contribution can be substantial due to its high spatial resolution and vertical accuracy. It can be used to define the initial state of the terrain and to monitor its evolution, both of which are necessary to calibrate forecast models about the behavior of coasts following increases in sea level, changes in storm systems or the construction of new coastal developments.

Despite the fact that it is still quite costly, this type of instrument is widely used on an international scale in coastal studies concerning risks of submersion, the impact of storms and monitoring developments. It is

---

Chapter written by Franck LEVOY, Franck GARESTIER, Laurent FROIDEVAL, Olivier MONFORT and Emilie POULLAIN.

also used more generally in the characterization of the coastal environment, sometimes in combination with other sensors such as hyperspectral imaging.

Though not exhaustive, this chapter uses case studies to present topographic LiDAR data's contribution to the study of the evolution of coastal sedimentary environments, notably in identifying and monitoring the coastline or in establishing a sediment overview. However, LiDAR also offers other possibilities for functional enhancement. One method proposed is the multi-temporal localization of channels in bays and estuaries based primarily on the use of elevation and intensity data. This method has been applied in the Mont-Saint-Michel bay, and is transposable to other sites.

On a more exploratory level, the second part of this chapter presents the processing of backscattered intensity data, which was the focus of recent developments with the aim of improving the characterization of coastal areas, bringing a notable added value to simple topographical data. First, this processing addressed the modeling of intensity signals necessary for an unbiased characterization of surfaces, followed by the use of the textural properties of intensity images to detect anisotropic surfaces, which are indicators of the means of sedimentary transfers, as well as an estimate of the moisture of sandy surfaces.

6.2. Characterization of coastal evolution

6.2.1. *Identification of coastlines for the study of kinematics on open coasts*

6.2.1.1. *Indicators of the position of the coastline: some reminders*

The shoreline is a significant geomorphological object on both a scientific and societal level. On many coasts, the shoreline is mobile for relatively short spans that vary from one season to a few years for loose coasts or those composed of cliffs made of soft rock. Its mobility, and especially its erosion, can lead to damages such as the destruction of goods (houses, roads, etc.) or, following the destruction of an embankment or a dune, lead to the submersion of low lands located immediately behind it. Identifying the shoreline on different dates may prove to be essential to better predicting coastal risks and managing their development.

Boak and Turner [BOA 05] state that there are several indicators available to characterize the position of a shoreline and its mobility. Generally, two broad categories of indicators have been identified: those which are visually discernable on the terrain, such as the ledge of a dune cliff, its foot, a border of vegetation or a structure (Figure 6.1), and those which rest on the tide gauge level and its intersection with the coastal topography. This level could be the mean level of the sea; theoretical higher levels (astronomical level) or any other level characterized by a specific return period (such as a centennial level).

After capturing the information, LiDAR technology is able to reproduce the terrain data with a large spatial density to characterize a detailed morphology of the shoreline whether it has natural or anthropized features. The topography generated by LiDAR can also locate a specific isohypse that reflects the reference tide gauge level, such as the astronomical high tide level, which is sometimes equated to a shoreline. This approach can also provide several indicators to better characterize the evolution of the system, each of which has advantages and disadvantages.

According to the coastal monitoring of the department of Manche (France), the shoreline is identified as the ledge of a dune when it is eroding (Figure 6.1(a)) or the vegetation boundary when it is stable or accreting (Figure 6.1(b)). These borders separate the continental area from the marine area, subject to the effects of the waves. This identification of the shoreline is planimetric and can be compared to historical coastlines from vertical aerial photographs. It does not have altimetric significance because, depending on the morphology of the dune bar, the two indicators could have very different altitudes even on a reduced geographical zone, with the ledge of the dune in erosion being altimetrically higher than the vegetation boundary of a dune in accretion.

6.2.1.2. *Identification of abrupt changes in slope as coastline indicators*

Several studies based on LiDAR technology identify zones as eroding, stable or accreting by comparing two studies conducted on different dates. However, this tool can also be used along with altimetric information such as the analysis of slopes at the land–sea interface in order to distinguish cliffs and microcliffs from the erosion of a dune and the topographical profiles of the convexo-concave morphology witnessed with accretion (Figure 6.2).

Figure 6.1. *Photos of coastlines and various visual indicators of their location: a) eroding shoreline and b) accreting shoreline*

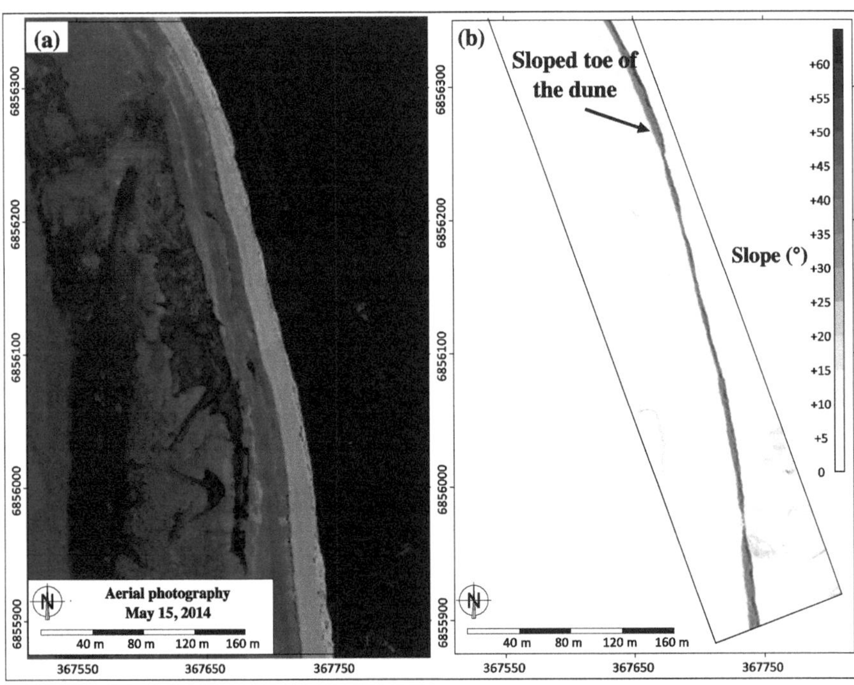

Figure 6.2. *a) Orthophography of the dune ledge in erosion in Saint-Jean-le-Thomas (Manche, May 2014) underlining the position of the coastline between vegetated dunes (in green) and beaches (in beige); b) map of slopes from LiDAR data presenting the toe of the dune (slopes from 20 to 35°) in May 2014 located between the foot of the dune and the ledge, two indicators of the position of the coastline. For a color version of this figure, see www.iste.co.uk/baghdadi/5.zip*

Information on the slope provided by the LiDAR data can be a useful basis for identifying shorelines represented by a ledge or crest of a dune, or by the foot of a dune for shorelines where erosion creates cliffs or microcliffs in the dunes (Figure 6.3(a)). However, in stable or accreting sectors, the slope variations between the beach and the dune system are not always very clear (Figure 6.3(b)). In these cases, the use of another indicator to integrate the presence of dune vegetation is essential.

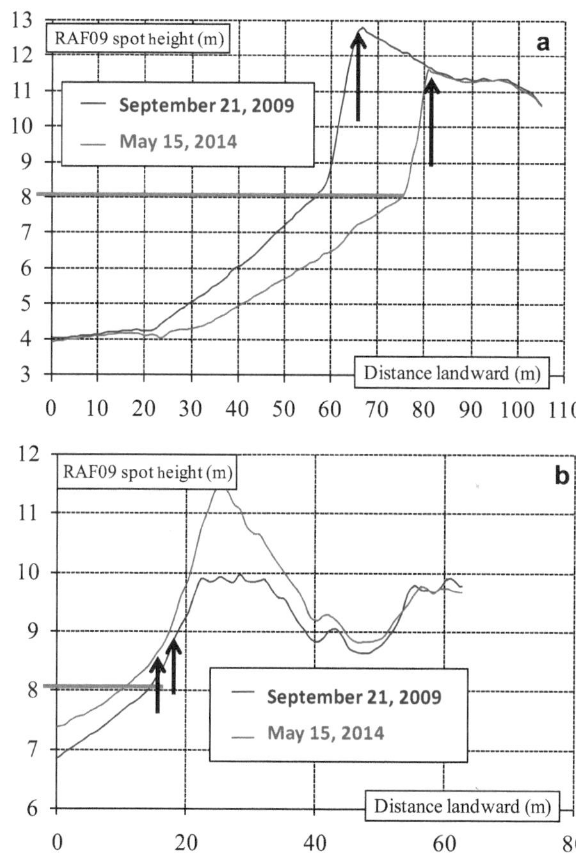

Figure 6.3. *a) Evolution of a topographic cross-shore profile generated by the LiDAR data between 2009 and 2014 in Saint-Jean-le-Thomas showing erosion of the coastline; b) evolution of the topographic profile from LiDAR data between 2009 and 2014 at Saint-Jean-le-Thomas showing an accretion of the coastline. In turquoise, the highest astronomical high tide level; the black arrows represent the position of the coastline from the GPS survey (dune ledges in Figure 6.3(a) and vegetation boundary for Figure 6.3(b)). For a color version of this figure, see www.iste.co.uk/baghdadi/5.zip*

6.2.1.3. *Identification of the coastline by extracting an isohypse of reference*

To retrieve the position of the coastline based on the characteristic water level, Stockdon *et al.* [STO 02] propose, in the case of LiDAR data, to fit a line on the profile points in a range of values around the retained water level, such as the level of high tide +/–0.5 m (constant to be adjusted according to the case), and to use the intersection between this line and the water level to determine the cross-shore position of the coastline. Figure 6.4 uses the Pointe d'Agon on the West Cotentin coast (49.01°N ; 1.59°W – France) as an example. It shows a gap in the longitudinal variable position of up to a dozen meters between the two selected shoreline identification indicators. The line of the highest astronomical tide is, in this particular sector, most often ahead of the shoreline provided by DGPS (differential GPS) which can be visually recognized by the identification of the dune ledge for the zones in erosion and the vegetation boundary for the zones that are stable or accreting.

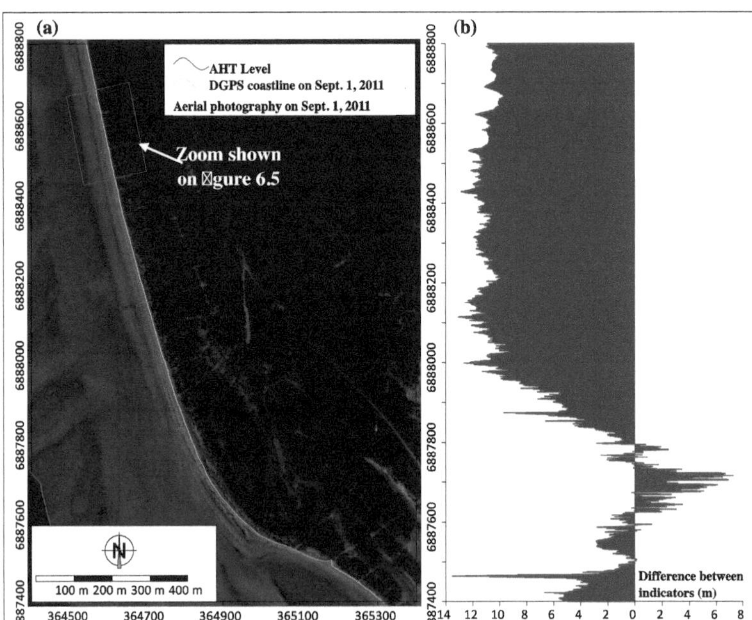

Figure 6.4. *a) Coastline provided by a DGPS survey compared to the isohypse of the level of the Astronomical High Tide (AHT) level from a LiDAR study (49.01°N ; 1.59°W). The red rectangle represents the zone of the zoom shown in Figure 6.5; b) comparison between the coastline from a DGPS survey and the coastline provided at the astronomical high tide level located with LiDAR data. For a color version of this figure, see www.iste.co.uk/baghdadi/5.zip*

6.2.1.4. *Identification of the coastline through an analysis of surface roughness at the land–sea interface*

This shaded image reveals the difference in terrain between the vegetated dune, with a complex topography, and a mostly smooth beach (Figure 6.5).

Figure 6.5. *On the left, a shaded image of a DEM on October 10, 2014 of the beach–dune interface at Pointe d'Agon. The location of the zone is indicated in Figure 6.4. On the right, a photo of the foredune and toe of a vegetated dune on the coast of Pointe d'Agon. For a color version of this figure, see www.iste.co.uk/baghdadi/5.zip*

There are clearly differences in terrain between the beach and the dune structure, as well as between the texture of these two areas. An analysis of the roughness between the beach and the dune can be used to determine a relatively clear border that can correspond to the delineation of the coastline by grouping both the variability of terrains (changes in slope and different shapes) and the vegetation cover, which is absent on the beach and dense on this stretch of dune (Figure 6.5). Similar to the method currently used in the study of materials, the roughness of the coastal area can be equated with the root mean square Rq assessed by:

$$Rq = \sqrt{\frac{1}{l}\int_0^l z(x)^2 dx} \qquad [6.1]$$

In this relation, x is the coordinate perpendicular to the general direction of the coast, $z(x)$ is the gap between the point of the DEM and its mean surface and l is the cross shore distance over which the roughness is integrated. For each point on the DEM, the slope of this mean surface perpendicular to the coast is calculated using linear regression on the DEM points on a transversal distance consistent with the dimensions of the relief under consideration. The vertical intercept resulting from this regression is calculated in such a way as to ensure that the surfaces of the coastal areas above and below the mean surface are equivalent. Once the roughness has been calculated on all points of the area, the cross-shore position of the coastline retained is the one with the maximum roughness in a band of 10 m from the point of contact between the beach profile and the local level of the highest astronomical high tides (Figure 6.6). The sensitivity of the final value chosen can be tested and modified based on a comparison of the results obtained with the reference sites on the ground.

6.2.1.5. *Discussion and conclusion*

On the coastline in question, the calculation of roughness between the dune structure that is mainly in erosion and the beach can be used to locate the position of the shoreline in a satisfactory way. This approach proves to be more effective than the use of an indicator based on the level of the astronomical high tides, a frequently used alternative (Figures 6.4(a) and (b)). In fact, considering the surface roughness, the difference between the location of the coastline in relation to data about the terrain gathered using DGPS is low, varying between 2 and 3 m. On the coastline in question, this gap corresponds to the distance between the ledge of the dune provided by DGPS and the foot of the dune provided by the roughness calculation (Figure 6.6(c)). However, the roughness indicator proves to be less satisfactory on the sectors of the bordering dunes in accretion composed of low dunes with small slopes, where the vegetation develops in the direction of the beach in an often discontinuous way (Figure 6.1(a)). For this type of land–sea interface, the use of an altimetric level of reference for the static water level proves more relevant. At this stage, LiDAR data are still not able to provide a universal indicator comparable with the delineation of coastlines produced by observing the terrain.

In conclusion, LiDAR data can be used to determine, in an automated way, the position of a shoreline that proves to be close to one that can be visually captured by DGPS or by other topometric methods. The use of the variation in roughness between the bordering dunes and the beach can locate the foot of the dune in a precise way when the dune is in erosion. For sectors of coasts in accretion, it is preferable to use a coastline with a tide gauge level of reference such as the level of astronomical high tides. A ground check remains essential to validate the relevance of a chosen indicator, to be adopted in the changing context of the toes of dunes to allow for automatic processing based on LiDAR data compared to a land survey.

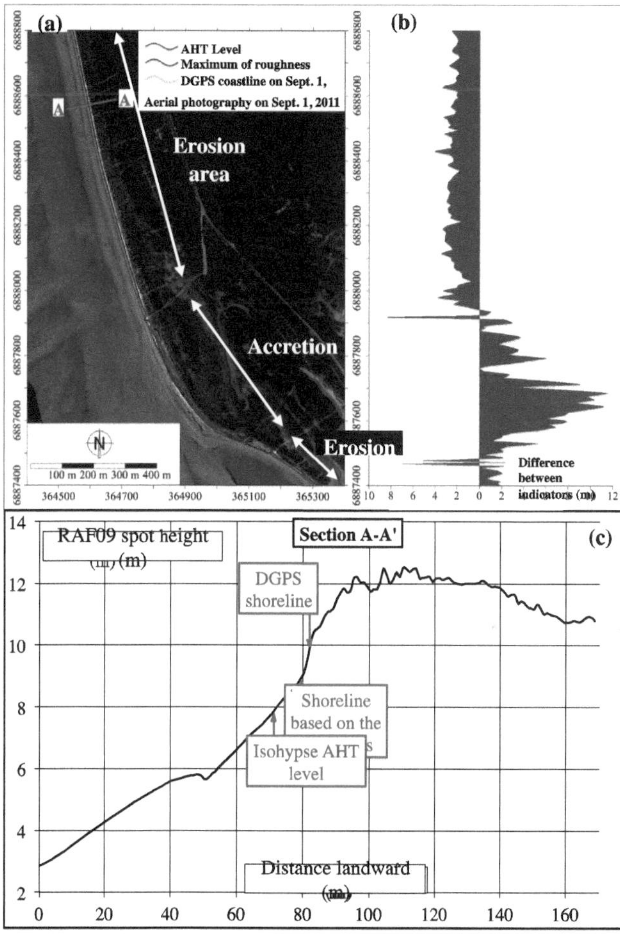

Figure 6.6. *a) Representation of the position of the coastline determined by two indicators calculated using LiDAR data, knowing the level of the astronomical high tides and the roughness of the land–sea interface, compared to the coastline provided by a DGPS study on the site Pointe d'Agon; b) comparison between the coastline from a DGPS study and the coastline from an analysis of the land–sea interface roughness given by LiDAR data; c) topographic section A-A' of the land–sea interface located Figure in 6.6(a) showing, on the same date, the location of the coastline revealed visually by DGPS, compared to indicators provided by LiDAR data, knowing the isohypse of the level of the astronomical high tides and the results of a roughness analysis. For a color version of this figure, see www.iste.co.uk/baghdadi/5.zip*

6.2.2. *Potential of airborne LiDAR for morphodynamic monitoring and the calculation of sedimentary budgets*

6.2.2.1. *Morphodynamic exploration at multiple levels*

LiDAR can be used to conduct topographical surveys of vast surfaces, as it can have a spatial resolution that is very high with densities surpassing 10 points per m^2. It is possible to distinguish types of coast, notably in sedimentary environments, which are a reflection of almost-instantaneous hydrodynamic processes such as sand megaripples, whose wave length is several meters, but also larger forms such as tidal deltas or sand bar systems at a kilometer to several kilometer-scale, whose establishment sometimes takes several decades (Figures 6.7 and 6.8).

The morphodynamic monitoring of the ebb tide delta of the Pointe d'Agon, which extends seaward more than 2 km, identified the existence of sandbars perpendicular to the shoreline whose north-facing migration, diametrically opposed to the sediment transports commonly seen in this zone, led to the formation of the Pointe d'Agon [LEV 13, MON 14].

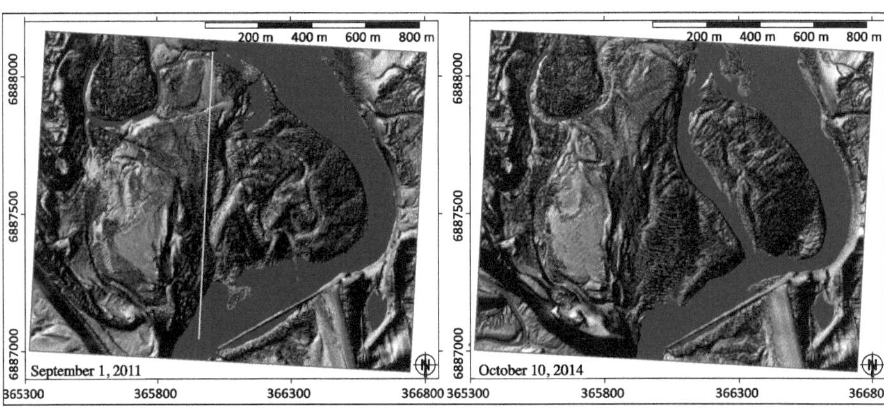

Figure 6.7. *Evolution of the Regnéville harbor delta (49.00°N ; 1.57°W – France) and its channels between September 2011 and October 2014. The channels in blue are visually bordered by the digitization of the intensity image associated with LiDAR topographic images. For a color version of this figure, see www.iste.co.uk/baghdadi/5.zip*

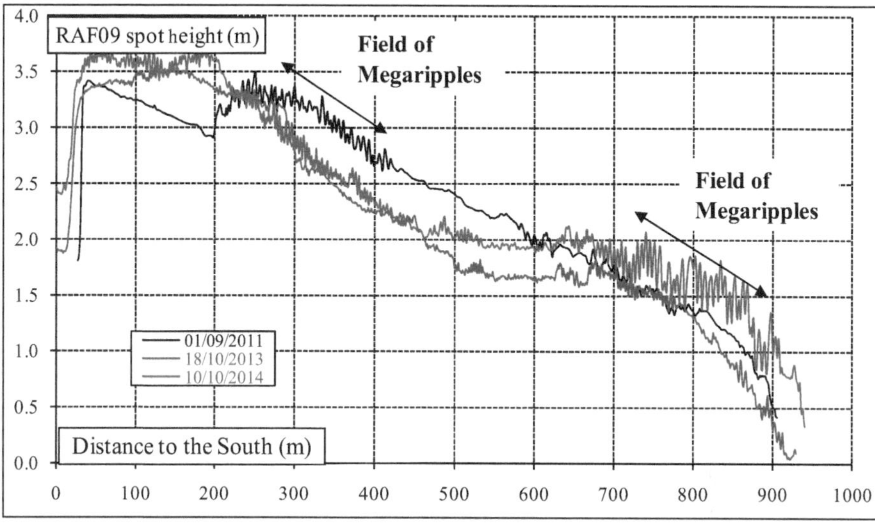

Figure 6.8. *Cross-section of the Regnéville harbor delta: visualization of the fields of megaripples in the surface characterizing the hydrodynamic force induced by the tidal runoff In green in Figure 6.7). For a color version of this figure, see www.iste.co.uk/baghdadi/5.zip*

6.2.2.2. *Calculating a sedimentary budget: relevance of estimates*

In addition to showing topographical changes, morphological monitoring of coastal sediment areas help in developing in sedimentary budgets and detecting major evolutionary trends in the volume of materials contributing to the changes. The accuracy of LiDAR data, especially vertically, for this type of study must be pointed out. It has been demonstrated that for coastal zones [SAL 03, ZHA 05, LIU 09], the altimetric accuracy of topographic LiDAR could not be much better than 0.10–0.15 m (root-mean-square error) and that it depended on the parameters of the flight and on environmental conditions. In the case of monitoring the Mont Saint-Michel bay, successive topographical studies the former approach route to Mont Saint-Michel (48.63°N ; 1.51°W – France, Figure 6.9) estimate a root-mean-square error of 0.087 m.

Systematically developed quality control sheets also assess the robustness of the results obtained using altimetry in order to best minimize uncertainty in comparisons between two morphological states. A comparison of the

LiDAR altimetry with DGPS data from a dozen control zones situated around the area under consideration can be used to determine relevant indicators about the flight quality (mean altimetric difference "DZ", root-mean-square error). These can also be compared to a large number of altimetric points (n.10^5–n.10^6 points) in the overlapping zones using the "measure match" tool in the Terra Match software. Finally, for topographic monitoring, a comparison between LiDAR altimetry on two different dates in the clearly defined control zones can be carried out (Figures 6.10(a) and (b)). In the chosen example, the root-mean-square errors are low, between 0.02 and 0.055 m.

This approach can be used to assess the quality of measurements made in order to draw conclusions about the consistency here of data gathered on different dates. It can also be used to improve understanding of uncertainties in the calculations of sedimentary budgets in the zone under consideration.

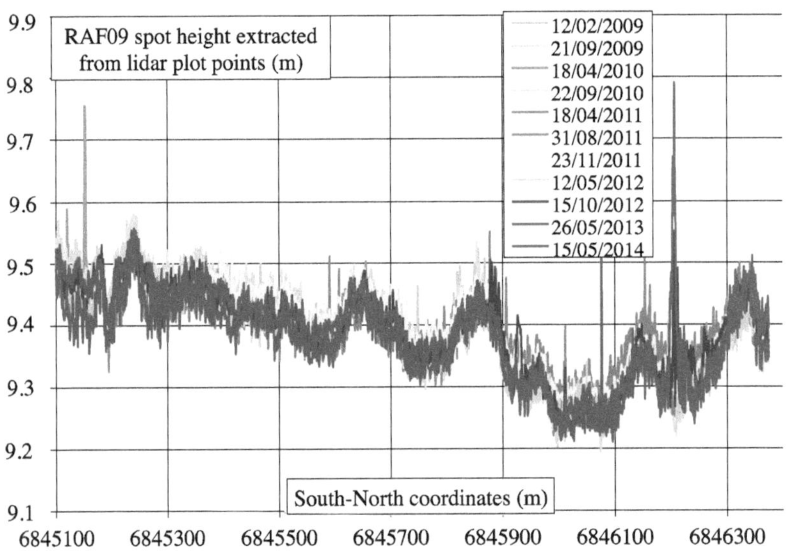

Figure 6.9. *Altimetric LiDAR profiles along the former embankment approach route to Mont Saint-Michel (48.63°N ; 1.51°W – France) made over 11 flights from February 2009 to May 2014 and calibrated according to the Référence Altimétrique Française RAF 09 grid. For a color version of this figure, see www.iste.co.uk/baghdadi/5.zip*

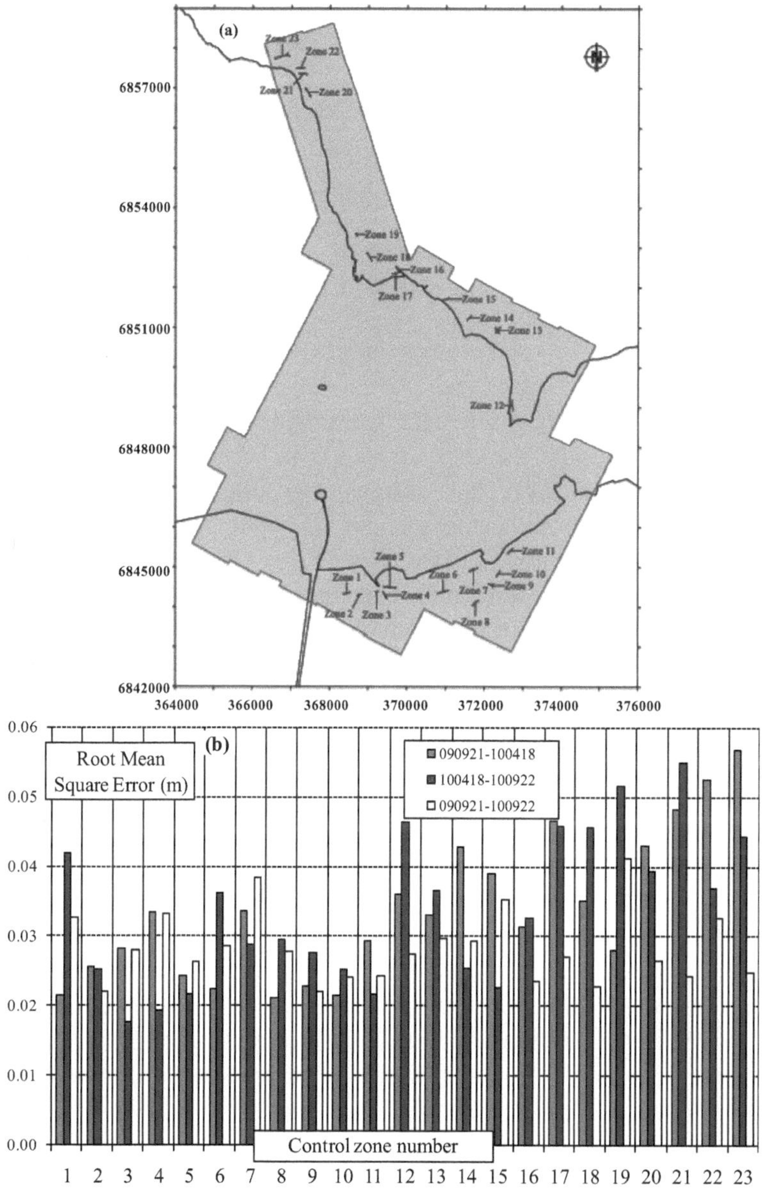

Figure 6.10. *a) Location of control zones around the Mont Saint-Michel bay (in blue); b) root-mean-square errors in 23 zones reviewed between September 21, 2009 and April 18, 2010, between April 18, 2010 and September 22, 2010, and between September 21, 2009 and September 22, 2010. For a color version of this figure, see www.iste.co.uk/baghdadi/5.zip*

Using seasonal LiDAR data from 2009 to 2012, maps of the differences are made by subtracting the altitudes from each node on the DTM to show the morphological changes (Figure 6.11). However, considering the uncertainties on DTMs is essential in order to identify and compare significant changes in the system under consideration. Recent studies [WHE 10] specified that DTM errors could spread in differential maps following this function:

$$\delta u_{DoD} = \sqrt{(\delta z_{new})^2 + (\delta z_{old})^2} \quad [6.2]$$

where δu_{DoD} is the error in the topographical differential and δz_{new} and δz_{old} are the errors in the new and old DTMs, respectively. In the monitoring of the Mont Saint-Michel bay, the mean uncertainty based on different controls is less than 0.10 m, with the error spreading to 0.15 m when comparing the two topographical states. The sedimentary budgets calculated must be interpreted with care, taking into account the value of the error and that the surface under consideration is more than 50 km^2. However, this type of bay often shows significant dramatic changes, both negative and positive, notably due to the shifting of the channels (Figure 6.11).

In this type of environment, sheltered from the effects of the waves, as in many estuaries, vegetated schorres occupy a large part of the edges of the intertidal zone. The extremely high density of vegetative cover on these schorres means that the laser generally does not penetrate the canopy well and it is difficult to get to the ground [SCH 11]. Furthermore, for periods of time in the order of a few years, the evolution of the ground in these zones is generally low [DET 11]. Airborne LiDAR is therefore not the most relevant equipment to monitor the topography of these spaces. It is better to exclude these areas from sedimentary budget calculations, which should be restricted to non-vegetated sediment spaces, at the risk of introducing substantial errors. Furthermore, the bathymetry of the water channels in tidal sediment environments cannot be found using LiDAR working in the near-infrared (1,064 nm), because this wavelength is not suitable for detecting the bottom under surfaces of water (see Chami *et al.*, Chapter 4, Bailly *et al.*, Chapter 5). The shallow depth of these channels at low normal tide (to the meter) and the turbidity also limit investigation with bathymetric LiDAR. As in the case of schorres, it is better to exclude the channels from the calculations of cubatures and the estimate of sedimentary budgets for this type of coastal environment.

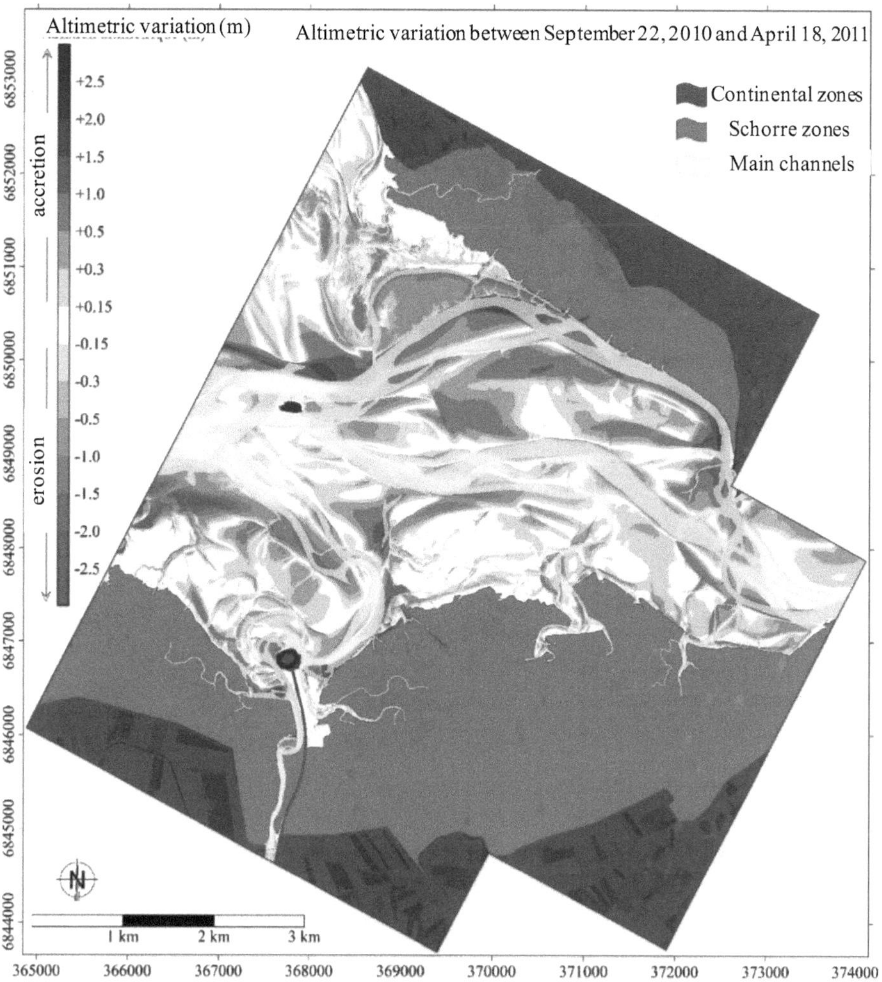

Figure 6.11. *Evolution of the small Mont Saint-Michel bay between September 22, 2010 and April 18, 2011*

To conclude, establishing sedimentary budgets in bays, or even estuaries, using topographical LiDAR data cannot be carried out with precision. Zones that are always under water, notably the bottom of channels, cannot be explored and require different measurement techniques. The floor in schorre areas with dense vegetative cover is equally difficult to reach. Calculations of development can therefore only be made on sedimentary non-vegetated zones outside of the water. Furthermore, the vertical errors of LiDAR measurements (RMSE from 0.1 to 0.15 m at best) cannot be used to find the orders of magnitude of sedimentary masses in movement. However, only airborne LiDAR technology offers topographical coverage of a coastal zone of several tens to several hundreds of km^2, such as the Mont-Saint-Michel bay, almost instantly – on the same date – with such a spatial resolution and absolute precision.

6.3. Method of identifying the main channels in the Mont Saint Michel bay combining topography and LiDAR intensity

Detecting the water surface using LiDAR data in coastal studies is an important step in order to deliver datasets that correctly render the topographical surface and not the surface of a body of water such as a channel or the sea. As previously indicated, considering water surfaces can bias the calculation of sedimentary budgets such as those conducted in a complex environment like the Mont Saint Michel bay. A multi-criteria method making it possible to easily determine the points belonging to the main channels in the bay has been developed so that these points can be removed from the calculations for the sedimentary budget.

6.3.1. *Hypotheses selected for the extraction of channels*

The approach consists of pairing a flatness criterion with an intensity criterion to identify the main channels. On the one hand, we assume that water contained in the main channels constitutes a flat surface. The relative flatness of the water in comparison with other surfaces is a criterion used by Smeeckaert *et al.* [SME 13]. To this effect, a method to calculate a slope adjusted to the complex environment of a bay is presented.

On the other hand, compared to other types of surfaces and in the near-infrared, water absorbs a large quantity of pulses emitted from the LiDAR [WOL 89]. A weaker intensity registers off water than off land. This hypothesis will be used to exclude high intensity points in our research area by fixing a threshold. Höfle and Pfeifer [HOF 07] correct the intensity according to the distance of the sensor from the illuminated surface and the variations of environmental parameters. Poullain *et al.* [POU 12] model the intensity according to the incidence angle on the surface illuminated by the laser. According to Song *et al.* [SON 02], the incidence angle constitutes a major part of the intensity noise registered because the returned energy depends greatly on the laser's angle of reflection on the surface. In this study, the approach is simpler. The intensity is not corrected and the variations in intensity do not take the incidence angle into account. The result at the level of the bay allows for an initial description of the position and shape of the channels. No parameters for angles of scan, distance or atmospherics are necessary, which is an advantage because these data are not always available. This method can be extended to other LiDAR data from various sources. However, this approach is an alternative to manual digitization, which is inherently subjective, based on an RGB image or an intensity raster.

6.3.2. *Description of data*

The LiDAR data used were acquired on May 12, 2012 with a Leica ALS 60 sensor by the CLAREC (*Contrôle par Laser Aéroporté des Risques Environnementaux Côtiers*) team at the Morphodynamique Continentale et Côtière laboratory at the University of Caen (France). This instrument operates with a laser that has a wavelength of 1,064 nm. The divergence of the laser is 0.15 mrad. The average altitude of flight was 1,500 m with a mean density of 2.2 points per square meter and a scan angle of ±40°. The overlap between two flight paths was 20% on average. The topographic survey was completed during a mean low tide with a coefficient of 60. This part of the bay is very exposed at low tide, even with the exception of the spring tides. The surface covered by the data is visible in Figure 6.12. The data was processed using Terrasolid software [SOI 05]. Corrections allowing for the trajectory of the plane and the internal structural movements of the instrument were applied, as well as an adjustment of several points provided by DGPS, assuring an accuracy of less than 10 cm as a root-mean-square error. The planimetric data (X,Y) are expressed in Lambert 93 and the Z dimension is computed with respect to the RAF09 geoid.

The ALS 60 is paired with an IGI system equipped with a Hasselblad camera. The flight plan was designed for the acquisition of high-resolution photographs, simultaneous to the LiDAR measurements. The photos were then corrected, orthorectified and assembled to create a mosaic with a 27 cm resolution representing the zone under consideration (Figure 6.12). This mosaic serves as a visual validation of the method presented.

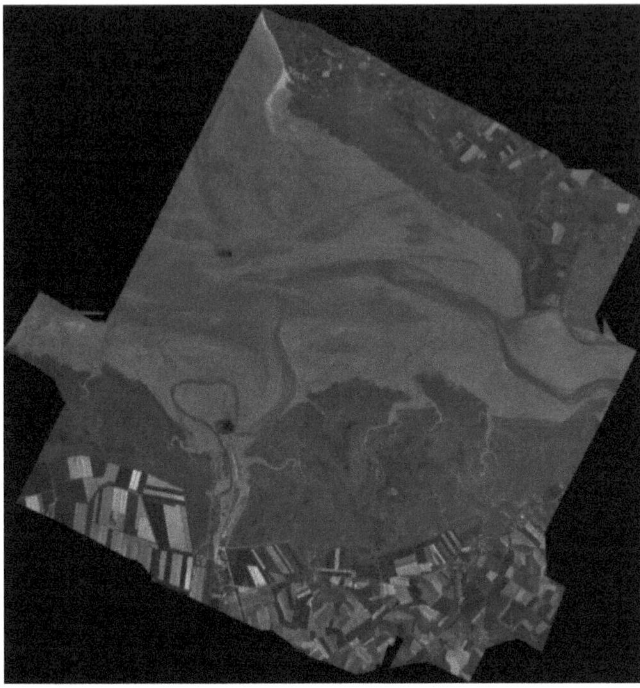

Figure 6.12. *Orthophotography representing the area captured during the LiDAR CLAREC flight on May 12, 2012 (Mont Saint-Michel bay – 48.63°N ; 1.51°W – France). For a color version of this figure, see www.iste.co.uk/baghdadi/5.zip*

6.3.3. *Description of the processing workflow of the channel extraction*

To reduce calculations, the zone of calculation was first restricted to an interval in Z of the absolute elevation corresponding to [–6 m, + 8 m]. These values were determined empirically and used to limit the research area while encompassing the whole bay and its channels.

A value associated with the slope was then generated for each point (X,Y,Z). The value is defined here as the absolute value of the gradient vector representing the variations in the three directions X, Y and Z. To reduce the noise inherent to this calculation's digital application on a point cloud, a Gauss filter must be applied beforehand. In each point, the Z coordinate is replaced by the mean of the values of the adjacent points, weighted by the distance with a standard deviation value of 1 m in accordance with Gauss's law. Furthermore, to improve the digital precision, the Z coordinate of each point is multiplied by an empirically determined factor (in this case, equal to 50). The points of the lowest slopes are extracted at the scale of the bay. The limit of the maximum slope is deduced with a sensitivity test, making it possible to remove regions with high slopes. In the interest of visibility, these points are then grouped in subclouds containing a minimum of 5,000 points and which cannot be spaced more than 10 m apart. They are represented in Figure 6.13. In practice, these calculations were applied in the Cloud Compare software [TEL 14] for an immediate visual depiction of the results.

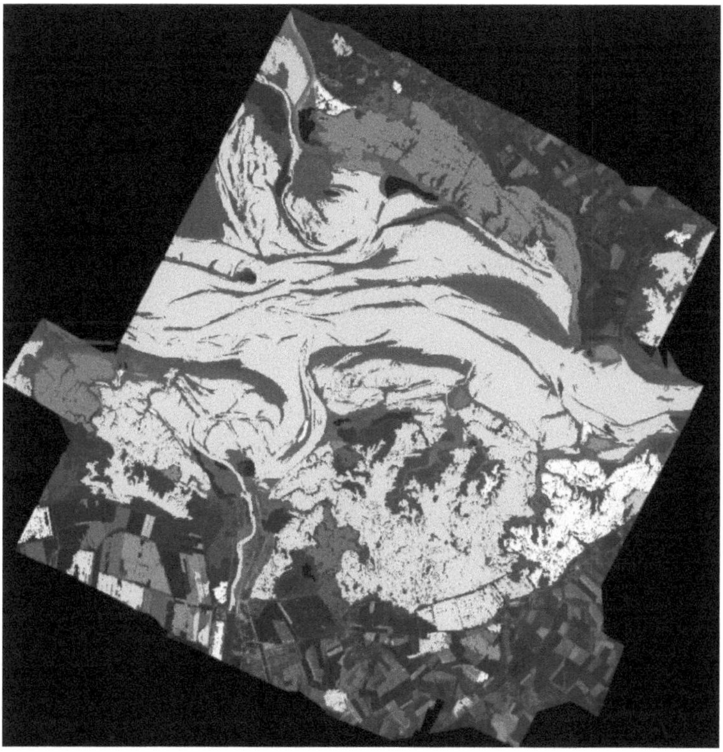

Figure 6.13. *Points of low slope grouped in subclouds. A random color is automatically attributed to each subcloud to distinguish between them. The background is an orthophotograph from the LiDAR CLAREC flight on May 12, 2012. For a color version of this figure, see www.iste.co.uk/baghdadi/5.zip*

The intensity associated with each point is used to differentiate the main channels. The ALS 60 sensor emits pulses in the near-infrared at 1,064 nm.

At this wavelength, a large part of the energy emitted is absorbed by water. However, turbid water can bounce back a certain amount. An arbitrary threshold of low intensity was used in this study. This criterion, coupled with the previous slope criterion, is sufficient to describe the position of the main channels in the bay in a satisfactory way. Figure 6.14 represents the low-intensity points grouped in the subclouds.

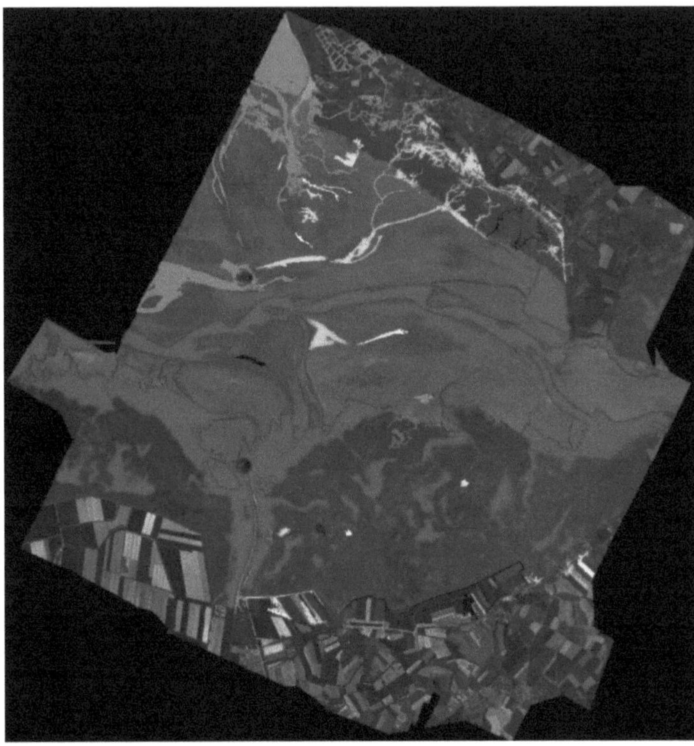

Figure 6.14. *Points of low-intensity grouped in subclouds. A random color was automatically attributed to each subcloud to distinguish between them. For a color version of this figure, see www.iste.co.uk/baghdadi/5.zip*

Figure 6.15 illustrates an area of superimposition of two criteria: points of low slope are represented in red and points of low intensity are represented in green. Certain low-intensity zones present a relatively high slope and can, for example, represent running water on the sedimentary surface of the bay, which will not be retained as a main channel in the water. Other areas with a low slope have an intensity that exceeds the tolerance threshold and can correspond to moist sand, for example. Only areas that combine a close proximity of points of low slope and low intensity will be selected as a channel. More precisely, starting from the cloud of raw points, we selected the points in a spherical area with a radius of 1 m around the points of low slope. From this selection, we considered only the points included in the spherical area with a radius of 1 m around the low-intensity points.

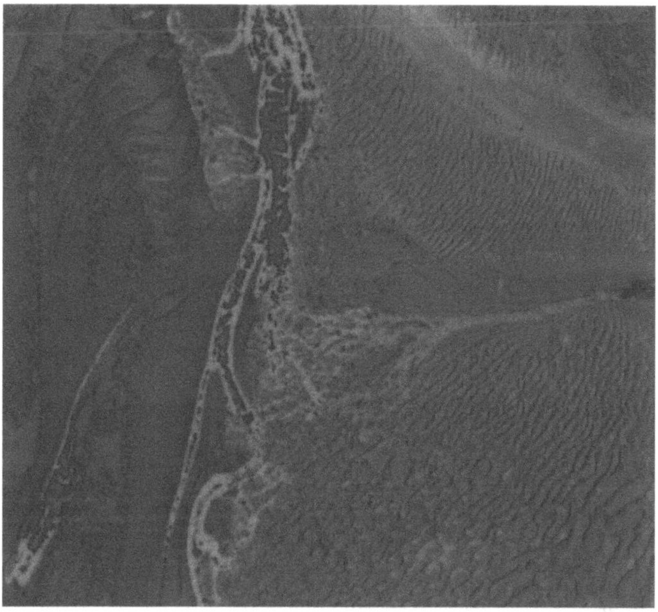

Figure 6.15. *Zoomed-in image of the superimposition of the two criteria. The low intensity points are green and the low slope points are red. The background is an orthophotograph from the LiDAR CLAREC flight on May 12, 2012. For a color version of this figure, see www.iste.co.uk/baghdadi/5.zip*

The resulting point cloud is then automatically partitioned into subclouds like the previous cases. A maximum distance of 10 m between each point belonging to the same subcloud is needed to introduce a continuity criterion to each channel and to avoid isolated points. To express a minimum channel size for the main channels, a minimum number of 5,000 points is required for each subcloud. An arbitrary color was randomly attributed to each subcloud determined in this way. The points that do not respect the continuity criterion do not belong to any subcloud. They are colored white. Figure 6.16 shows the subclouds as well as the white points that do not belong to any subcloud. The white points will not be considered as part of a channel. All of the points included in a subcloud are retained. The white points are automatically excluded. Visually, only the linear forms will be retained (Figure 6.17).

6.3.4. *Results and discussion*

In relation to the mosaic of images, the position and size of the main channels are generally respected. The semi-automatic method presented allows for easy extraction of the main channels of the Mont Saint-Michel bay. However, reflections of the water to the nadir should be analyzed to improve the channel modeling. The laser is reflected in a specular manner to a greater or lesser extent according to the condition of the channel surface to the nadir. These high-intensity returns are not taken into account by the low-intensity criterion. However, the sensor does not receive any returned energy in certain areas, according to the emission angle of the laser in relation to the nadir and the condition of the water surface. This results in a density of low, or zero, points that could be modeled to improve the description of the channels.

Figure 6.16. *Result of combining the elevation, slope and intensity criteria divided into subclouds. A random color was automatically attributed to each subcloud to distinguish between them. For a color version of this figure, see www.iste.co.uk/baghdadi/5.zip*

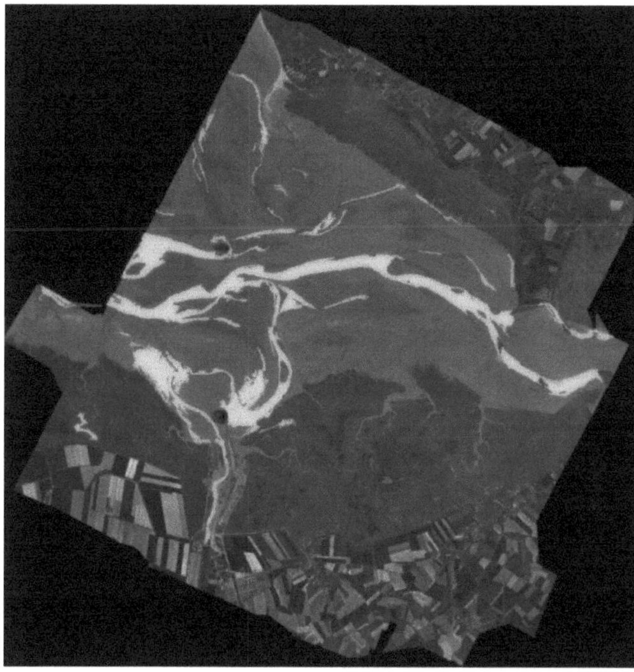

Figure 6.17. *Result of combining the three criteria after cleaning. The color blue is associated with points considered to belong to a channel. For a color version of this figure, see www.iste.co.uk/baghdadi/5.zip*

6.4. Backscattered signal intensity applications

6.4.1. *Modeling of the backscattered intensity as a function of the incidence angle*

The intensity of the backscattered signal is registered simultaneously with each distance measurement and, as such, constitutes complementary information for the characterization of natural surfaces. The investigation of the system parameters that influence LiDAR intensity has been addressed by several works in the past years, references of which can be found in [KAA 11] and [POU 16a]. As shown by the image of intensity, which mainly covers a channel and its banks made up of sand saturated by water (Figure 6.18), obtained by the rasterization of point clouds, the effect of the incidence angle presents a fundamental influence on the signal intensity. The incidence angle is formed by the laser beam and the normal of the surface. It is estimated using the LiDAR system scan angle, which corresponds to the angle separating the laser beam emitted from the nadir of the plane and the DEM. The incidence angle is null close to the nadir (yellow line) on water surfaces, as well as in the vicinity of low topography banks, meaning the intensity is reaching its maximum. As a coefficient of a Fresnel equation, the intensity return takes the minimum values for the highest incidence angles (in the order of 20° for this example). The considerable intensity dependence at the incidence angle was modeled to be used in characterizing surfaces and estimating parameters.

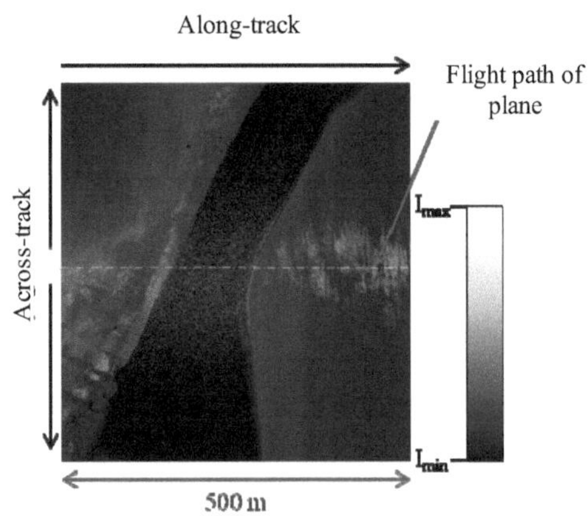

Figure 6.18. *Intensity map on a water zone. For a color version of this figure, see www.iste.co.uk/baghdadi/5.zip*

In order to investigate very different intensity behaviors reflected by different surfaces, a comparative study of several reflection models used for image synthesis and virtual reality was conducted on different types of surfaces [POU 12, POU 16a]. The objective was to extend previous laboratory analyses of coastal surfaces conducted by Kaasalainen *et al.*. [KAA 11] in order to determine a general reflection model to describe the intensity behavior as a function of the incidence angle on all of the surfaces.

The exhaustive review of reflection models aimed to elaborate a model:

– based on the physics of interaction, taking into account the diffuse and specular components of reflection;

– that optimizes the number of variables, which is necessary for parameter inversion.

The Torrance–Sparrow (TS) [TOR 67] and Lommel–Seeliger [HAP 93] models were selected to be evaluated on the images of intensity backscattered by different types of surfaces.

The TS model (equation [6.3]), used with the Beckmann distribution D [BEC 63], considers the surface as a distribution of microfacets that are randomly oriented and reflect light in a purely specular way. The D factor depends on the θ_i incidence angle and the m surface roughness. The reflectance of the surface is defined by the F Fresnel factor. The G factor defines the function of geometric attenuation; this factor depends on the microfacet distribution and therefore on surface roughness.

$$f_{TS} = \frac{FD(\theta_i,m)G^2(\theta_i,m)}{cos^2\theta_i} \quad [6.3]$$

where

$$G(\theta_i) = \frac{cos\theta_i}{cos\theta_i - kcos\theta_i + k} \text{ with } k = \sqrt{\frac{2m^2}{\pi}} \quad [6.4]$$

and

$$D(\theta_i) = \frac{1}{m^2 cos^4\theta_i} e^{-\left\{\frac{tan^2\theta_i}{m^2}\right\}} \quad [6.5]$$

The Lommel–Seeliger model is based on a theory of radiative transfer. The Lommel–Seeliger function combined with Lambert's cosine law was used to establish the behavior of the intensity backscattered by an airborne laser scanner [KAA 11]. The intensity is expressed as:

$$I(\theta_i) = a(w, g_s)(1 - b(w, g_s)\,(1 - cos\theta_i)) \quad [6.6]$$

In equation [6.6], ω represents the albedo and g_s represents the granulometric size of the material. When the incidence angle is zero, the intensity is equal to a. If b is equal to 1, then the intensity becomes Lambertian. If b is greater than 1, the diffusion of light becomes more specular than Lambert's law. The intensity behavior described by the combined model is more diffuse than a Lambertian surface when the b factor is less than 1.

6.4.2. *Characterization of scanned surfaces*

The study of Poullain *et al.* [POU 13] surveyed different coastal surfaces (sea, of various levels of moistures and surfaces of various levels, salt marshes or schorres, etc.) located along the edge of the Mont-Saint-Michel, as well as urban zones, in order to consider all surface types.

The models selected in section 6.4.1 (TS and Lommel–Seeliger) are applied to the intensity data associated with the different surfaces (Figure 6.19).

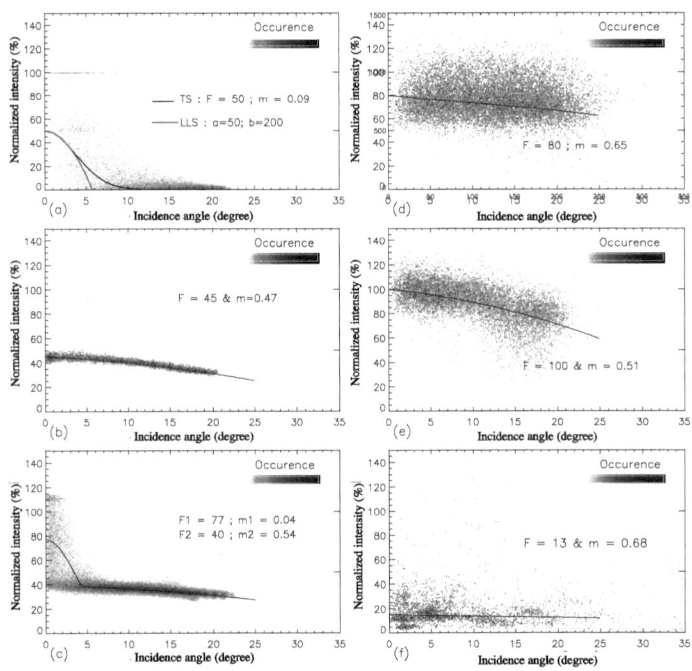

Figure 6.19. *Intensity as a function of the incidence angle over a water surface: a) a sandy surface containing little water; b) a very moist sandy surface; c) an area of salt marsh that is slightly moist; d) an area of salt marsh containing more water; and e) an urban area; f) the intensity was normalized to a common value on all the surfaces and the blue scale represents the abundance of each intesity value as a function of the incidence angle. The legend of each graph indicates the values of parameters associated with the Torrance-Sparrow model: the F factor represents the Fresnel term and m represents the associated roughness parameter (the two values of Figure 6.19 (c) corresponding to two components). For Figure 6.19(a), the Lommel–Seeliger curve is modeled on equation [6.6] (TS = Torrance-Sparrow and LLS = Lommel–Seeliger. For a color version of this figure, see www.iste.co.uk/ baghdadi/5.zip)*

The specular signature of the water surfaces (open water and saturated sand) can only be modeled by the microfacet distribution, since the Lommel–Seeliger law is not suitable for describing the intensity behavior on the entire interval of the incidence angle, as shown by the red curve in Figure 6.19(a). The TS model associated with the Beckmann microfacet distribution models the intensity behavior observed on all the surfaces, from specular behavior, with a very low *m* roughness factor, to diffuse behavior, associated with higher values of this parameter. To describe the intensity behavior on any surface type, it is possible to generalize the TS model as a sum of n-components of microfacet distributions:

$$I(\theta_i) = \frac{1}{cos\theta_i} \sum_{n=1}^{k} F_{(m_n)}\, D_{(m_n)} G^2{}_{(m_n)} \quad [6.7]$$

For most of the surfaces, the intensity behavior as a function of the incidence angle corresponds to a unique component microfacet distribution, the sandy surface only requiring two components. On the other hand, the decomposition of the relation of the intensity-incidence angle into n-components allows an unsupervised spectral analysis for the characterization of the surfaces, as a wavelet decomposition approach. Then, from the resulting spectrum, it is possible to reconstruct the intensity-incidence angle relation through summation of the different components.

The spectrum of the decomposition in Figure 6.20 includes the information of the dispersion of data (width of the shore), the *F* Fresnel coefficient associated with different components, and each *m* roughness parameter. Then, the spectral decomposition of different surfaces (Figure 6.20) constitutes a tool for unambiguous characterization of the surfaces.

6.4.3. *Anisotropic surface detection by texture analysis*

As shown above, LiDAR intensity presents a high sensitivity to the surface type, through the Fresnel coefficient *F* and incidence angle relation, whose local variations depend on small scale changes in topography. The important texture information observable on the intensity images reveals the transitions between the different surface types and the variations of the topography whose amplitude may be lower than the LiDAR altimetric accuracy. For coastal surface monitoring, the textural information of the intensity images appears complementary to the simultaneously acquired DEMs, as they have the same spatial sampling.

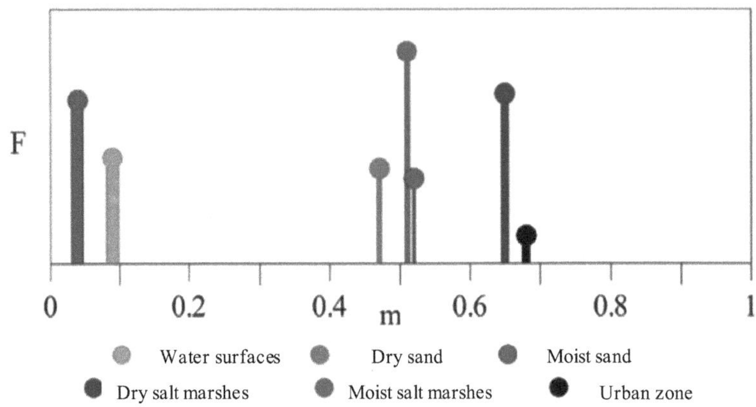

Figure 6.20. *Decomposition into Beckmann n-distributions for each surface type. The spectrum is represented in the Fresnel factor F – roughness space m and the stem width is proportional to the standard deviation of the distribution. Each color is associated with a surface. For a color version of this figure, see www.iste.co.uk/baghdadi/5.zip)*

From the perspective of coastal geomorphology, the intensity images are able to distinguish (1) the discontinuities between surfaces with different lithology, roughness and moisture, and (2) the current ripples attesting to the hydrodynamic sediment transport, the same as wind ripples formed by the wind. From a textural perspective, the common feature of these types of surfaces is obtaining spatial anisotropy properties.

Recent algorithms for characterization, orientation estimation and texture segmentation have used different decomposition approaches (spectrum, wavelets and S transforms [DRA 09, ABD 09, ACH 08, LEF 11]). More specific to coastal areas, an estimator has been developed to define the degree of spatial anisotropy on coastal surfaces included in a sliding window in order to monitor the spatial distribution of natural discontinuities and ripple fields over time [GAR 15]. Because the size of the window optimizes the spatial resolution that limits the complexity of the surface sample and the periodic nature of more complex surfaces (ripples), the estimator uses a correlation formalism in the Fourier space.

The following two matrices contain the Fourier Transforms (FT) of lines and columns of the surface sample $\boldsymbol{Z}$ of size $N \times N$:

$$\boldsymbol{W} = \begin{bmatrix} TF\left\{(z_{1j})_{1\leq j\leq N}\right\} \\ \vdots \\ TF\left\{(z_{Nj})_{1\leq j\leq N}\right\} \end{bmatrix} \quad [6.8]$$

and

$$\boldsymbol{W}_{\perp} = \begin{bmatrix} TF\{(z_{i1})_{1\leq i\leq N}\} \\ \vdots \\ TF\{(z_{iN})_{1\leq i\leq N}\} \end{bmatrix} \quad [6.9]$$

Out of concern for normalization, the energy associated with each line and column is contained in these vectors:

$$\boldsymbol{W}_{\Sigma} = \begin{bmatrix} \|(w_{1k})_{1\leq k\leq N}\|_2 \\ \vdots \\ \|(w_{Nk})_{1\leq k\leq N}\|_2 \end{bmatrix} \quad [6.10]$$

and

$$\boldsymbol{W}_{\perp\Sigma} = \begin{bmatrix} \left\|(w_{\perp_{1k}})_{1\leq k\leq N}\right\|_2 \\ \vdots \\ \left\|(w_{\perp_{Nk}})_{1\leq k\leq N}\right\|_2 \end{bmatrix} \quad [6.11]$$

So, it is possible to form a correlation matrix of all of the combinations between lines and columns as follows:

$$\boldsymbol{C}_{\rho} = \sqrt{\left(Re\{\boldsymbol{W}\boldsymbol{W}_{\perp}^{\dagger}\}^{2} + Im\{\boldsymbol{W}\boldsymbol{W}_{\perp}^{\dagger}\}^{2}\right)} \oslash \boldsymbol{W}_{\Sigma}\boldsymbol{W}_{\perp\Sigma}{}^{T} \quad [6.12]$$

and

$$\rho = \frac{1}{N^2}\sum_{i,j=1}^{N} C_{\rho_{i,j}} \quad [6.13]$$

where Ø symbolizes the Hadamart division, Re symbolizes the real part and Im symbolizes the imaginary part. The mean correlation of the two orthonormal dimensions ρ is found by averaging the matrix $\boldsymbol{C_p}$. Concerning real data, subject to the thermal noise of the system, an unbiased estimate must include a correction taking into account the signal-to-noise ratio (SNR), $\rho_{SNR} = (1 + SNR^{-1})^{-1}$. By dividing by this term, as explained in Garestier *et al.* [GAR 15], the estimate of the degree of spatial anisotropy will remain independent of the reflectivity of surfaces and will only depend on their texture.

For a relevant characterization of anisotropic surfaces, it is therefore possible to consider the amplitude of the texture because, in the case of current ripples, it denotes the energy involved in the processes of sediment transfer. To do this, the surface sample has to be modified in the following way before calculating $\rho(\boldsymbol{Z}_\Delta)$, as shown in Garestier *et al.* [GAR 15]:

$$\boldsymbol{Z}_\Delta = \frac{\boldsymbol{Z} - min(vec(\boldsymbol{Z})) - \Delta_Z}{\Delta} + 1 \qquad [6.14]$$

where Δ_Z is the amplitude of the surface sample included in the given window and Δ is the maximum amplitude observed in the image. The correction related to SNR is modified in the following way:

$$SNR(\alpha z + \beta) = SNR(z) + \frac{2\alpha\beta<z>+\beta^2}{\alpha^2<|n|^2>} \qquad [6.15]$$

with:

$$\alpha = \Delta^{-1} \qquad [6.16]$$

and

$$\beta = \frac{-min(vec(\boldsymbol{Z}))}{\Delta} - r + 1 \qquad [6.17]$$

where $r = \Delta_z/\Delta$ and $|n|^2$ are the power of the noise. Practically, the SNR of the surface included in the sliding window is estimated first. The SNR of the surface that is modified to take into account the texture amplitude will be computed using equation [6.15], which accounts for the adjustment parameters of the window and the power of the noise. The actual SNR assessed then can be used to calculate the degree of corrected anisotropy of the SNR: $\rho_{corrected} = \rho(\boldsymbol{Z}_\Delta)/\rho_{SNR}(\boldsymbol{Z}_\Delta)$.

The image on the left in Figure 6.21 covers a typical coastal surface containing areas of water, current ripples and discontinuities separating sands of different moistures and granulometries. In the image on the right, we can first verify the independence of the estimator of the degree of spatial anisotropy, estimator accounting for the amplitude of the texture to the SNR, since the coherence value $\rho(\boldsymbol{Z}_{\Delta})$ is totally independent of the intensity value. Its coherence appears minimal over areas of current ripples with maximum amplitude and strong discontinuities. The coherence would be maximal over surfaces of sand and water that are spatially isotropic, and takes intermediary values on anisotropic surfaces with less amplitude.

In a complementary way, the investigation of the sole degree of spatial anisotropy can prove to be relevant in cases of anisotropic surfaces with more complex geometries, without considering the amplitude information. To do this, we can estimate the coherence of each surface sample after normalization in the following way:

$$\boldsymbol{Z}_n = \boldsymbol{Z} - min(vec(\boldsymbol{Z})) \qquad [6.18]$$

The correction of SNR described by equation [6.15] will take the values $\alpha = 1$ and $\beta = -min(vec(\boldsymbol{Z}))$.

In Figure 6.22, the coherence value only depends on the degree of anisotropy, which appears highly variable on the intensity image in Figure 6.21, gathered along a section of sea on a coastline. The estimator also detects three sea states by distinguishing the plane waves from refracted waves as well as the surf zone.

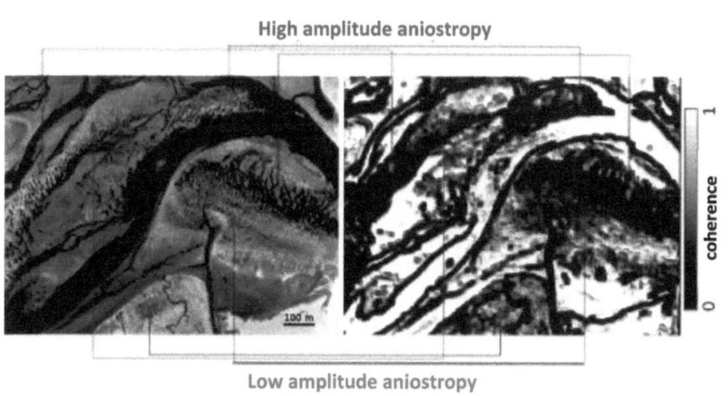

Figure 6.21. *Intensity image of a coastal zone (left) and map of the degree of spatial anisotropy taking into account the texture in a window of 15 x 15 pixels (right). For a color version of this figure, see www.iste.co.uk/baghdadi/5.zip)*

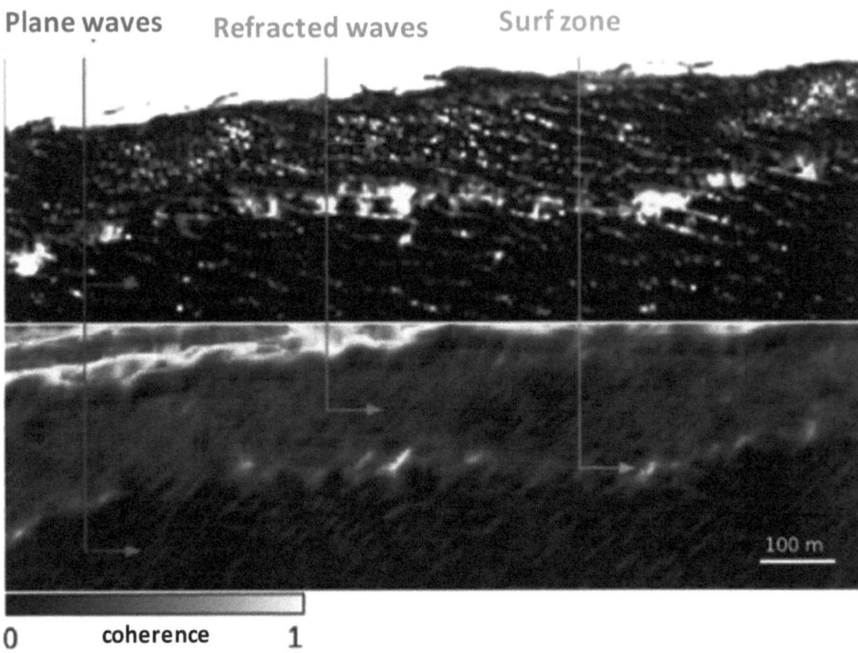

Figure 6.22. *Intensity image of different sea regimes in the coastal zone (top) and a map of the degree of spatial anisotropy not taking into account the amplitude of texture (bottom). From the open sea to the coast: refracted incident waves and, at the coast, a line of breakers, then saturated waves in the surf zone. For a color version of this figure, see www.iste.co.uk/baghdadi/5.zip)*

In summary, the two approaches are best used simultaneously for unambiguous characterization of the anisotropic surfaces, which are widely represented on the coast. A representation in a two-dimensional space of coherences $\rho(\mathbf{Z}_\Delta)$ and $\rho(\mathbf{Z}_\Pi)$ can not only be used to distinguish anisotropic surfaces from isotropic ones, but also from other anisotropic surfaces with different amplitudes.

6.5. Quantification of the sandy surface moisture of Ls

After having determined a general model describing the behavior of backscattered intensity as a function of the incidence angle, the influence of the physical parameters of coastal surfaces, such as the granulometry and the surface moisture, was studied in Poullain *et al.* [POU 16b]. The measurements of these two parameters were carried out in a laboratory using samples processed *in situ*, simultaneously to the LiDAR flights.

On the given site, which is composed of large areas of dry sand, the granulometric class ranges from fine to very coarse sand (size of grains between 0.125 and 2 mm). It appeared that the factor *F*, representing the intensity at the nadir, remained independent of the granulometry of the sediment (Figure 6.23). The variation of the *m* roughness factor is low and remains included in the interval of the data dispersion.

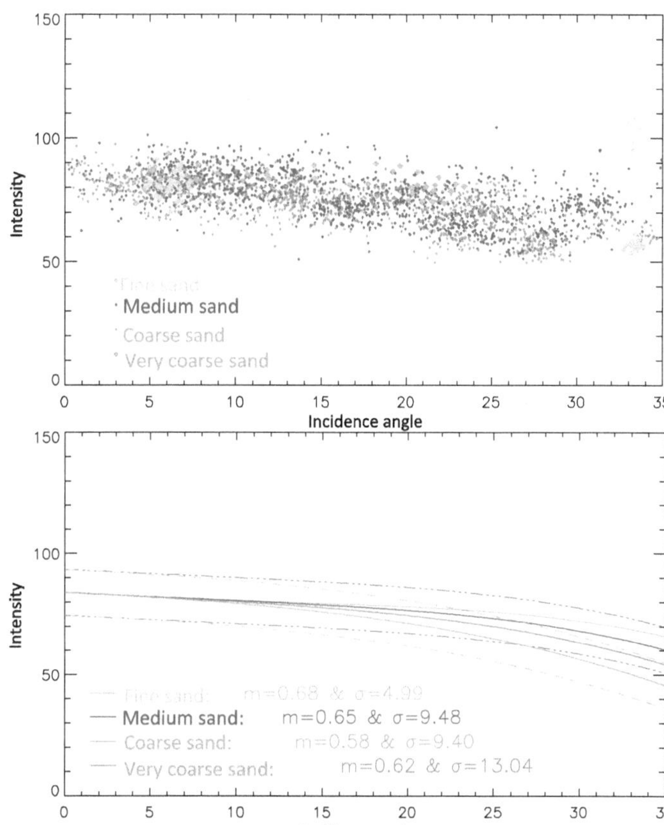

Figure 6.23. *Intensity data backscattered by medium and coarse sand as a function of the incidence angle (top) and modeling using m microfacet distribution (bottom, in points ± standard deviation σ). For more visibility, the standard deviation of the fine and very coarse sands is not represented because their statistical sample is too small. For a color version of this figure, see www.iste.co.uk/baghdadi/5.zip)*

After observing of the independence of granulometry to the intensity–incidence angle relation, the influence of the moistured sandy surfaces' was analyzed at constant granulometry. For the different categories of mass wetness varying from 0 to 25% (Figures 6.24(a–e)), the model of the intensity–incidence angle with the law of microfacet distribution showed that:

– the roughness factor m remained constant with the same value, to the nearest uncertainty, as the one estimated during the granulometric study ($m = 0.60 \pm 0.05$);

– the Fresnel coefficient F, representing the intensity returned to the nadir, continually decreases in an exponential manner as a function of the moisture factor (Figure 6.24(f)).

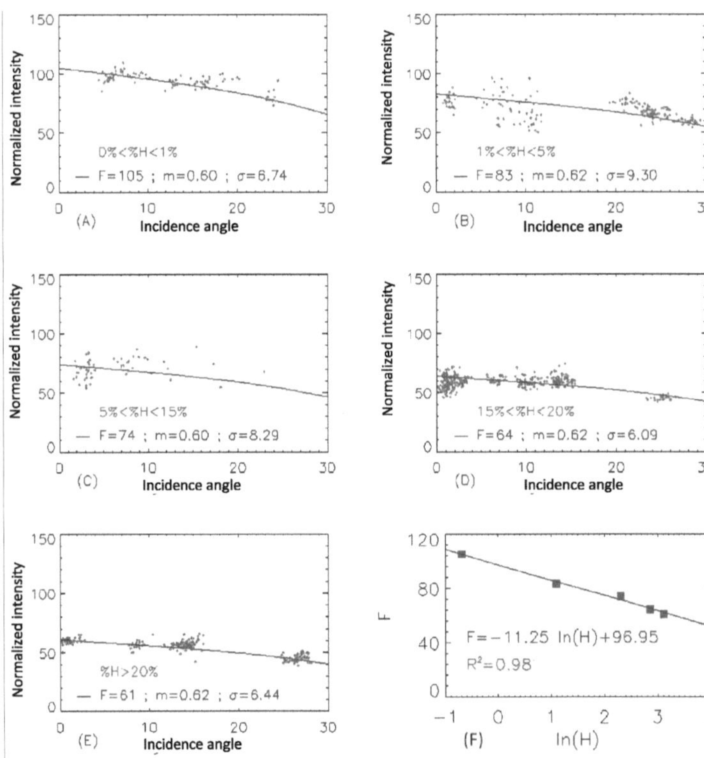

Figure 6.24. *a)–e): study of intensity as a function of the angle for zones with different mass wetnesses; f): modeling of the behavior of the Fresnel factor F as a function of moisture H. For a color version of this figure, see www.iste.co.uk/baghdadi/5.zip)*

After observing the independence of the m roughness factor (it remains constant in its interval of uncertainty for the fine, medium, coarse and very coarse sand, as the granulometric study shows), it is possible to determine the mass wetness ($\%H_m$) of a surface using backscattered intensity $I(\theta)$ and the associated incidence angle θ, the latter being estimated from the scan angle and the topography obtained by a distance measurement:

$$\%H_m = exp\left[8.62 - \frac{I(\theta)cos^5(\theta)exp\left(\frac{tan^2(\theta)}{0.36}\right)}{11.25}\left(1 + \frac{0.48tan\,(\theta)}{2-erfc\left(\frac{1}{0.85tan\,(\theta)}\right)}\right)^2\right] \quad [6.19]$$

Therefore, it is theoretically possible to estimate the rate of mass wetness for each point of the image. In practice, the use of an interval of minimal angles, leading to a light degradation of resolution, would be necessary in order to simultaneously detect sandy surfaces on which it would be possible to apply the inversion methodology.

6.6. Prospects

Airborne topographic LiDAR is a tool suited for determining the terrain of large surfaces, especially when they are difficult to access, with a potentially high spatial resolution (>10 points/m²). This capacity to cover large areas is due to a good absolute altimetric precision, which other technologies can match, although mostly on smaller surfaces than those covered by LiDAR.

The use of an intensity signal prompts interesting discussions on the subject of automatic or semi-automatic identification of objects and physical characteristics of surfaces. With the repeatability of LiDAR flights becoming more accessible, it will be possible to study the spatio-temporal developments of surface properties (anisotropy, moisture, roughness, etc.) and, if necessary, make corrections such as the one for the scan angle. The use of the intensity combined with the other types of data (topographical, color imagery, etc.) could improve this knowledge.

Joining together the two different approaches presented in this chapter could lead to an improvement on the accuracy, coherence and automatization of sedimentary budgets or the monitoring of other developments over time. Furthermore, for certain types of very pronounced terrain (cliffs, coastal slopes or mountain complexes), the combination of data gathered using an airborne topographic LiDAR and a land-based laser scanner with complete backscattering offers the prospect of fusing complementary data.

6.7. Key points

In coastal zones, the data gathered from airborne topographic LiDAR have traditionally been used for reference to determine digital records of the terrain of an interface zone whose topography is still not well known. Increasingly, monitoring is being conducted with this tool because it can measure real developments on a morphological and volumetric level with a high accuracy over vast surfaces. The extraction of objects such as channels in the water of a bay or an estuary using only intensity data coupled with topographic data, or a shoreline using a multi-criteria approach, represents a significant added value to the source data. A deeper analysis of the intensity data will improve knowledge of scanned surface characteristics and also lead to improving the monitoring of their evolution.

6.8. Bibliography

[ABD 09] ABDESSELAM A., "Texture image retrieval using Fourier transform", *Proceedings of International Conference on Communication, Computer and Power (ICCCP)*, Sultan Qaboos University, Oman, pp. 343–348, 15–18 February 2009.

[ACH 08] ACHARYYA M., MALAY K., "Extraction of noise tolerant, gray scale transform and rotation invariant features for texture segmentation using wavelet frames", *International Journal of Wavelets Multiresolution and Information Processing*, vol. 6, no. 3, pp. 391–417, 2008.

[BEC 63] BECKMANN P., SPIZZICHINO A., *The Scattering of Electromagnetic Waves from Rough Surfaces*, Pergamon Press, New York, 1963.

[BOA 05] BOAK E.H., TURNER I.L., "Shoreline definition and detection: a review", *Journal of Coastal Research*, vol. 21, no. 4, pp. 688–703, 2005.

[DET 11] DÉTRICHÉ S., SUSPERREGUI A.S., FEUNTEUN E. *et al.*, "Interannual (1999-2005) morphodynamic evolution of macro-tidal salt marshes in Mont-Saint-Michel bay (France)", *Continental Shelf Research*, vol. 31, pp. 611–630, 2011.

[DRA 09] DRABYCZ S., STOCKWELL R., MITCHELL J.R., "Image texture characterization using the discrete orthonormal S-transform", *Journal of Digital Imaging*, vol. 22, no. 6, pp. 696–708, 2009.

[GAR 15] GARESTIER F., BRETEL P., MONFORT O. *et al.*, "Anisotropic surface detection over coastal environment using near-IR LiDAR intensity maps", *IEEE Journal of Selected Topics in Applied Earth Observations and Remote Sensing*, vol. 8, no. 2, pp. 727–739, 2015.

[HAP 93] HAPKE B., *Theory of Reflectance and Emittance Spectroscopy*, Cambridge University Press, Cambridge, UK, 1993.

[HOF 07] HÖFLE B., PFEIFER N., "Correction of laser scanning intensity data: data and model-driven approaches", *ISPRS Journal of Photogrammetry and Remote Sensing*, vol. 62, no. 6, pp. 415–433, 2007.

[KAA 11] KAASALAINEN S., JAAKKOLA A., KAASALAINEN M. *et al.*, "Analysis of incidence angle and distance effects on terrestrial laser scanner intensity: search for correction methods", *Remote Sensing*, vol. 3, no. 10, pp. 2207–2221, 2011.

[LEF 11] LEFEBVRE A., CORPETTI T., HUBERT-MOY L., "Estimation of the orientation of textured patterns via wavelet analysis", *Pattern Recognition Letters*, vol. 32, no. 2, pp. 190–196, 2011.

[LEV 13] LEVOY F., ANTHONY E.J., MONFORT O. *et al.*, "Formation and migration of transverse bars along a tidal sandy coast deduced from multi-temporal LiDAR datasets", *Marine Geology*, vol. 342, pp. 39–52, 2013.

[LIU 09] LIU H., "Shoreline mapping and coastal change studies using remote sensing imagery and LIDAR data", in YANG X. (ed.), *Remote Sensing and Geospatial Technologies for Coastal Ecosystem Assessment and Management*, Springer, 2009.

[MON 14] MONTREUIL A.L., LEVOY F., BRETEL P. *et al.*, "Morphological diversity and complex sediment recirculation on the ebb delta of a macrotidal inlet (Normandy, France): a multiple LiDAR dataset approach", *Geomorphology*, vol. 219, pp. 114–125, 2014.

[POU 12] POULLAIN E., GARESTIER F., BRETEL P. *et al.*, "Modeling of ALS intensity behavior as a function of incidence angle for coastal zone surface study", *Proc. IEEE International Geoscience and Remote Sensing Symposium (IGARSS)*, Munich, Germany, pp. 2849–2852, 2012.

[POU 13] POULLAIN E., Exploitation de l'intensité du signal LASER d'un LIDAR topographique aéroporté pour des environnements littoraux sableux, PhD thesis, University of Caen Basse-Normandie, p. 188, 2013.

[POU 16a] POULLAIN E., GARESTIER F., LEVOY F. *et al.*, "Analysis of ALS intensity behavior as a function of incidence angle in coastal environments", *IEEE Transactions on Geoscience & Remote Sensing*, vol. 9, no. 1, pp. 313–325, 2016.

[POU 16b] POULLAIN E., GARESTIER F., LEVOY F., "Soil moisture estimation using near IR LiDAR intensity", *IEEE Transactions on Geoscience & Remote Sensing*, 2016.

[SAL 03] SALLENGER A.H., KRABILL W.B., SWIFT R.N. *et al.*, "Evaluation of airborne topographic LiDAR for quantifying beach changes", *Journal of Coastal Research*, vol. 19, no. 1, pp. 125–133, 2003.

[SCH 11] SCHMID K.A., HADLEY B.C., WIJEKOON N., "Vertical accuracy and use of topographic LIDAR data in coastal marshes", *Journal of Coastal Research*, vol. 27, pp. 116–132, 2011.

[SME 13] SMEECKAERT J., MALLET C., DAVID N. *et al.*, "Large-scale classification of water areas using airborne topographic LiDAR data", *Remote Sensing of Environment*, vol. 138, pp. 134–148, 2013.

[SOI 05] SOININEN A., "TerraScanUser's Guide", Terrasolid, 2005.

[SON 02] SONG J.H., HAN S.H., YU K.Y. *et al.*, "Assessing the possibility of land-cover classification using LiDAR intensity data", *International Archives of Photogrammetry Remote Sensing and Spatial Information Sciences*, vol. 34, no. 3/B, pp. 259–262, 2002.

[STO 02] STOCKDON H.F., SALLENGER A.H., LIST J.H. *et al.*, "Estimation of shoreline position and change using airborne topographic LiDAR data", *Journal of Coastal Research*, vol. 18, no. 3, pp. 502–513, 2002.

[TEL 14] TELECOM PARISTECH EDF R&D, CloudCompare (version 2.5.5.2) (GPL software), available at: http://www.cloudcompare.org/, 2014.

[TOR 67] TORRANCE K.E., SPARROW E.M., "Theory for off-specular reflection from roughened surfaces", *Journal of the Optical Society of America*, vol. 57, no. 9, pp. 1105–1114, 1967.

[WHE 10] WHEATON J.M., BRASINGTON J., DARBY S.E. *et al.*, "Accounting for uncertainty in DEMs from repeat topographic surveys: improved sediment budgets", *Earth Surface Processes and Landforms*, vol. 35, pp. 136–156, 2010.

[WOL 89] WOLFE W., ZISSIS G.J., *The Infrared Handbook*, The Infrared Information Analysis Center, Environmental Research Institute of Michigan, Detroit, 1989.

[ZHA 05] ZHANG K., WHITMAN D., LEATHERMAN S. *et al.*, "Quantification of beach changes caused by hurricane Floyd along Florida's Atlantic Coast using airborne laser surveys", *Journal of Coastal Research*, vol. 21, no. 1, pp. 123–134, 2005.

7

Mangrove Forest Dynamics Using Very High Spatial Resolution Optical Remote Sensing

7.1. Introduction

Assessing the role of tropical forests in biogeochemical cycles is one of the greatest scientific challenges of the century [NAT 15]. The impact of climate change could increase the mortality of tropical forests and decrease their ability to store atmospheric carbon dioxide [BRI 15]. Efforts to model the forest dynamics in these regions are hindered by the lack of ground measurements [PAC 96] and the spatial-temporal overlap of function management processes [CLA 99]. Some of the models developed attempt to show the environment's influence on forest dynamics, particularly by studying the capacity for canopy deformation (plasticity of the crown), as explained in Purves *et al.* [PUR 07]. The models allow us to better understand the development of intertree competition [VIN 08], the capacity of species to adapt to changes [GRU 14] and how the forest functions as a whole [BOH 12].

The models of forest dynamics, which connect the forest parameters measured on the ground and the properties of the canopy (defined here as the upper layer of the forest cover), have much to gain by joining together with

Chapter written by Christophe PROISY, Jean-Baptiste FÉRET, Nicolas LAURET and Jean-Philippe GASTELLU-ETCHEGORRY.

spatial approaches that provide information about the canopy. The first representative studies on the subject of such a combination were conducted with radar remote sensing [KAS 90, RAN 97]. The idea was to use the proven ability of radar signals to discover forest structures in low frequencies (bands L and P). That said, even at these frequencies, the radar signal saturates at above-ground biomass levels of 150–200 tons of dry matter per hectare (tMS/ha), and above-ground biomass in tropical forests can reach 500 tMS/ha [IMH 95]. To this day, the full potential of the results of ALOS-2 (band L) and the future BIOMASS (band P) satellite for the study of tropical forests has not been entirely explored [MER 15, LET 12]. Studies of tropical forest dynamics cannot ignore optical remote sensing approaches which, even if obscured by the cloud cover, provide information about the function of vegetation cover [FER 08, FER 11] and the spatial organization of canopies.

With the aid of scanned aerial photographs, Couteron *et al.* [COU 05] have developed a method of analyzing canopy grain, called FOTO (Fourier-based Textural Ordination; see section 7.4.1). They showed that the properties of canopy organization and the structure of the forest cover below could be described using texture gradients. Due to the use of satellite images with a very high spatial resolution (VHR), these results have been validated and suggest interesting correlations between the textural signatures and the above-ground biomass of forest stands in terra firma tropical forests [PLO 12]. In the case of the Guianese mangrove forests, which can reach biomass levels of over 400 tMS/ha, Proisy *et al.* [PRO 07] confirmed a direct link between the texture of VHR satellite images and the biomass of mangrove forest stands. This last point is interesting because the FOTO texture index does not show saturation at high biomass levels, contrary to what was found using radar remote sensing on the same mangrove forests [PRO 00, PRO 12]. When combined with the increasing availability of VHR optical images like those provided by the Ikonos, Formosat-2, Geoeye-1, Pléiades 1A/1B, Spot 6/7, Quickbird or Worldview-1/2/3satellites, these promising results allow us to consider a consolidation of research on tropical forest dynamics using the analyses of the texture of the canopy images. In certain regions, the availability of historical aerial photographs combined with VHR satellite images going back to 2001 could be used to plan a study of changes in the forest over several decades.

In order to take full advantage of the large number of VHR optical images and their potential, it is necessary to physically interpret the solar radiation propagation mechanisms inside forest covers before inverting the parameters that describe the structural organization of forests [SOU 14]. Neglecting the significance of the image acquisition parameters can lead to erroneous interpretations of the forests' variation in spectral responses [MOR 14]. One challenge for research in optical remote sensing would be to succeed in simulating VHR optical images that are sufficiently realistic of the range:

– of image acquisition modes in terms of sensor responses and angular configurations (position of the sun and viewing geometry);

– of forest characteristics in terms of geometric and optical properties [SCH 14].

In this chapter, French Guiana mangrove forests were selected as the subject of study due to their exceptional spatial and temporal dynamics. First, we explain the concerns connected with the study of mangrove forests by VHR optical remote sensing and the specifications encountered in French Guiana. Next, we describe the signal processing chain required to simulate VHR images of the forest canopy using ground measurements, the allometric models describing the variations in above-ground biomass as a function of the diameter of trunks and a model of solar radiation propagation (radiative transfer) in the forest cover. In the final section, the principles and the potential of the FOTO method are presented through a texture analysis conducted on a temporal series of VHR images. This method is used to test the realism of VHR images of mangrove forests simulated at 50 cm of spatial resolution. Finally, the set of results obtained with VHR optical remote sensing is discussed with the aim of better predicting the dynamics of mangrove forests.

7.2. Dynamics of mangrove forests

7.2.1. *General context*

Mangrove forests grow in the intertidal zone of the tropical and intertropical regions. They are representative of the concern to monitor the dynamics of forest ecosystems because, although they only contribute to about 1% of carbon dioxide stored by all forests, they can be responsible for at least 14% of the carbon dioxide stored by the oceans [ALO 12]. When mangrove forests are destroyed or degraded, there are significant greenhouse gas emissions [SID 13]. However, the growth of these forests continues to decline at an annual rate of 1–2% [DUK 07]. For countries that want to be eligible for funding from the REDD+ (Reducing Emissions from Deforestation and Forest Degradation, with conservation, sustainable management of forests, and carbon storage [OLA 12]) and Blue Carbon (carbon captured by oceans and coastal ecosystems [HER 12]) programs, there is an urgent need to use spatial methods capable of monitoring the characteristics of forest structure and the specific composition in large regions of mangrove forests at a fine spatial scale (typically 1 hectare).

7.2.2. *The case of Guianese mangrove forests*

French Guiana has a particularly interesting coast on which to test and develop remote radar sensing methods [PRO 00], passive optics [PRO 07] and LiDAR [PRO 09]. Changes in the features of the Guianese coastal landscape are wide-ranging and continuously observed at all points along the coast (Figure 7.1 [FRO 04]).

Figure 7.1. *Top: coastal dynamics illustrated using excerpts of 6 km-wide Spot images on a portion of the Guianese coast between 1986 and 2001. These excerpts show an expansion until 1995 and then the treatment and destruction of the surface occupied by the mangrove forests (zones in bright red). Bottom: forest dynamics illustrated by land photographs at various stages of development. For a color version of this figure, see www.iste.co.uk/baghdadi/5.zip*

Phases of erosion (destruction of the mangrove) and muddy accretion (appearance of new mangrove forests) occur all along the French Guiana coast [ANT 10] in accordance with the shifting of gigantic mud banks to the north-west. The juvenile mangrove forest changes quickly. In a few years, the mangroves reach more than 12 m in height. The "adult" stages of trees taller than 25 m are attained after interindividual competition and mortality (Figure 7.2). The above-ground biomass of mature mangrove forest stands exceeds 300 tMS/ha [FRO 98]. During phases of erosion, the sea-mangrove shoreline can retreat several hundreds of meters per year. These erosive phases are displayed in the remote sensing images by a sawtooth shoreline (Figure 7.1, see Spot images from 2000 and 2001). Readers can refer to Chapter 8 to find a list of works and results obtained by monitoring the coastline evolution through remote sensing. This chapter focuses on the monitoring of the structural development of mangrove forest stands (Figure 7.2).

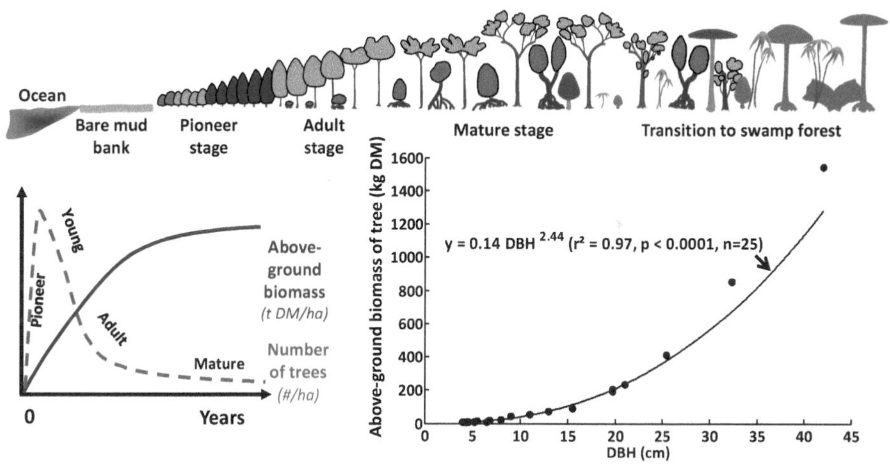

Figure 7.2. *Top: sea–land transect of the progression of the development stages of Guianese mangrove forests. Bottom left: typical temporal evolution of the number of trees and the above-ground biomass of a mangrove forest stand. Bottom right: allometric relation between the DBH and the total individual biomass (expressed in kg of dry matter) for the mangrove Avicennia germinans [FRO 98] found by cutting and weighing* in situ *of 25 individuals (the black dots)*

7.2.3. *Modeling forest dynamics in mangrove forests*

By definition, forest dynamics describe the changes of structural factors, such as the diameter of the trunk at breast height (hereafter DBH to represent the acronym diameter at breast height), the height of the tree, the number of trees per hectare and the temporal and spatial crown dimensions (Figure 7.2).

Forest dynamic models developed to study the mangrove forests are mostly "individual-based" [BER 08]. The models can be used to understand the evolution of the DBH of several individuals by describing the interactions (competition and facilitation) of the neighbors and the influence of environmental factors such as salinity. The physical characteristics of trees, difficult to measure with a laser scanner [FEL 14], are still only somewhat integrated, despite being recognized as valuable for analyzing the evolution of interactions between neighboring trees over time [GRU 14]. In addition, the DBH growth equation does not indicate where the biomass produced will be allotted in the tree between crown, trunk and roots [PET 14].

Separate from the forest dynamic models, the above-ground biomass of mangroves is assessed using allometric equations (Figure 7.2, bottom right) established empirically as a function of DBH [KOM 05] from tree cutting and weighing. The use of these allometric equations on these sites in French Guiana or elsewhere, where the growth conditions and the appearance of the forest stands are different from those encountered on the sample sites, can be contaminated with errors. Extrapolating these equations outside of their domain of validity (the maximum DBH cut) should be considered with caution [OLA 16].

7.2.4. *Research concerns in VHR optical remote sensing of mangrove forests*

In a visual analysis of VHR satellite images of French Guiana mangrove forests, the eye is able to distinguish the openings and gaps in the forest cover as well as different heights and densities of crowns corresponding to different stages of development (Figure 7.3). Some definitive advances in the prediction of mangrove development can be achieved if we can make use of the synoptic vision offered by spatial observations to describe the set of forest situations encountered. However, the visually attractive side of the VHR optical images does not presuppose the reliability of observations nor the robustness of the methods developed to invert the forest parameters [ADA 06]. Simulating images of the canopy can provide the physical bases needed to interpret the variability of spectral and textural signatures observed, depending on the sensor parameters and forest structures. But can we simulate realistic images of the canopy at VHR with the current state of knowledge and data gathered in the mangrove forests?

Figure 7.3. *Excerpt of an Ikonos image in which the four blue, green, red and near infra-red channels have been fused on the panchromatic channel (pixel size = 1 m). For a color version of this figure, see www.iste.co.uk/baghdadi/5.zip*

7.3. Methods

7.3.1. *Field experiments*

Field experiments are conducted in order to understand the reality and the variability of forest structures and to provide information that is useful for image analysis and the calibration of a radiative transfer model. These tests determine the robustness and the range of the methods developed with remote sensing.

7.3.1.1. *Characterization of forest structures*

The characterization of forest structures in a region of mangrove forests captured at a very high spatial resolution occurs through the meticulous positioning of a sufficient number of sample zones. This positioning, which must be representative of the dynamic of a mangrove region, has been facilitated by the advent of VHR satellite images since 2001, and has also been distributed freely via Internet platforms like Google Earth®. In each sample zone, the user must define one or more sections whose size of inventory is to be adjusted as a function of the homogeneity of the forest stands and the number of trees per surface unit. For French Guiana, the defined areas usually vary from 100 m², for very dense pioneer facies, to 1 hectare for the mature stages. Once the section has been chosen, the user measures the DBH of each individual (Figure 7.4) with a trunk diameter greater than the set threshold of its development stage.

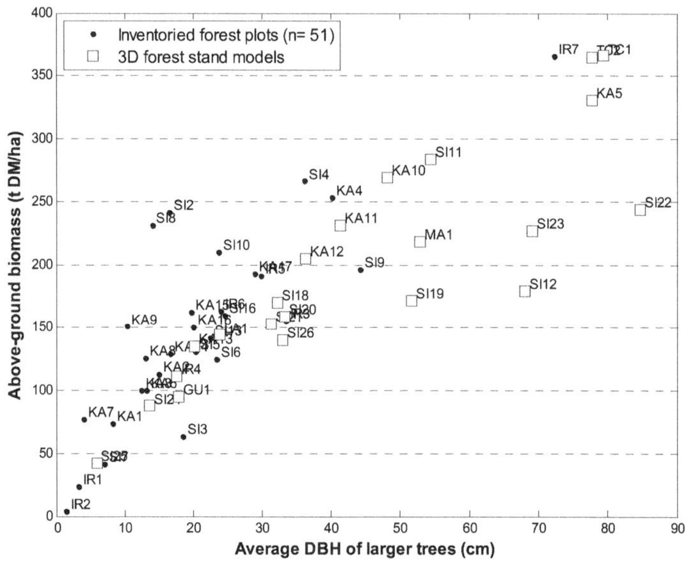

Figure 7.4. *Estimate of above-ground biomass for 51 forest stands inventoried since 1996 on different mangrove forest sites in French Guiana and differentiated by the codes GU, IR, KA, MA, SI and TC. The average DBH of the largest trees is averaged from a threshold at 20% over the total number of trees. The 22 forest stands that were the subject of the production of models are indicated by yellow squares. This graph can be compared to the graph in Figure 7.2, bottom left. It relates some variability in the trajectories of the development of above-ground biomass in Guianese mangrove forests. For a color version of this figure, see www.iste.co.uk/baghdadi/5.zip*

A laser range finder can be used to reveal the X, Y positions of each inventoried tree and to evaluate the total height, the insertion height of the first branch and the diameter of each individual's crown. Structural parameters like the density of branches by DBH class or the above-ground biomass of the forest stand (if the allometric relations are available) can then be calculated. As an illustration of the forest dynamic observed in French Guiana, Figure 7.4 shows the range and variability of the biomasses of each section according to the average DBH of the largest trees in sections characterized by a significant gradient of development stages. These forest measurements are used to create forest scenes, based on which physical modeling can be conducted.

7.3.1.2. *Characterization of optical properties*

To simulate VHR optical remote sensing images of forests, it is also necessary to measure the optical properties of the vegetation and soil in the spectral domain used by satellite sensors, between 0.4 and 1.1 μm. This can be done using a ground spectroradiometer. The measurement protocols of the optical properties vary depending on the equipment used [SCH 06]. In French Guiana, a portable spectroradiometer (model PSR-1100 F from the brand Spectral Evolution) was used. It works in a spectral domain ranging from 300 to 1,150 nm. This instrument is composed of a 5 W halogen tungsten light source, an optical fiber, a leaf clip and a lens opening at 25°. The optical spectrum was obtained by repeating the measurements of a coherent sample (between 30 and 50 measurements) of leaves, barks and types of soil.

Sampling the sun and shade leaves on several individuals of each species gives the best indication of intraindividual and intraspecies variability (a cross-section of the branches may be necessary). The leaf reflectance is found using optical fiber, a leaf clip and a light source. The calibration is done quickly, before each measurement is made, using a small reflective target provided by the manufacturer placed in the interior section of the clip. The radiation reflected by a leaf is then evaluated according to the Lambertian hypothesis (homogeneous diffusion in all directions). The transmittance measurement requires the use of an integrating sphere, which is difficult to implement in a mangrove forest and requires an additional budget. In the absence of this measurement, it is possible to assess the transmittance using reflectance measurements (see section 7.3.2.2). Figure 7.5 presents the average foliar spectrums of reflectance and transmittance obtained for three species of mangroves in French Guiana and used for our simulations.

For the soil, the range of possible situations is large because it depends on flood conditions, water salinity and the presence or absence of a layer of pneumatophores. This measurement is taken with the lens positioned at the nadir and in such a way as to maximize direct illumination. Before each measurement and in order to limit inaccuracies due to variations in illumination, a calibration is conducted by placing a Spectralon® target in the sample's place.

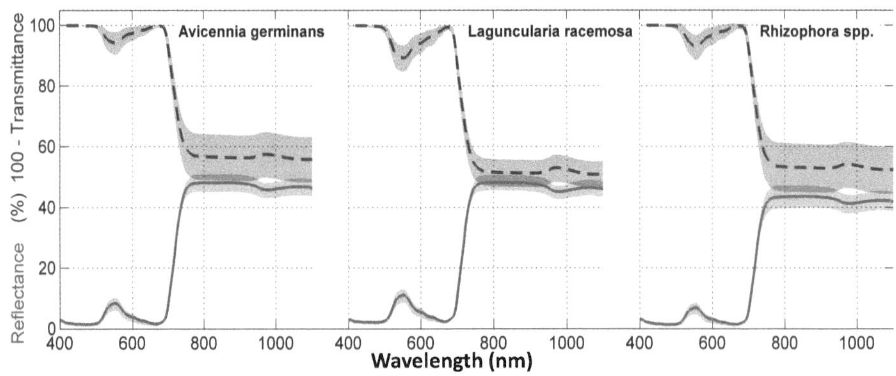

Figure 7.5. *Spectrums of reflectance and average foliar transmittance for three species of mangroves. The reflectance values were measured on the ground (between 30 and 50 samples per species, the standard deviations are represented by the colored perimeter around the mean curves) and the transmittance values were obtained by inverting the PROSPECT foliar radiative transfer model*

7.3.2. *Modeling 3D radiative transfer with DART*

In this section, the principles and essential steps for the configuration of the Discrete Anisotropic Radiative Transfer (DART) model are presented. To this day, DART is the most complete model for 3D radiative transfer in terms of both electromagnetic calculations of complex scenes and the development of graphic interfaces for the configuration and analysis of results. DART was developed at the UMR CESBIO, in Toulouse, beginning in 1992 (patent PCT/FR 02/01181, 2003). The code is continually improving and benefits from CPU parallelization. There are 32 bit or 64 bit versions available for Windows or Linux. For research and teaching activities, free licenses are available at: http://www.cesbio.ups-tlse.fr/dart/license/newDartWebSite. The version of DART used for this work was version 5.5.2 from March 17, 2015 (Windows, 64 bit).

7.3.2.1. *Principles of modeling with DART*

DART is a model of radiative transfer (radiation propagation) in the Earth–atmosphere system [GAS 15]. It simulates the radiation budget and the satellite or airborne spectroradiometer images at every visible wavelength of the optical domain of the thermal infrared using different methods to monitor the beam or photon. DART can operate in any conditions, whether experimental (atmosphere, terrain and solar direction) or instrumental (spectral range, look direction, spatial resolution, etc.). The model also simulates ground, airborne and satellite LiDAR measurements (shape of wave and photon count) [YIN 15].

The radiative transfer calculation is made using three-dimensional (3D) scenes, which is to say a 3D matrix of parallelepipedic cells (called voxels) composed of natural elements (trees, crop plants, soil, water, etc.) and/or urban elements (houses, roads, cars, etc.) with or without terrain. The elements that make up a scene can be created directly in DART or imported. They are described either in the form of translucent facets (which define the perimeter of the object) or by the juxtaposition of cells or "turbid" volume elements. A facet generates diffusions and emissions that are isotropic or anisotropic. A volume element produces diffusions and emissions that can be calculated knowing the density, the cross-section and the diffusion function of the turbid matter contained in the element. The choice of a certain mode of representation has different implications in terms of storage volume and calculation time as well as the level of detail provided for forest scenes (elaborated on in the next section). The turbid approach (cloud of diffusing elements distributed according to the Poisson law) is most commonly used since we are far from being able to describe the position, form and dimensions of the leaves, branches and trunks that make up a natural tropical forest stand.

Documentation for both the physical principles underpinning the radiative transfer calculation and the use of the DART model is available for download on the website mentioned above.

7.3.2.2. *Parameterization of experimental and instrumental image acquisition configurations*

Parameterization is conducted in the *Editor* window which is divided into three sections (Figure 7.6, top). The left section details the parameters necessary to complete the radiative transfer calculations, including, from top to bottom: the different modes and spectral intervals of the calculation, the geometry of observation (solar angles and look directions), the optical properties of the building blocks of the scene, the characteristics of the scene (*maket*) and the stand of trees up to the description of the atmospheric model to use. The central section represents the simulation scene viewed from above. In the right section, linked to the left section, it is possible to adjust the values of the listed parameters (or use the default values).

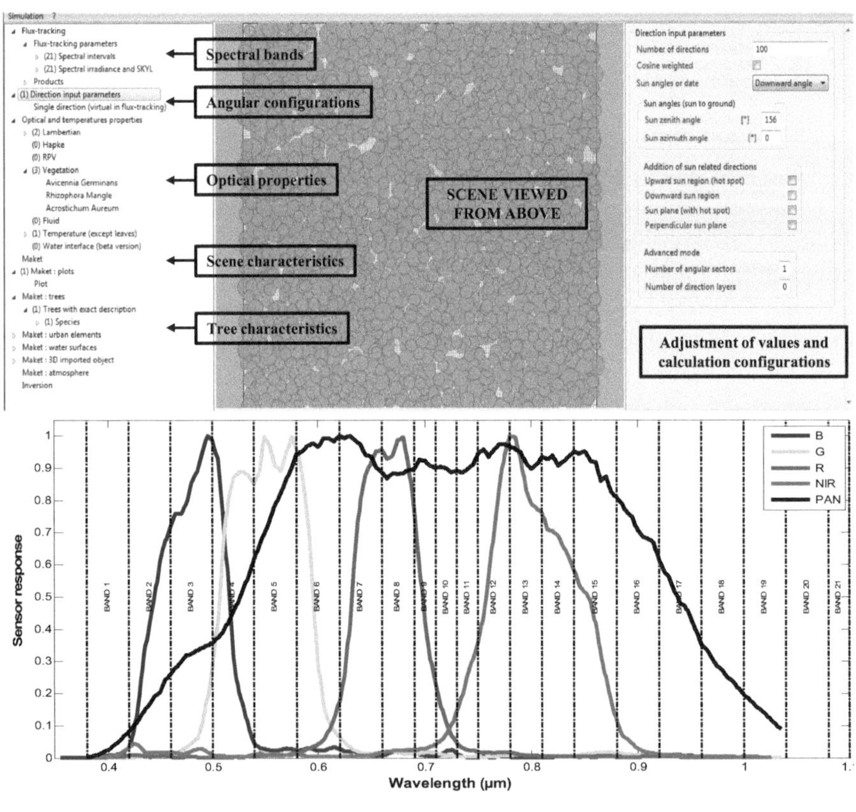

Figure 7.6. *Top: illustration of the* Editor *window content and the three sections used for configuration. Bottom: responses from sensors onboard the satellite Ikonos with placement of 21 spectral bands defined and used in DART to calculate "large band" images corresponding to multi-spectral channels of current VHR satellite sensors*

Twenty-one spectral bands of widths varying between 0.02 and 0.04 μm were defined (Figure 7.6, top). These spectral bands, in which the radiative transfer calculations will be conducted, were adjusted in order to accurately break down each of the large bands of the aforementioned satellite sensors (Figure 7.6, bottom).

A DART simulation produces as many images as there are observation directions and solar illuminations specified. Below, the zenith angles will be denoted as θ_s and θ_v, to indicate the directions of solar illumination and observation, respectively. The azimuth difference is denoted as $\varphi_{s\text{-}v}$.

The optical properties of the soil and the leaves are to be determined. At the current stage of the analyses, the identical optical properties of leaves and bark are assigned to all individuals of the same species. These individuals are thus only distinguishable using different geometric dimensions (trunk and crowns).

With regard to leaves, DART uses properties of reflectance and foliar transmittance. If no measurement is available, the transmittance of each species can be estimated from the reflectances measured using the method proposed by Féret *et al.* [FER 08]. That method gathers all of the foliar biochemical and structural variables by inverting the PROSPECT-5 model of foliar radiative transfer [JAC 90]. Inverting the PROSPECT model using the reflectance measurements gathered with the leaf clip (which are not directional-hemispherical measurements) is conducted by assuming that the foliar surface is Lambertian and that the measurements are comparable to directional-hemispherical measurements. The levels of chlorophyll, carotenoids, water and dry matter, as well as a structure index showing the complexity of the foliar anatomy deduced can, through DART (after activating the PROSPECT module), generate the directional-hemispherical foliar transmittance (Figure 7.5). An inversion procedure for the PROSPECT model has been integrated in the most recent versions of DART (DART 5) [GAS 15].

7.3.2.3. *Construction of forest scenes*

An important step in the image simulation process involves the geometric construction of forest stand models. The dimensions and the content of these forest scenes are to be defined by the user and can be seen in the center section of the DART *Editor* window (Figure 7.6, top). The size of the cells that make it up is also to be determined by the user. In this case, DART was configured with 100 m × 100 m forest scenes and a repetition factor of 2, which allowed us to artificially increase the size of the images produced while managing the side effects.

This forest scene is described in a text file that lists all of the trees present, noting each tree's species number, its XY position, its diameter, its trunk height below and in the crown, and the geometry and dimensions of its crown. The forest inventories cannot completely populate this file because they were not all conducted on 100 m × 100 m surfaces with a position report and the measurements of the heights and widths of the crowns for each tree in the inventoried section. If not using a forest dynamic model capable of simulating both the DBH growth and the crowns as well as the position of the trees, then an empirical approach based on the development of equations can relate the diameter of the trunk, heights (Table 7.1) and widths of the crowns using available measurements. For the crown dimensions, a terrestrial laser scanner was used on a sample of 36 mangroves [OLA 16] of the species Avicennia germinans to establish equation [7.1] that relates the surfaces projected from the ground to the crowns, denoted as *SH* (expressed in m²), in rows *R1* or *R2* of the equivalent ellipse.

$$SH = \pi.R1.R2 = 236\, DBH_{\mathrm{cm}}^{4/3} + 7.5 \qquad [7.1]$$

The relation between *R1* and *R2*, which varies randomly between 1 and 1.3 for each tree, can give a less circular appearance to surfaces projected to the ground of the ellipsoids. However, this does not have any ecological accuracy.

$H = a X^2 + b X + c$ ($X = DBH_{cm}$)	H_{1b} (m)					H_{tot} (m)				
	A	*b*	*c*	r^2	*e*	*a*	*b*	*c*	r^2	*e*
***Avicennia germinans* (n=1004)**	0.74	2.9	13.2	0.35	3.9	-0.7	6.7	20.4	0.70	4.0
***Rhizophora spp.* (n=474)**	0.25	3.9	9.5	0.60	2.9	0.28	5.9	15.9	0.74	3.6

Table 7.1. *Equations used to find the insertion heights of the first branch H_{1b} and total heights H_{tot} using the DBH measurements for the 2 dominant species of mangrove in Guiana. The coefficients of the correlation r^2 and the root-mean-square error e (in meters) are also indicated*

The distribution of measured trunks' diameters was extrapolated proportionally to the relation between the real surface of the inventory and a hectare. In the same way, the distribution of the positions of trees on a 100 m × 100 m surface was obtained using an iterative method of the "birth and death" model [MAT 86]. This method chronologically orders the largest trees then the smallest according to the minimum neighbor distance associated with the height of the crown [BAR 12]. Figure 7.7 presents an overview of the forest model renderings produced this way.

Finally, the DART configuration requires the user to indicate a leaf area index (LAI), either on the same section or on a group of trees of the same species. If lacking measurements, the mean LAI values are set by section at three for juvenile forests and five for mature forests.

7.3.2.4. *Simulations of satellite images of the canopy at VHR*

The DART model produced images of the canopies for each of the 21 "fine" spectral bands and for all of the specified look directions. It takes between 1 and 2 h of calculating and about 20 GB of RAM on 64-bit Windows with an i7 2.8 GHz processor to analyze images at 50 cm in 10 observation configurations. To find satellite sensor images (Figure 7.8) made up of four multi-spectral channels (or eight for Worldview 2/3) and a panchromatic channel, the reflectance of each pixel of a "large band sensor" image is calculated proportional to the influence of the sensor response on the "fine" spectral bands defined in DART.

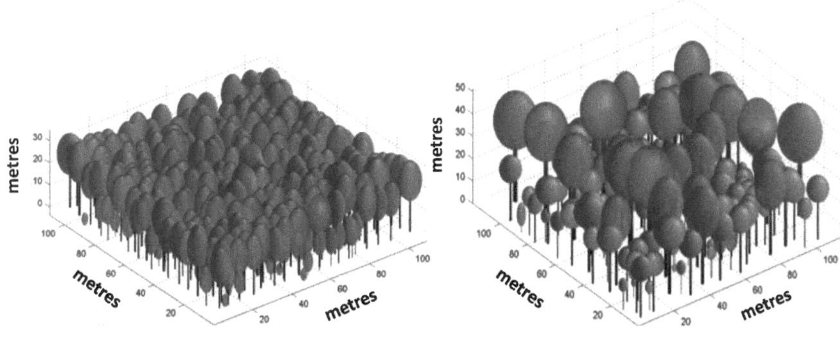

Figure 7.7. *Examples of forest stand models of sections KA12 (left) and TC1 (right) produced at 100 m × 100 m*

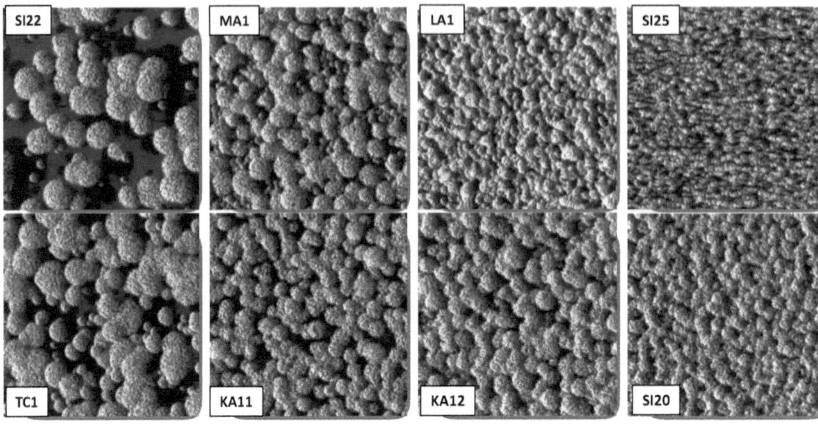

Figure 7.8. *Multi-spectral images of 100 m × 100 m simulated in the configuration of the Geoeye image acquisition (Θ_s = 24°, Θ_v = 27°, $\varphi_{s\text{-}v}$ = 27°) with 50 cm pixels on eight sections of mangrove forest characteristic of the variability of the observed forest dynamics. Each image is an RGB-colored composite of the four Geoeye channels ((NIR+Red)/2, V, B). For a color version of this figure, see www.iste.co.uk/baghdadi/5.zip*

7.4. Application to the monitoring of Guianese mangrove forest dynamics

7.4.1. *Principles, potential and limits of the FOTO method*

Rao and Lohse [RAO 96] explained that the human perception of "textured" images was primarily sensitive to the presence or absence of pattern repetition. Couteron *et al.* [COU 05] demonstrated that pattern repetition could be observed in the aerial photographs of *terra firme* tropical forests and proposed a method to measure it. The method developed was named FOTO.

The FOTO method works in two steps using a grid that divides the VHR image of the forest into identically sized windows (Figure 7.9, left). The first step consists of calculating a Fourier 2D spectrum (denoted as r-spectrum) and averaging all of the azimuth directions included in each window on the grid (there are as many spectrums as there are windows). The r-spectrum graphs are frequential spectrums that reveal the number of times (cycles/kilometer) that a forest pattern repeats inside a 100 m × 100 m window. An r-spectrum with a peak clearly indicated in the low spatial frequencies (expressed in cycles/km) indicates a rough canopy grain (Figure 7.10). On the contrary, if the peak is located at high frequency, there is a fine canopy grain (fine texture). These r-spectrums are compiled in a table to which a principal component analysis (PCA) is applied in order to reduce the number of variables (or spatial frequencies) to three principal components.

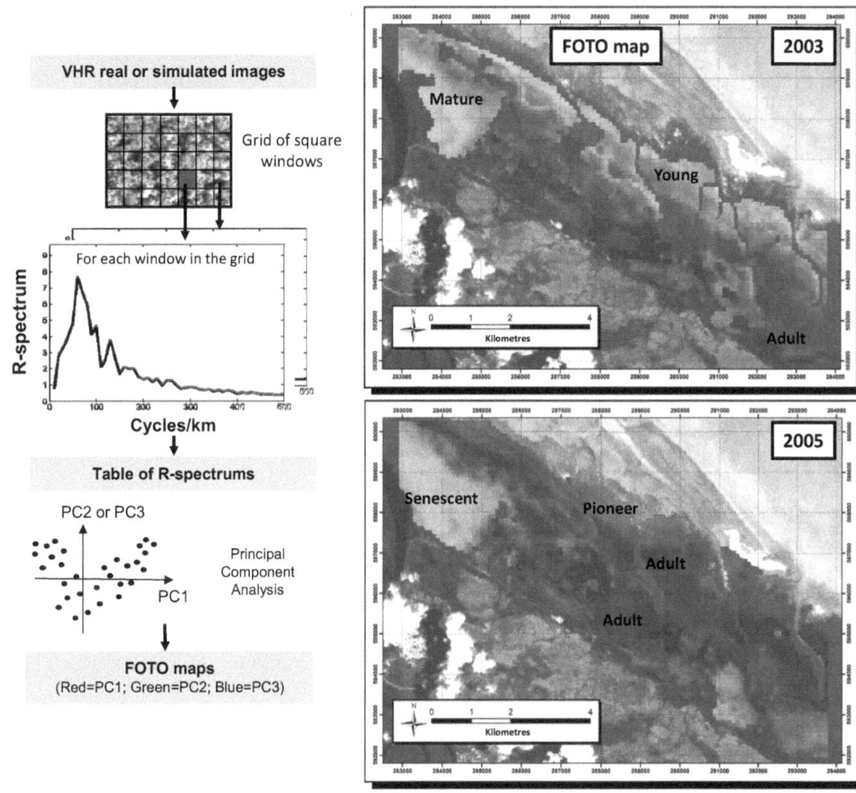

Figure 7.9. *Left: principles of the FOTO (*Fourier-based textural ordination*) method. Right: FOTO maps from 2003 and 2005. The map background is a Landsat image from 2002. The extension (colonization) of the mangrove forest toward the sea (red color) from 2003 to 2005 should be noted. For a color version of this figure, see www.iste.co.uk/baghdadi/5.zip*

The height of the windows on the grid must be sufficient to account for a repetition of a forest pattern (reflecting both the spatial distribution and variability of the dominant shapes of crowns in the canopy).

For the Guianese mangrove forests that present a significant texture gradient between the pioneer stages and the mature stages, window sizes between 75 and 125 m per side are usually appropriate [PRO 07]. However, if the spatial resolution of the images degrades (spatial resolution > 2 m), the pioneer facies will no longer be distinguishable because the textural signature becomes smooth.

In order to verify that the sensitivity of the FOTO method for spatial variations in texture is sufficient to account for the growth of mangrove forest stands monitored over several years, a FOTO analysis of a 100 m window was conducted on three VHR spatial images acquired in the Sinnamary region between 2003 and 2009 (Table 7.2). The PCA was carried out on a single table that compiled the r-spectrums of three images. The operation produced FOTO indices (the coordinates PC1, PC2 and PC3, Figure 7.9, left) for each analysis window on different dates. The respective coordinate distribution in each window on the first three axes of the display channels in red, green and blue produces FOTO images or maps whose pixel size is that of the analysis window. These FOTO maps (Figure 7.9, right) show the zones of juvenile stages of development (shaded from orange to red), adult facies (shaded in blue) and the mature facies open or with gaps (shaded in green).

Platform	Date of acquisition	Angles (°)		
		Look angle (θ_v)	Solar angle (θ_v)	Azimuth difference (φ_{s-v})
Ikonos	12 October 2003	16	30	69
IGN	October 2005	NA	NA	NA
Geoeye	6 September 2009	27	24	27

Table 7.2. *Dates and acquisition parameters of the analyzed images. It was not possible for us to access the angular configurations of the acquisition of IGN photographs. They vary within a same photograph and for all of the sections considered*

About 20 zones that correspond to the location of forest inventories typical of a gradient of forest development stages were selected in the 2003 Ikonos image. The texture indices of each zone are expressed on the PC1 axis and grouped by three observation dates in Figure 7.10. The spread of the point coordinates along the PC1 axis translates a canopy texture gradient ranging from a coarse grain (the lowest values of PC1) to a finer grain (the highest values).

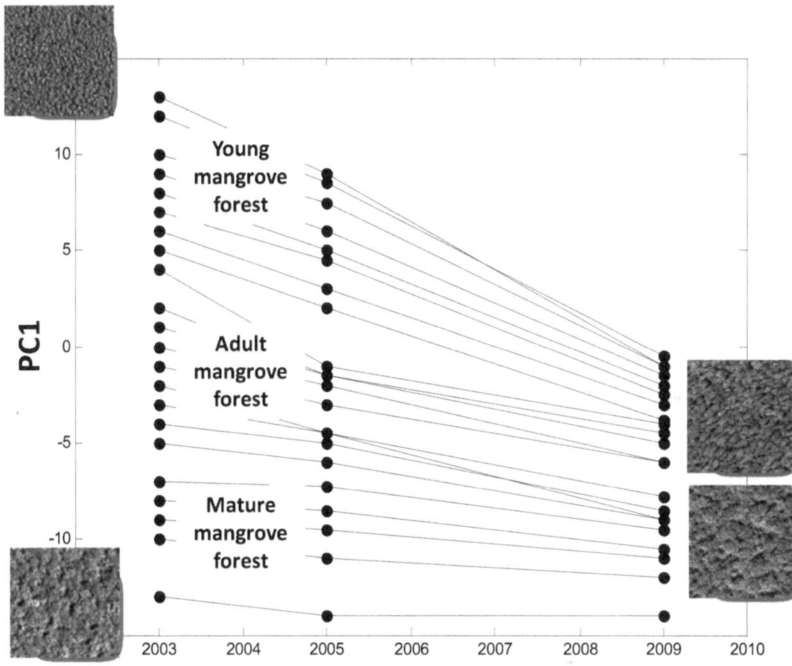

Figure 7.10. *Temporal variations between 2003 and 2009 of the FOTO PC1 texture index of the mangrove stands at different stages of development, using the FOTO analysis of the Ikonos image (2003), the BDORTHO IGN (2005) and the Geoeye image (2009). For a color version of this figure, see www.iste.co.uk/baghdadi/5.zip*

Figure 7.10 reveals that the fine textures of the juvenile mangrove forests evolved more quickly than the coarser textures, suggesting a faster transformation in juvenile mangrove stands (decreasing number of individuals and widening of crowns). The result is in line with a classical trajectory of forest development (Figure 7.2) and reflects the rapidity at which the Guianese mangrove stands develop [FRO 04]. The restriction of the point clouds in 2009 indicates that the stands considered to be juvenile in 2003 show, 6 years later, textures comparable to those observed on the adult forest stands, which suggests a quick evolution in the appearance of the juvenile mangrove forests.

These preliminary results are promising. The challenge is to develop the capacity to physically interpret these temporal variations in texture in such a way as to ensure that the inversions of forest parameters using FOTO texture indices are robust.

7.4.2. *Potential and limits of simulated images*

Tests are needed to verify the realism of the DART model calibration, which produces a forest dynamic model that generates forest models at different ages in the form of a stand of trees with crowns represented by ellipsoids. With this in mind, the FOTO method was used to compare the textural properties of real and simulated images of the canopy at VHR on 22 forest sections (Figure 7.4). The first step was to evaluate the difference between the r-spectrums of the excerpts of real images and those of simulated DART images (Figure 7.11).

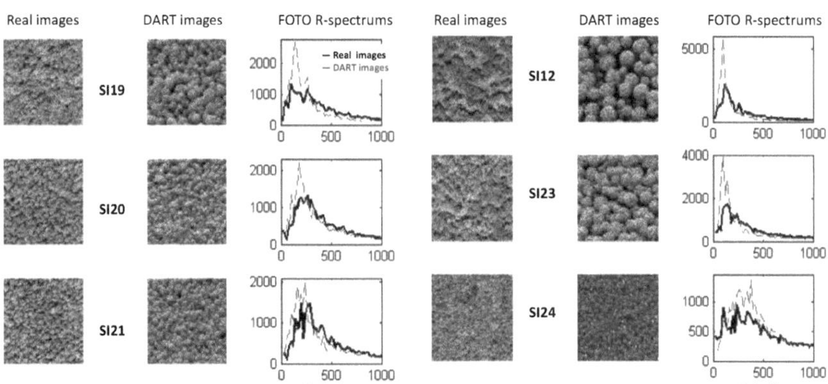

Figure 7.11. *Comparison of the visual rendering of real images (Ikonos panchromatic channel) and simulated images for six typical forest sections of the gradient observed in Guianese mangrove forests. The r-spectrums are also shown*

When the dominant frequency borders on 50 cycles/km (sections SI12 and SI23), the canopy grain is coarse and has a wavelength equivalent to λ=20 m. For a frequency of 200 cycles/km (λ=5 m), the canopy grain is fine (sections SI19 and SI24).

An analysis of Figure 7.11 shows that the dominant frequencies of r-spectrums of simulated and real images coincide well enough. That said, the frequency peaks observed in the simulated images are more distinct than in real images. Similarly, in the high frequencies (>400 cycles/km), the values of the r-spectrums of simulated images are slightly lower than those given by the r-spectrum of real images. The use of forest models made up of trees represented by ellipsoids produces less fine textures. The r-spectrum of simulated images is denser around a dominant frequency, though it remains consistent with the frequency peak observed in the r-spectrum of real images.

The PCA will enhance the consistency of the dominant frequencies across the distribution of coordinates on the PC1 and PC2 axes. The r-spectrums of the 22 simulated images were added to the table of r-spectrums of the Geoeye image before conducting a PCA of the whole table that shows the r-spectrums of the three real images and the 22 simulated images (Figure 7.12).

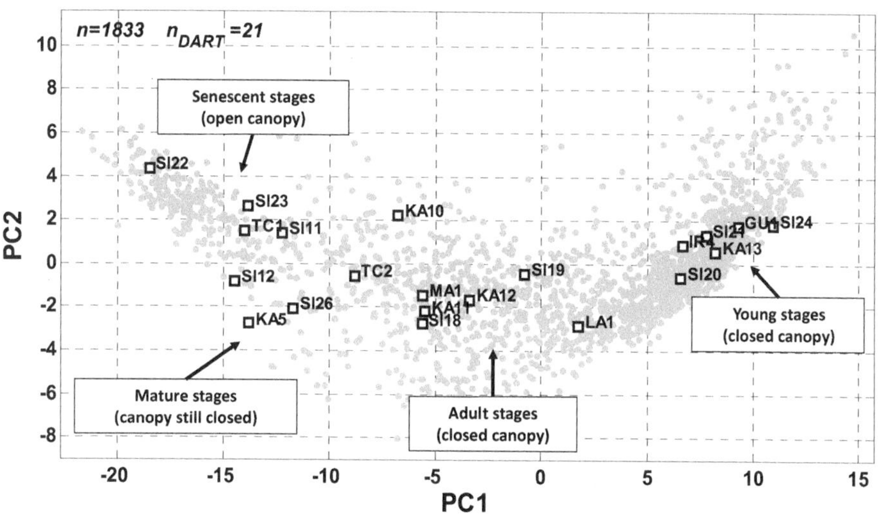

Figure 7.12. *Variability of the FOTO texture indices on the first two PC1 and PC2 axes of the principal component analysis as observed in the Geoeye image from September 2009 (1 833 windows of 1 hectare) and in the 22 simulated images (the yellow squares). For a color version of this figure, see www.iste.co.uk/baghdadi/5.zip*

Figure 7.12 shows the typical texture gradient expressed by the PC1 axis (as already presented in [PRO 07]). The lowest values of PC1 are very open, representing the sections of senescent mangrove forests like SI22, as opposed to the sections of young mangroves like SI24 and SI25. Furthermore, the coordinates on the PC1 and PC2 axes of 22 simulated images are suitably distributed in the point cloud representing the textures observed in the whole Geoeye image (about 18 km^2). However, none of the simulated images present PC2 values less than –3, which likely suggests too great of a geometric homogeneity in the forest models produced following the conclusions of Barbier *et al.* [BAR 12], made on simulations of VHR images at 2 m of spatial resolution.

7.5. Conclusion and prospects

This chapter presented and discussed a way to analyze very high spatial resolution optical images of the canopy in order to study the dynamics of the Guianese mangrove forests. This approach entails the simulation of multi-spectral images at a spatial resolution between 50 cm and 1 m. This study represents an important step for research in VHR optical remote sensing because the simulated images, obtained from forest models where the tree crowns were represented by ellipsoids, demonstrate textural characteristics comparable to those of real images. Combined with the FOTO method, which proved its ability to describe the textural variability of the mangrove forest canopies, this opens up a new research approach geared toward the combination of forest dynamics models with 3D radiative transfer models such as DART.

The research described here provides the fundamental elements for conducting a theoretical study on the contribution of the multi-spectral and hyper-spectral to the characterization of the function and biodiversity of mangrove forests.

Finally, this chapter shows that the development of a robust method of spatial remote sensing dedicated to the study of forests cannot be achieved without a double approach that combines the experimental (acquisition of terrain data and analyses of several real images in different acquisition configurations) and the theoretical (simulation of realistic VHR images for future sensitivity studies of all parameters susceptible to decrease the reproducibility of the developed method).

7.6. Key points

– The use of very high resolution (VHR) optical remote sensing is promising for the study of mangrove forest dynamics.

– The challenge is to develop methods of remote sensing capable of describing the variability of the forest parameters independent of the variability of the image acquisition configurations: the FOTO method of canopy grain analysis has confirmed its potential for this undertaking.

– It is possible to simulate VHR optical images using the DART 3D radiative transfer model. This tool has proven to be relatively simple to use and well suited for studying forests using remote sensing.

– The VHR-simulated images based on forest models, in which the trees are represented by ellipsoids, have realistic textural characteristics like the real images acquired on a large gradient of stages of development in the Guianese mangrove forests.

– Research in remote sensing can now embark upon an analysis of the potential of VHR multi-spectral and hyper-spectral images for the characterization of the function and biodiversity of tropical forests.

7.7. Bibliography

[ADA 06] ADAMS J.B., GILLESPIE A.R., *Remote Sensing of Landscapes with Spectral Images, A Physical Modeling Approach*, Cambridge University Press, 2006.

[ALO 12] ALONGI D.M., "Carbon sequestration in mangrove forests", *Carbon Management*, vol. 3, no. 3, pp. 313–322, 2012.

[ANT 10] ANTHONY E.J., GARDEL A., GRATIOT N. *et al.*, "The Amazon-influenced muddy coast of South America: a review of mud-bank-shoreline interactions", *Earth-Science Reviews*, vol. 103, no. 3–4, pp. 99–121, 2010.

[BAR 12] BARBIER N., COUTERON P., GASTELLY-ETCHEGORRY J.P., "Linking canopy images to forest structural parameters: potential of a modeling framework", *Annals of Forest Science*, vol. 69, no. 2, pp. 305–311, 2012.

[BER 08] BERGER U., RIVERA-MONROY V.H., DOYLE, T.W. *et al.*, "Advances and limitations of individual-based models to analyze and predict dynamics of mangrove forests: A review", *Aquatic Botany*, vol. 89, no. 2, pp. 260–274, 2008.

[BOH 12] BOHLMAN S., PACALA S., "A forest structure model that determines crown layers and partitions growth and mortality rates for landscape-scale applications of tropical forests", *Journal of Ecology*, vol. 100, no. 2, pp. 508–518, 2012.

[BRI 15] BRIENEN R.J.W., PHILLIPS O.L., FELDPAUSCH T.R. *et al.*, "Long-term decline of the Amazon carbon sink", *Nature*, vol. 519, no. 7543, pp. 344–348, 2015.

[CLA 99] CLARK D.A., CLARK D.B., "Assessing the growth of tropical rain forest trees: issues for forest modeling and management", *Ecological Applications*, vol. 9, no. 3, pp. 981–997, 1999.

[COU 05] COUTERON P., PÉLISSIER R., NICOLINI E. *et al.*, "Predicting tropical forest stand structure parameters from Fourier transform of very high-resolution remotely sensed canopy figures", *Journal of Applied Ecology*, vol. 42, pp. 1121–1128, 2005.

[DUK 07] DUKE N.C., MEYNECKE J.O., DITTMANN S. *et al.*, "A world without mangroves?", *Science*, vol. 317, no. 5834, pp. 41–42, 2007.

[FEL 14] FELICIANO E., WDOWINSKI S., POTTS M.D., "Assessing mangrove above-ground biomass and structure using terrestrial laser scanning: a case study in the Everglades National Park", *Wetlands*, vol. 34, no. 5, pp. 955–968, 2014.

[FER 08] FÉRET J.B., FRANÇOIS C., ASNER G.P. *et al.*, "PROSPECT-4 and 5: advances in the leaf optical properties model separating photosynthetic pigments", *Remote Sensing of Environment*, vol. 112, no. 6, pp. 3030–3043, 2008.

[FER 11] FÉRET J.B., ASNER G.P., "Spectroscopic classification of tropical forest species using radiative transfer modeling", *Remote Sensing of Environment*, vol. 115, pp. 2415–2422, 2011.

[FRO 98] FROMARD F., PUIG H., MOUGIN E. *et al.*, "Structure, above-ground biomass and dynamics of mangrove ecosystems: new data from French Guiana", *Oecologia*, vol. 115, no. 1, pp. 39–53, 1998.

[FRO 04] FROMARD F., VEGA C., PROISY C., "Half a century of dynamic coastal change affecting mangrove shorelines of French Guiana. A case study based on remote sensing data analyses and field surveys", *Marine Geology*, vol. 208, no. 2–4, pp. 265–280, 2004.

[GAS 15] GASTELLU-ETCHEGORRY J.P., YIN T., LAURET N. *et al.*, "Discrete anisotropic radiative transfer (DART 5) for modeling airborne and satellite spectroradiometer and LIDARacquisitions of natural and urban landscapes", *Remote Sensing*, vol. 7, no. 2, pp. 1667–1701, 2015.

[GRU 14] GRUETERS U., SELTMANN T., SCHMIDT H. *et al.*, "The mangrove forest dynamics model mesoFON", *Ecological Modelling*, vol. 291, pp. 28–41, 2014.

[HER 12] HERR D., PIDGEON E., LAFFOLEY D. (eds), "Blue carbon policy framework: based on the discussion of the international blue carbon policy working group", *International Union for Conservation of Nature (IUCN)*, Gland, Switzerland, p. 36, 2012.

[IMH 95] IMHOFF M.L., "Radar backscatter and biomass saturation: ramifications for global biomass inventory", *IEEE Transactions on Geoscience and Remote Sensing*, vol. 33, no. 2, pp. 511–518, 1995.

[JAC 90] JACQUEMOUD S., BARET F., "PROSPECT: a model of leaf optical properties spectra", *Remote Sensing of Environment*, vol. 34, no. 2, pp. 75–91, 1990.

[KAS 90] KASISCHKE E.S., CHRISTENSEN JR N.L., "Connecting forest ecosystem and microwave backscatter models", *International Journal of Remote Sensing*, vol. 11, no. 7, pp. 1277–1298, 1990.

[KOM 05] KOMIYAMA A., POUNGPARN S., KATO S., "Common allometric equations for estimating the tree weight of mangroves", *Journal of Tropical Ecology*, vol. 21, pp. 471–477, 2005.

[LET 12] LE TOAN T., VILLARD L., LASNE Y. *et al.*, "Assessment of tropical forest biomass: a challenging objective for the BIOMASS mission", *IEEE International Geoscience and Remote Sensing Symposium*, Munich, Germany, pp. 7581–7584, 22–27 July, 2012.

[MAT 86] MATÉRN B., *Spatial Variation*, 2nd ed., Springer-Verlag, New York, 1986.

[MER 15] MERMOZ S., RÉJOU-MÉCHAIN M., VILLARD L. *et al.*, "Decrease of L-band SAR backscatter with biomass of dense forests", *Remote Sensing of Environment*, vol. 159, pp. 307–317, 2015.

[MOR 14] MORTON D.C., NAGOL J., CARABAJAL C.C. *et al.*, "Amazon forests maintain consistent canopy structure and greenness during the dry season", *Nature*, vol. 506, no. 7487, pp. 221–224, 2014.

[NAT 15] NATURE GEOSCIENCE, "Ecology in a changing climate", *Nature Geoscience*, vol. 8, no. 6, pp. 417–417, 2015.

[OLA 12] OLANDER L.P., GALIK C.S., KISSINGER G.A., "Operationalizing REDD+: scope of reduced emissions from deforestation and forest degradation", *Current Opinion in Environmental Sustainability*, vol. 4, no. 6, pp. 661–669, 2012.

[OLA 16] OLAGOKE A., PROISY C., FÉRET J.B. *et al.*, "Extended biomass allometric equations for large mangrove trees from terrestrial LiDARdata", *Trees*, vol. 30, no. 3, pp. 935–947, 2016.

[PAC 96] PACALA S.W., CANHAM C.D., SILANDER J.A.J. *et al.*, "Forest models defined by field measurements: estimation, error analysis and dynamics", *Ecological Monographs*, vol. 66, no. 1, pp. 1–43, 1996.

[PET 14] PETERS R., VOVIDES A.G., LUNA S. *et al.*, "Changes in allometric relations of mangrove trees due to resource availability – a new mechanistic modelling approach", *Ecological Modelling*, vol. 283, pp. 53–61, 2014.

[PLO 12] PLOTON P., PELISSIER R., PROISY C. *et al.*, "Assessing aboveground tropical forest biomass using Google Earth canopy images", *Ecological Applications*, vol. 22, no. 3, pp. 993–1003, 2012.

[PRO 00] PROISY C., MOUGIN E., FROMARD F. *et al.*, "Interpretation of polarimetric radar signatures of mangrove forests", *Remote Sensing of Environment*, vol. 71, no. 1, pp. 56–66, 2000.

[PRO 07] PROISY C., COUTERON P., FROMARD F., "Predicting and mapping mangrove biomass from canopy grain analysis using Fourier-based textural ordination of IKONOS images", *Remote Sensing of Environment*, vol. 109, no. 3, pp. 379–392, 2007.

[PRO 09] PROISY C., GRATIOT N., ANTHONY E.J. *et al.*, "Mud bank colonization by opportunistic mangroves: a case study from French Guiana using LiDARdata", *Continental Shelf Research*, vol. 29, no. 3, pp. 632–641, 2009.

[PRO 12] PROISY C., BARBIER N., GUÉROULT M. *et al.*, "Biomass prediction in tropical forests: the canopy grain approach", in FATOYINBO T.E. (ed.), *Remote Sensing of Biomass: Principles and Applications / Book 1*, INTECH Publisher, pp. 59–76, 2012.

[PUR 07] PURVES D.W., LICHSTEIN J.W., PACALA S.W., "Crown plasticity and competition for canopy space: a new spatially implicit model parameterized for 250 North American tree species", *PLoS ONE*, vol. 2, no. 9, p. e870, 2007.

[RAN 97] RANSON K.J., SUN G., WEISHAMPEL J.F. *et al.*, "Forest biomass from combined ecosystem and radar backscatter modeling", *Remote Sensing of Environment*, vol. 59, no. 1, pp. 118–133, 1997.

[RAO 96] RAO A.R., LOHSE G.L., "Towards a texture naming system: Identifying relevant dimensions of texture", *Vision Research*, vol. 36, no. 11, pp. 1649–1669, 1996.

[SCH 06] SCHAEPMAN-STRUB G., SCHAEPMAN M.E., PAINTER T.H. *et al.*, "Reflectance quantities in optical remote sensing—definitions and case studies", *Remote Sensing of Environment*, vol. 103, no. 1, pp. 27–42, 2006.

[SCH 14] SCHNEIDER F.D., LEITERER R., MORSDORF F. *et al.*, "Simulating imaging spectrometer data: 3D forest modeling based on LiDAR and in situ data", *Remote Sensing of Environment*, vol. 152, pp. 235–250, 2014.

[SID 13] SIDIK F., LOVELOCK C.E., "CO2 efflux from shrimp ponds in Indonesia", *PLoS One*, vol. 8, no. 6, p. e66329, 2013.

[SOU 14] SOUDANI K., FRANCOIS C., "Remote sensing: a green illusion", *Nature*, vol. 506, no. 7487, pp. 165–166, 2014.

[VIN 08] VINCENT G., HARJA D., "Exploring ecological significance of tree crown plasticity through three-dimensional modelling", *Annals of Botany*, vol. 101, no. 8, pp. 1221–1231, 2008.

[YIN 15] YIN T., LAURET N., GASTELLU-ETCHEGORRY J.P., "Simulating images of passive sensors with finite field of view by coupling 3-D radiative transfer model and sensor perspective projection", *Remote Sensing of Environment*, vol. 162, pp. 169–185, 2015.

8

Remote Sensing-based Monitoring of the Muddy Mangrove Coastline of French Guiana

8.1. Introduction

8.1.1. *The state of mangrove coastlines: information derived from remote sensing*

Although mangrove forests constitute only 0.7% of tropical forests, they provide important ecological services, such as the protection of coasts against erosion by waves, and the storage of carbon (see Chapter 7 of this Volume). They also provide nursery and spawning zones for aquatic species. The development of remote sensing since the 1970s has greatly expanded our understanding of mangrove environments. NOAA, Landsat, SPOT and ERS satellite images have enabled the mapping of mangrove forests on a global scale [SPA 97, SPA 10]. In 2000, an estimated 130,000 km of coastline was covered by mangrove forests, or about 17% of coastlines worldwide and 65% of tropical coastlines (Figure 8.1). Several studies using remote sensing have been dedicated to monitoring changes in mangrove area and have been used to draw up statistics by country [SPA 97, SPA 10].

Chapter written by Romain WALCKER, Nicolas GRATIOT and Edward J. ANTHONY.

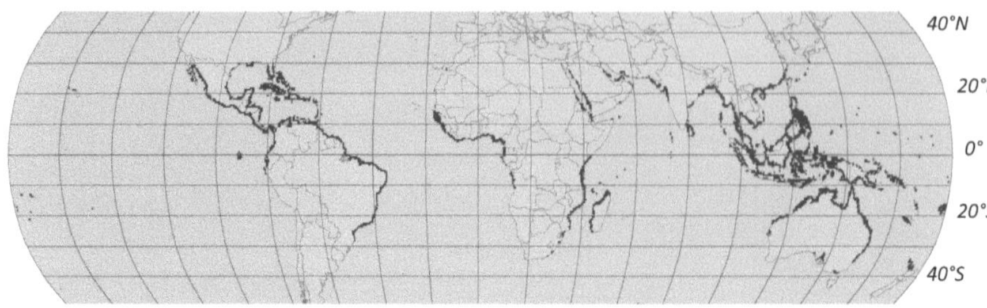

Figure 8.1. *Global distribution of mangrove forests: Map created using data from Giri et al. [GIR 11]*

8.1.2. *The Guianas' mangrove coastline: a challenge for coastal applications of remote sensing*

The 1,500 km-long mangrove coastline of the Guianas, between the mouths of the Amazon and the Orinoco Rivers in South America, is the most variable coastline in the world in terms of intensity and frequency of changes [ANT 10]. About 150 million tons of sediment from the Amazon, or about 15–20% of the annual Amazon sediment load, is transported along this coastline every year. A combination of hydrodynamic mechanisms involving the mixing of fresh and salt water, and interactions with waves, tides and currents, produces, just north of the mouth of the Amazon, large coastal mud banks on which mangrove forests thrive. Remote sensing has shown that these Amazon-derived banks, spaced several tens of kilometers apart and which can be up to 60 km long and 20 km wide, migrate toward the mouths of the Orinoco at a speed of 0.5–3 km per year [GAR 05]. The coastal system associated with these mud banks and their alongshore migration constitutes a unique natural laboratory for studying the interactions between ocean waves, muddy coasts and mangrove forests. This coastline has been monitored through remote sensing observations since the middle of the 2000s, notably following the installation of a receiving station (called SEAS) for SPOT satellite images in French Guiana specifically aimed at producing data on the Amazon-influenced environment. The use of remote sensing in the Amazon coastal region has undoubtedly improved our understanding of this environment by unraveling sedimentary and ecological processes in coastal zones that are otherwise largely inaccessible.

8.1.3. *Chapter outline*

With this regional and thematic context in mind, we present four examples of applications of coastal dynamics monitoring with remote sensing:

– the contribution of remote sensing to the quantification and monitoring of the mineral and organic composition of coastal waters;

– the contribution of remote sensing to the cartography and monitoring of the migration and morphodynamics of mud banks;

– remote sensing as the preferred tool to monitor the shoreline and its evolution at a regional level;

– innovative methods for topographic modeling of mud banks.

In conclusion, we will see how the different monitoring methodologies based on remote sensing provide data for coastal planning and management, while enabling a better understanding of the impact of mangrove degradation on coastal sedimentary and ecological processes.

8.2. Monitoring coastal water color with remote sensing

8.2.1. *The concentration of phytoplankton and the biogeochemical composition of coastal waters*

Coastal waters are sites of remarkable biogeochemical cycles and processes at a global level, especially carbon cycles. These coastal processes are especially important for the planet because they affect not only the global carbon balance, but also directly affect fishing and microbial biodiversity and the primary production of the region [ABR 14, ALL 06, ALL 10]. With the launch of the first satellite dedicated to the study of water color in 1978 (Nimbus 7 and its sensor *Coastal Zone Color Scanner*), monitoring the abundance, distribution and dynamics of phytoplankton at a global level became possible [MAR 09a]. Only recently, however, has remote sensing proven to be a tool capable of analyzing the biogeochemical composition of coastal surface waters, especially their particulate and dissolved organic carbon load [ALV 12].

8.2.2. *Water color as an indicator of sediment dynamics*

Water color is related to the presence of colored dissolved matter and suspended particulate matter (SPM), whether organic, such as phytoplankton and the associated degradation products, or mineral, such as the silts and clays that make up 90% of the SPM off the mouth of the Amazon [DAG 04]. These particles control whether the waters are more or less clear or colored. In coastal waters, the SPM usually comes from nearby estuaries, but can also travel long distances as a result of coastal currents, which is the case along the Guianas coast. Through an analysis of archived spatial measurements of water color, Martinez *et al.* [MAR 09b] quantified the annual sediment discharge of the Amazon at 754 million tons (Mt) (Figure 8.2), and estimated its seasonal and interannual variability between 1995 and 2001. Vantrepotte *et al.* [VAN 12] linked the concentration of SPM detected by satellite on the French Guiana coast to the morpho-sedimentary dynamics, and showed that not only can water color be used to monitor the subtidal sedimentary dynamics (below the lowest tide level) but also erosion of the foreshore (intertidal zone). There still remains a lot of work to be done in order to better explain the interactions between SPM, oceanic forcing and the dynamics of the Guianas coast.

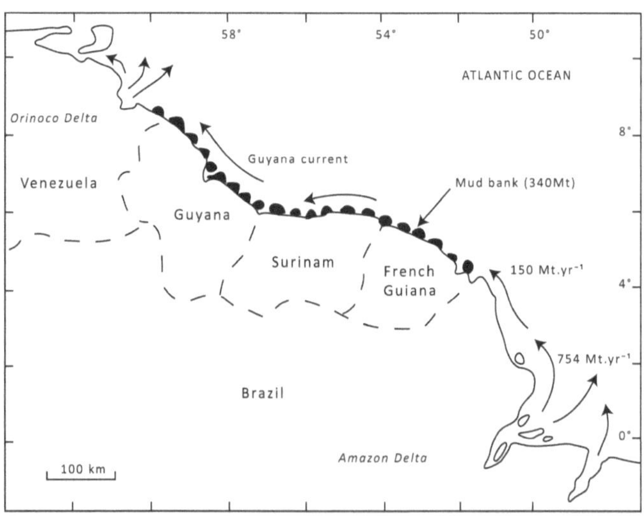

Figure 8.2. *15–20% of the suspended particulate matter (SPM) discharged annually by the Amazon forms a plume along the 1,500 km of mangrove coasts, westward to the Orinoco delta in Venezuela. Much of the SPM is contained in mud banks that can contain the equivalent, and even more, of the Amazon's annual sediment discharge*

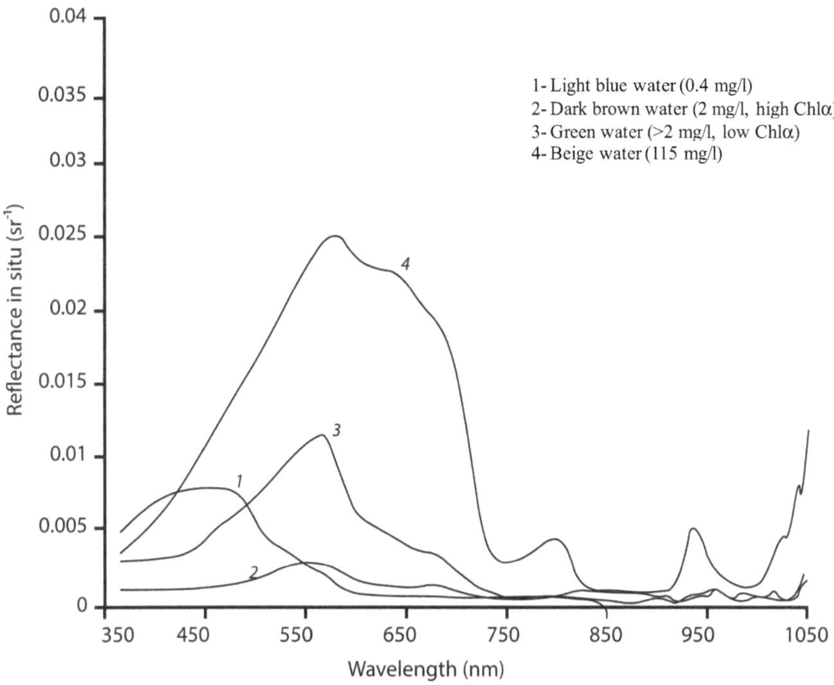

Figure 8.3. *Characterization of the four dominant types of spectral signatures of the water surface acquired at sea by a spectroradiometer. These types correspond to colored waters, the SPM concentrations of which vary from 0.4 to 115 mg/l [FRO 02]*

8.2.3. *Estimating the concentration of SPM using remote sensing*

Remote sensing measures the water color in the first few centimeters below the surface (see Chapter 4 of this Volume). The principle of estimating SPM concentrations using remote sensing consists of inverting the measurement of the water surface color, expressed by a physical quantity, and the marine reflectance, via empirical relations that enable prediction of the SPM concentration (Figure 8.3). Although data from remote sensing are less precise than those obtained by *in situ* measurements, they offer the advantage of covering large areas in a repetitive manner. Vantrepotte *et al.* [VAN 11] showed for the coastal waters of French Guiana that the marine reflectance in the red domain varied mainly as a function of the SPM content (Figure 8.4). Before being used in an inversion algorithm, the reflectance captured and recorded in a satellite image must first be corrected for the effects of diffusion and absorption of the signal in the atmosphere, essentially due to air molecules and aerosols. We consider that 90% of the blue light captured by the satellite sensor is caused by the atmosphere and is not related to the terrestrial or marine surface observed. The particularly complex quality of the atmospheric correction significantly influences the use of images to study water color. Further efforts are needed to improve these algorithms (see Chapter 4 of this Volume). The SPM prediction models as a function of marine reflectance are obtained via simultaneous luminance measurements conducted *in situ* using a spectroradiometer onboard a ship and measurements of the SPM content in the surface waters. The spectral signatures obtained *in situ* are then resampled to the resolution of the chosen satellite's spectral bands. Finally, empirical models are tested in order to link the concentrations measured in the sea and the reflectances observed by satellite. The validity of the final products is established via analyses comparing the satellite products to the ground-truth data acquired in the course of the passage of the satellite [CAR 11].

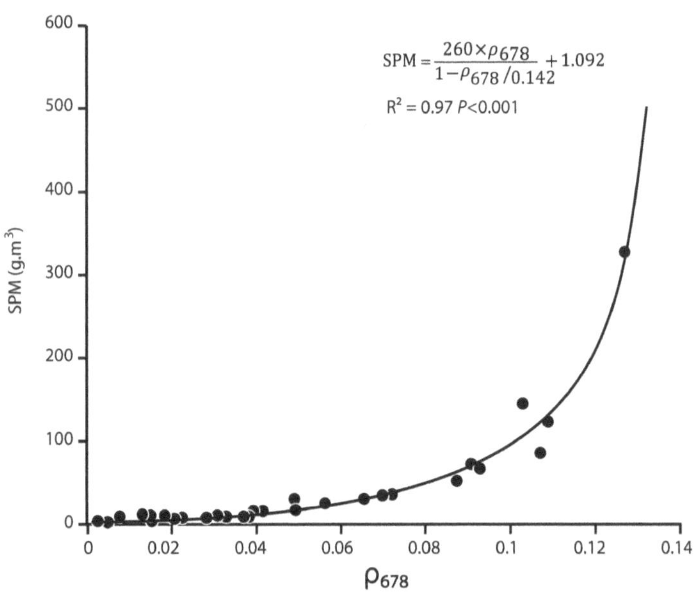

Figure 8.4. *Algorithm developed for French Guiana coastal waters relating the in situ reflectance (ρ) of water in the red band (678 nm) and the SPM concentration [VAN 11]*

8.2.4. *Satellites used to study water color*

The first satellite dedicated to water color (Nimbus 7 and its sensor CZCS) was launched in 1978. The sensor clearly demonstrated the effectiveness of remote sensing for monitoring phytoplankton at a global scale, which was previously not feasible using ships. Since then, several satellites dedicated to this domain have followed, notably SeaWiFS (1997–2010), MERIS (2002–2012) and MODIS (still active). In principle, all optical sensors could be used for the study of water color. Although optical sensors onboard the satellite MODIS have a low spatial resolution (250 m) compared to SPOT or Landsat sensors (20–30 m), they still have several advantages that make them ideal tools to monitor SPM. Firstly, they have been active since 2002 and cover the Earth's ocean surfaces in 1–2 days. Secondly, with 36 channels, MODIS is a high spectral resolution tool for the study of water color [LOI 13].

8.3. Remote sensing of coastal mud banks

Sediments transported downstream by rivers are deposited in estuary mouths, usually sheltered from waves. In the case of the Guianas coast, part of the sediment load discharged by the Amazon is swept by waves and wave-induced currents nearly 1,500 km alongshore toward the mouths of the Orinoco in Venezuela. 90% of this alongshore migrating mud is composed of silts and clays [DAG 04]. These circulating particles move along the coast as distinct mud banks. We estimate that some mud banks can contain more than the equivalent of the annual sediment discharge of the Amazon, or about 754 Mt of sediments [ANT 10]. The contribution of remote sensing to monitoring the intertidal part of these mud banks is presented below.

8.3.1. *Problem of delineation of mud banks linked to tides*

In the course of different studies on the Guianas coast, it has been shown that aerial photographs and high spatial resolution satellite images (shooting scale < 1:100,000[e] or pixel size < 900 m^2) are well suited to define the intertidal mudflats characterizing the surfaces of mud banks over large stretches [PRO 93, CHA 92]. The comparison of boundaries obtained on different dates has enabled the quantification of the volume and speed of migration in conjunction with factors of environmental forcing [GAR 05, GRA 07]. Although the reflectance of emergent muddy surfaces is easily identifiable on satellite images in relation to other neighboring surfaces [PRO 07], their definition still presents some difficulties in terms of multi-date comparisons. The main difficulty resides in the precise qualification of the water level at the time of image acquisition. The direct detection of the intertidal zone of a mud bank is impossible when its surface is covered by a few centimeters of water, whatever the sensor used, whether optical or a radar. This is due to the high sediment load in suspension [TRE 05]. As such, the different water levels at different times of acquisition are problematic. The sun-synchronous trajectory of high spatial resolution satellites makes it difficult to obtain images at the same water level, because they always pass at the same time each day whereas, the tide times shift by several minutes each day. In the equatorial zone, this difficulty is paired with that of obtaining an optical image during the day and without clouds. Satellite image acquisitions must be repeated to maximize the chances of obtaining useful information.

To make use of multiple observations at different tide times, Gardel and Gratiot [GAR 05] proposed a method to correct the water level, thus enabling comparison of the boundaries provided by images taken at different tide levels (Figure 8.5). The authors deduced the average slope of a mud bank (α) using two images obtained at nearby dates with a strong amplitude in water levels. They then used this information to correct the difference of the water line position in images as a function of the tide level at the time of acquisition (obtained by a tide model such as the one used by the French National Hydrographic Service). Thus:

$$IE(x,t) = Intertidal\,(x,t) + \frac{h_{tide}(t) - 1}{\alpha} \qquad [8.1]$$

where *Intertidal* is the water line position identified on the satellite image at a point x of the coast and at a time t (Figure 8.5), *IE* is the corrected water line, h_{tide} is the water level and α is the average slope.

8.3.2. *Monitoring the migration of mud banks*

In French Guiana, mud banks migrate from the southeast to the northwest, driven by waves and coastal currents [AUG 04]. Sediments are eroded at the trailing edge of the bank (east) and accumulate at the leading edge of the bank (west). Gardel and Gratiot [GAR 05] calculated the migration speed of mud banks by monitoring the position of the barycenter overt time, a point of the x-axis delimiting equal surface distance between the leading and the trailing edges of the bank (Figure 8.5). Using this technique, the authors documented highly variable migration speeds, ranging from 0.5–3 km.year^{-1} (Figure 8.6). This variability allows us to better understand the link between mud bank dynamics and oceanic forcing.

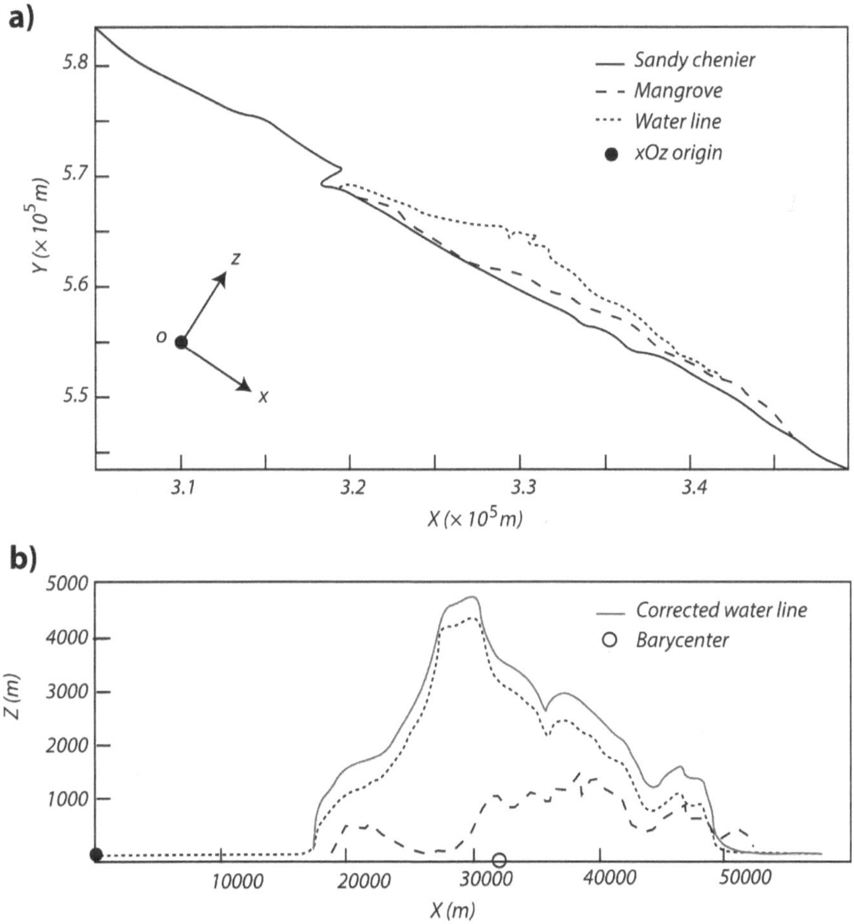

Figure 8.5. *Example of correction of the effects of the tide level on the positioning of mud bank boundaries: a) projected map of the mud bank boundaries (water line) obtained from a georeferenced satellite image; b) change of the coordinate system (X,Y) as indicated in b) to a new coordinate system (X,Z). Application of a correction to the position of the shoreline and the position of the mud-bank barycenter [GAR 05]*

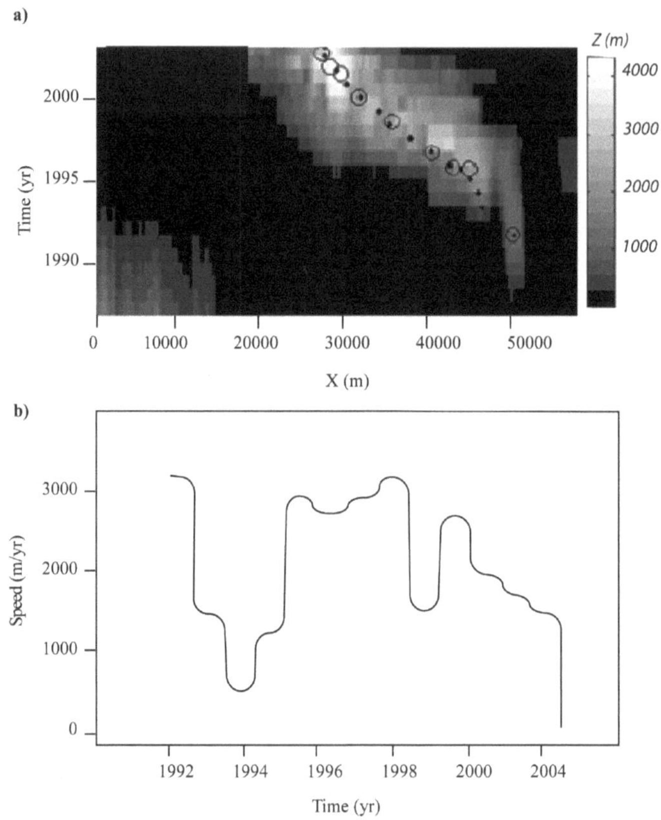

Figure 8.6. *Quantification of mud bank migration: a) representation in matrix form of the displacement of the water line in distances along the coast. The displacement is evaluated by monitoring the monthly position of the water line within coastal sections of 1,000 m (ordered) by width. The farther (closer) the water line is from the coast, the lighter (darker) the color of the matrix cell. The position of the barycenter of the mud bank along the coast is indicated by the circles (measured positions) and the points (interpolated positions); b) evolution of the migration speed of the barycenter of a mud bank between 1992 and 2004 [GAR 05]*

8.3.3. *Radar remote sensing of mud banks*

Because mangrove coastlines occupy an area of the Guianas coast where the latitudinal position of the intertropical convergence zone (ITCZ) fluctuates seasonally, optical remote sensing is usually impeded by the presence of clouds during the rainy season when coastal changes are the most apparent. The possibility of using radar images was tested in order to mitigate this problem. Baghdadi *et al.* [BAG 04] showed that radar images can effectively detect the surface parts of banks, and in particular very smooth surfaces called "mud lakes" (Figure 8.7), where the water depth is only a few centimeters. There is a total dampening of ocean waves and surface tension waves, resulting in the formation of a very smooth body of water (the "lake"), characterized by a very low radar backscatter coefficient (between –21 and –25 dB for a RADARSAT image with horizontal polarization with an incidence angle of 28°). In Figure 8.7, the mud lake is represented by a black zone with distinct contours due to the dampening of surface tension waves.

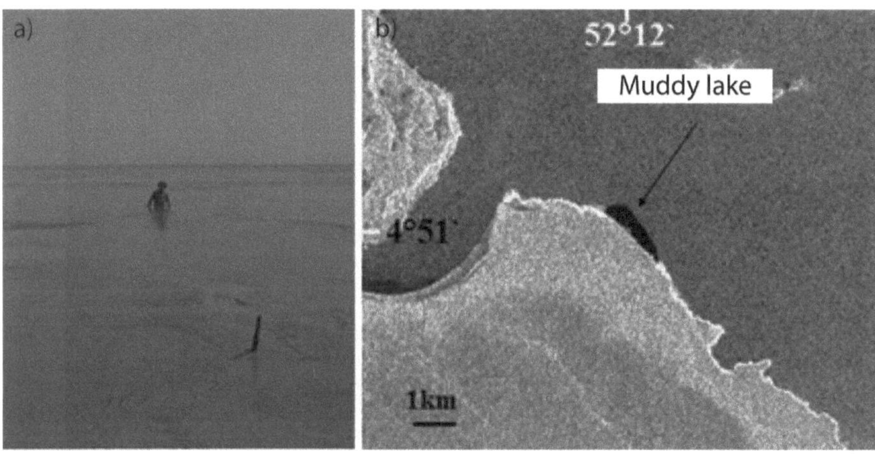

Figure 8.7. *Detection of "mud lakes": a) photograph of a mud lake at low tide; b) detection of a mud lake on a RADARSAT image [BAG 04]*

8.4. Monitoring the shoreline with remote sensing

8.4.1. *The shoreline: definition and issues posed by a multi-faceted feature*

The shoreline usually marks the separation between the land and sea. The concept of a fixed line is not the best tool to describe this interface because the passage between the two zones is gradual over space and mobile over time [BER 05]. There are several definitions of the shoreline: foot of a dune, foot of a cliff, average line attained by spring tides, highest astronomical tides, etc. [ROB 02]. For the *Conseil national de l'information géographique*

(CNIG, the French National Council for Geographic Information), a shoreline is the "physical boundary where the ground stops", which is the boundary attained by the highest astronomical tides. Mapping this area with remote sensing is problematic because the satellite's passage must coincide with the schedule of the highest astronomical tides, the cycles of which can be quite long (tidal cycle linked to the Saros lunar cycle of 18.6 years), and such timing is rarely possible. The very high turbidity of the water also makes the boundary between the submerged/emergent parts of mud banks extremely unclear. The geomorphological shoreline was therefore conceptualized by the CNIG as "an area of morphological discontinuity characteristic of the probable limit of the sea along the coast. Examples: erosional bluff, foot of a dune, vegetation boundary on a beach, exceptional spring tide debris lines, storm berms… This morphological boundary is important for the monitoring of the physical evolution of the coast".

The vegetation boundary is a relevant biological indicator to demarcate the limits of the land zone. Mangrove forests, like salt marshes, are directly adapted to coastal hydrodynamic and sedimentary conditions and are significant for monitoring the physical evolution of the coast. These environments are easily identifiable by their radiometric and textural contrasts on optical remote sensing images [GRE 98]. Vegetation is also a valid indicator of the land boundary for sandy coasts. In the Guianas, the top of the beach is colonized by herbaceous species, especially ipomea, which can be considered as a shoreline marker. Anthony *et al.* [ANT 15] have used remote sensing to determine the highly changeable dynamics of the sandy beaches of Cayenne. Even higher on these sand beaches and cheniers (beaches of sand and/or shells formed in muddy environments), there is often a forest characterized by species such as courbaril. In rocky zones, the foot of the rocky outcrop is the geomorphological element to consider.

8.4.2. *Exceptional dynamics of the Guianas coastline*

In the Guianas, the coastal dynamics are closely tied to the migration of mud banks (Figure 8.8). The seaward fringe of the mangrove forests is an essential indicator of coastal changes [BLA 96]. Gratiot *et al.* [GRA 08] compiled 39 SPOT satellite images between 1986 and 2006 and mapped the mangrove shoreline in each image. In order to obtain a synoptic view of the shoreline dynamics over time, they adopted the methodology of Gardel and Gratiot [GAR 05] and interpolated the data to obtain a complete temporal view in the form of a diagram (Figure 8.9(a)). This diagram allowed them to represent the temporal evolution of the shoreline in relation to its position in 1988, in the first image acquired by the authors. Time (years) is represented on the x-axis, distance along the coast is represented on the y-axis and expresses the distance to the source of the coastal sediments (the mouth of the Amazon), and the position of the shoreline in meters in relation to 1988 is indicated by color. Estuaries of rivers debouching on the coast northwestward of the Amazon are depicted by triangles on the y-axis. Warm colors express accretion, while the cool colors express erosion.

Figure 8.8. *Dynamics of the Guianas coastline: a) zone of mangrove and shoreline. The retreat of the mangrove forest is due to ocean waves undercutting the substrate. The mangrove trees fall and the coastline retreats; b) zone of shoreline accretion and mangrove development. Some of the sediments transported by mud banks weld onto the coast, promoting accretion, and forming mudflats that emerge at low tide and that are extensively colonized by juvenile mangroves in a few weeks. The older, higher mangrove forest forming the previous shoreline is visible in the background*

Using these data, the authors showed the existence of a disparity in the sediment balance at a multi-decadal scale (Figure 8.9(b)). They revealed that between 1988 and 1999, 37 Mt of sediment disappeared from the intertidal zone of the coast each year. This trend changes after the year 2000 with an annual increase of 35 Mt. The authors highlighted an important connection with the 18-year and 6-month tidal cycles. This cycle of tides corresponds to the low frequency modulation of the tide due to the Saros lunar rotation cycle. For the authors, the imbalance is explained by a cyclical fluctuation of the highest levels of the tides.

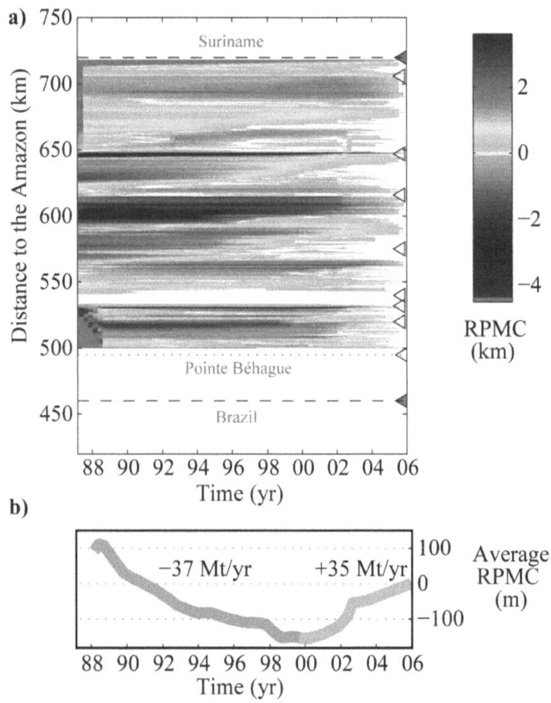

Figure 8.9. *Relative position of the mangrove coastline (RPMC) from 1988 to 2006 in the Guianas: a) spatio-temporal fluctuation of the coastline from Pointe Béhague to Suriname; b) average position over time and sediment budget in millions of tonnes per year (Mt y^{-1}) [GRA 08] . For a color version of this figure, see www.iste.co.uk/baghdadi/5.zip*

In 2015, Walcker *et al.* [WAL 15] extended this study to the period 1950–2010. By analyzing the changes in the extents of mangrove forests together with the changes in the heights, periods and average directions of waves in the North Atlantic, the authors showed a link between the coastal changes and the wave climate on a multi-decadal scale. They highlighted the role of the North Atlantic Oscillation on the modulation of wave impact, and showed that its effect is more significant on the cycle of tides than previously suggested.

8.5. Intertidal topography

In the context of rising average sea levels and an increase in the frequency and intensity of storm events, the geomorphological monitoring of coasts generally responds to a societal need for surveillance against coastal erosion, the silting of ports and navigation channels, and marine submersions near inhabited areas. Topographical modeling of coastal forms and the monitoring of their dynamics require recourse to spatial remote sensing tools due to the inaccessibility of the terrain, particularly the mudflats of the Guianas coast. Digital terrain models (DTMs) of these mudflats at low tides are traditionally acquired by topographical surveys using total electronic or laser stations. These instruments measure spot heights that are then interpolated to obtain a topographic map of the terrain. In unstable environments such as mud banks, this terrain monitoring technique is particularly difficult to implement and can only provide a basic approximation of the terrain trends over short distances. This approach is presented below. Other approaches using remote detection tools have been tested in French Guiana to provide information on the topography of mud banks.

8.5.1. *Ground surveys using a laser tacheometer or differential GPS*

Surveys using a laser station or total electronic station, or a high-resolution GPS, represent the classic method of DTM creation. Several authors [ANT 08, LEF 04] have shown that this ground-survey method provides an understanding of the general topography of the terrain, but the spatial coverage of such monitoring is limited to only a few tens of hectares (Figure 8.10). The accessibility of mudflats and the inherent instability of the muddy substrate (poorly consolidated sinking mud) considerably limit the efficacy of these techniques on the Guianas coast. These techniques are also limited by the number of points from which the DTM is constructed.

Although more precise than the total electronic station and the GPS due to the very large number of points it measures per scan, the terrestrial laser scanner, another topographical monitoring tool, has two disadvantages. Its scan radius is relatively restricted, and to cover a large surface it needs to be redeployed several times, which is difficult to accomplish on a muddy substrate. It is also much more costly than the other two techniques.

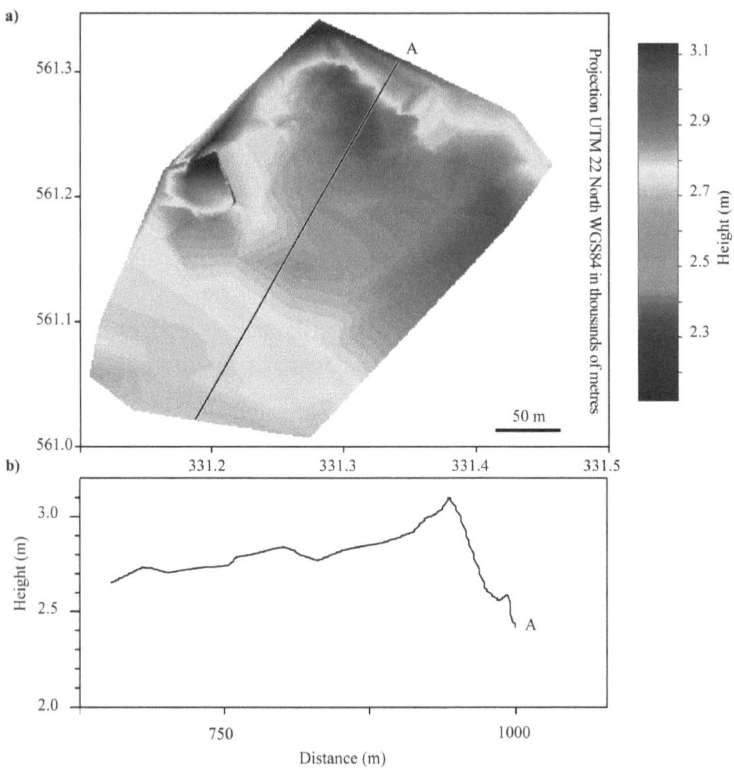

Figure 8.10. *Digital terrain model (DTM) obtained by GPS and total electronic station surveys in the region of Macouria, French Guiana: a) terrain map; b) cross-section of DTM at the location indicated in (a). Modified from Anthony et al. [ANT 08]. For a color version of this figure, see www.iste.co.uk/baghdadi/5.zip*

8.5.2. *Interpolation of water level isolines on satellite images*

One method using remote sensing tools was progressively developed by several authors [MAS 95, AAR 03] and adapted to the Guianas coast based on the use of SPOT satellite images [GRA 05]. The principle consists of mapping the water line of several successive images taken at different heights of the tide. The images must be acquired in a time interval which is sufficiently short so that no significant topographical changes can take place, ideally within a few weeks of each other. The set of water lines constitutes an isobath map, or topographical isoline, which is then interpolated in a geographical information system (GIS) in order to obtain the DTM of the intertidal part of the bank (Figure 8.11). The success of the method depends on the acquisition of images with a significant tidal amplitude over a short time period. The limitation is its ability to precisely identify the water line in waters with high turbidity, especially during low swells.

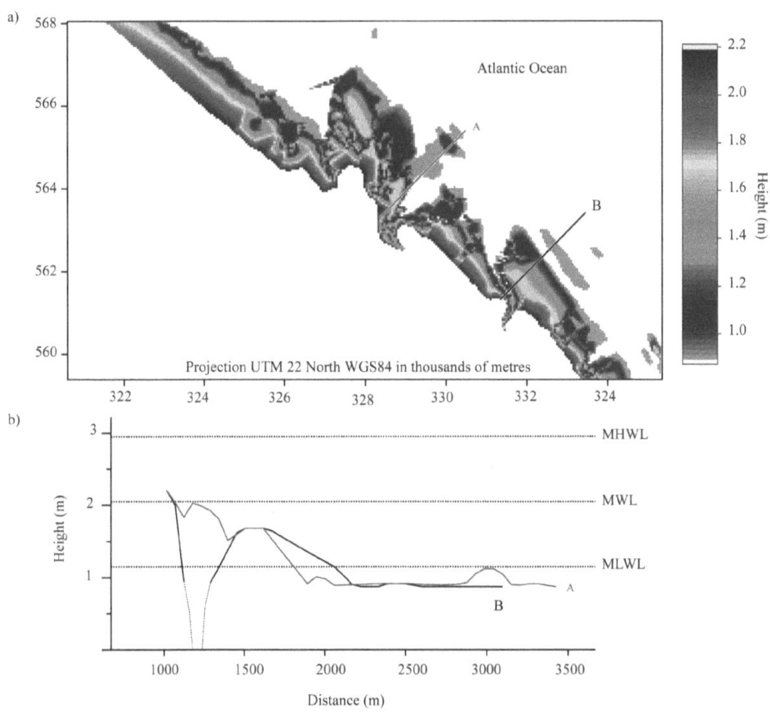

Figure 8.11. *Digital terrain model (DTM) obtained by interpolating the successive water lines of several SPOT images in the region of Macouria, French Guiana. Modified from Anthony et al. [ANT 08]: a) terrain map; b) cross-sections of DTM the locations of which are indicated in (a). For a color version of this figure, see www.iste.co.uk/ baghdadi/5.zip*

8.5.3. *Airborne LiDAR*

One important possibility that circumvents the problems associated with the methods described above is that of airborne laser scanning (light detection and ranging, or LiDAR). This technology combines the advantages of very high resolution remote sensing, since it is based on ground measurements of an infinitely larger number of points than those of a total electronic station or a GPS, with the possibility of covering very large surfaces (several km^2). The DTMs derived from LiDAR are more precise than those provided by the other two techniques (Figure 8.12). LiDAR is being increasingly used in physical studies of coasts [MON 14].

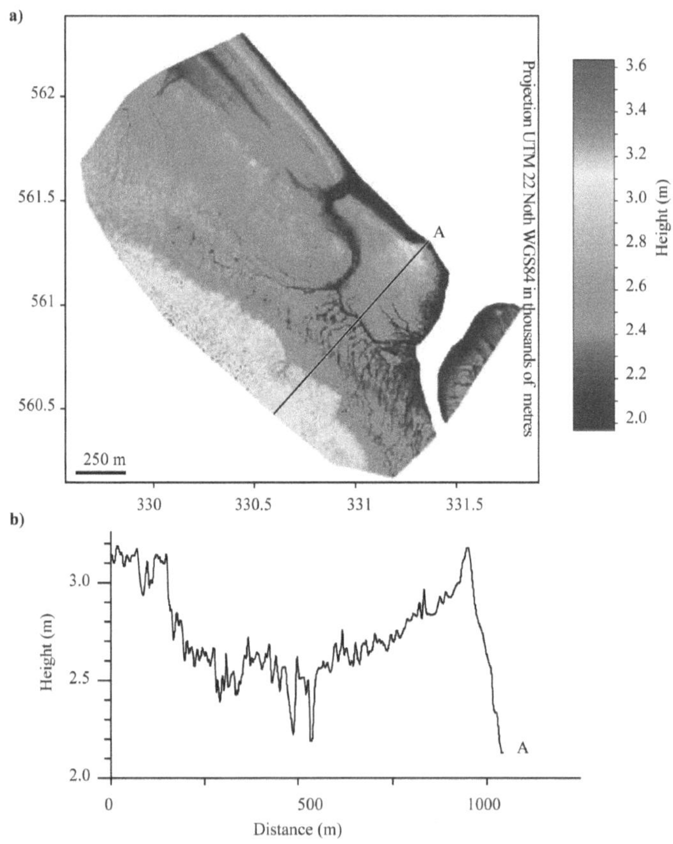

Figure 8.12. *Digital terrain model (DTM) obtained by airborne LiDAR in the region of Macouria, Guiana: a) terrain map; b) cross-section of the DTM with location indicated in (a) Modified from Anthony et al. [ANT 08]. For a color version of this figure, see www.iste.co.uk/baghdadi/5.zip*

8.5.4. *Photogrammetry*

In the past few years, there has been a considerable development of low-cost but very high-resolution photogrammetry using small aerial vehicles (airplanes, drones and ULMs) linked to the development of algorithms for processing adjusted images, notably the Structure-from-Motion technique (SfM [WES 12]). These new developments, made possible by advances in photographic quality combined with data processing, enable the generation of DTMs at a very high resolution (comparable to that of LiDAR and terrestrial laser scanners) by non-specialists and at low cost [WES 12, BRU 16]. The SfM photogrammetry technique is currently being applied to the muddy coasts of the Guianas where it presents considerable advantages for very high-resolution topographic monitoring of mud bank surfaces. Notwithstanding, airborne photogrammetry requires control points on the ground that must be obtained by GPS. This is one limitation of the method, given the difficulty of accessing the surfaces of mud banks and their instability.

8.6. Conclusion

Today, remote sensing is an indispensable tool for monitoring coastal dynamics on a large scale, especially on mangrove coasts. The instruments and methods derived from remote sensing enable us to observe, quantify and analyze biogeochemical, hydrodynamic, sedimentary and ecological interactions. This knowledge can then be used to provide environmental assessments that are useful for planners, developers and managers of such coasts. By comparing the French Guiana coast, with its still highly pristine mangrove forests, to the heavily urbanized coast of Guyana, where mangroves have been largely removed, researchers have recently shown the harmful impact to the coast caused by the massive and costly construction of dykes to defend against coastal erosion [ANT 12]. Using satellite data to characterize the advances and retreats of the shoreline on a multi-decadal scale, researchers have determined the width of the segment of mangrove forests that assures good coastal resilience following episodes of erosion. All along this very dynamic system, a good understanding of the mud banks and mangrove forests shows how this ecosystem has an eco-engineering role in coastal protection that is much more efficient than that provided by costly dykes, which concentrate wave energy and reduce the ability of the mud banks to absorb this energy.

It is certain that in the years to come, spatial remote sensing will continue to play a leading role in the evaluation of the state of mangrove coasts on a global scale, especially with the development of more and more powerful sensors capable of acquiring images of very high spatial, spectral and temporal resolution. For example, the images acquired by the Sentinel-2 satellite have recently offered a new tool to monitor land and coastal surfaces unparalleled by previous satellites (Landsat, SPOT, etc.). This mission provides systematic coverage with a spatial resolution of 10 m, passing over the site every 10 days. This performance will surely encourage improvement of, and innovation in, methods of time-series processing used for monitoring high-frequency changes in surfaces, which was previously only possible with average to low spatial-resolution images such as those used in meteorology. In 2020, the spatial radar mission *Biomass* will provide a biannual map in three dimensions of the heights of forests. Combined with allometric models, these data will enable us to estimate the standing biomass at a large spatial scale, and also airborne carbon stores located in mangrove forests. To conclude, we must not forget the importance of reanalyzing satellite data archives which, since the 1970s, represent an important accumulated mass of information on terrestrial and oceanic surfaces. Spatial remote sensing will always be an essential tool for monitoring and acquiring a better understanding of natural environments.

8.7. Key points

– mangroves are a coastal forest ecosystem in tropical regions, bordering coasts and waterways subject to the influence of the tides. Mangrove forests represent about 17% of global coasts and 65% of tropical coasts;

– since the 1980s, more than 35% of mangroves have disappeared due to deforestation. Today, some rare mangroves are still protected from the direct actions of humans and constitute the last remaining laboratories enabling scientists to understand the natural interactions between climate, sedimentary processes and ecosystem dynamics;

– mangrove coasts are often large and difficult to access, so monitoring them requires the use of remote sensing;

– the mangrove coast in the Guianas between the Amazon and Orinoco rivers in South America is very dynamic spatially, since it is subject to the combined influences of the Amazon's sediment discharge and waves and currents from the Atlantic. These interactions produce migrating mud banks at the heart of this ecosystem's unique dynamics;

– the contribution of remote sensing to the monitoring and understanding of coastal dynamics has been proven. Remote sensing is a tool for observation, quantification and analysis of coastal waters, the area and topography of coastal mud banks, and mangrove coasts. The spatial information can be combined in the form of time series to provide data that will facilitate decision-making by coastal planners, developers and managers.

8.8. Bibliography

[AAR 03] AARNINKHOF S.G.J., TURNER I.L., DRONKERS T.D.T. *et al.*, "A video-based technique for mapping intertidal beach bathymetry", *Coastal Engineering*, vol. 49, pp. 275–289, 2003.

[ABR 14] ABRIL G., MARTINEZ J.M., ARTIGAS L.F. *et al.*, "Amazon River carbon dioxide outgassing fuelled by wetlands", *Nature*, vol. 505, pp. 395–398, 2014.

[ALL 00] ALLISON M.A., LEE M.T., OGSTON A.S. *et al.*, "Origin of Amazon mudbanks along the northeastern coast of South America", *Marine Geology*, vol. 163, pp. 241–256, 2000.

[ALL 06] ALLER R.C., BLAIR N.E., "Carbon remineralization in the Amazon-Guianas tropical mobile mudbelt: a sedimentary incinerator", *Continental Shelf Research*, vol. 26, pp. 2241–2259, 2006.

[ALL 10] ALLER J.Y., ALLER R.C., KEMP P.F. *et al.*, "Fluidized muds: a novel setting for the generation of biosphere diversity through geologic time", *Geobiology*, vol. 8, pp. 169–78, 2010.

[ALV 12] ALVAIN S., VANTREPOTTE V., UITZ J. *et al.*, "Use of global satellite observations to collect information in marine ecology", in LE GAILLARD J.-F., GUARINI J.-M., GAILL F. (eds), *Sensors for Ecology*, CNRS INEE, Paris, 2012.

[ANT 08] ANTHONY E.J., DOLIQUE F., GARDEL A. *et al.*, "Nearshore intertidal topography and topographic-forcing mechanisms of an Amazon-derived mud bank in French Guiana", *Continantal Shelf Research*, vol. 28, pp. 813–822, 2008.

[ANT 10] ANTHONY E.J., GARDEL A., GRATIOT N. *et al.*, "The Amazon-influenced muddy coast of South America: a review of mud-bank–shoreline interactions", *Earth-Science Reviews*, vol. 103, pp. 99–121, 2010.

[ANT 12] ANTHONY E.J., GRATIOT N., "Coastal engineering and large-scale mangrove destruction in Guyana, South America: averting an environmental catastrophe in the making", *Ecological Engineering*, vol. 47, pp. 268–273, 2012.

[ANT 14] ANTHONY E.J., GARDEL A., GRATIOT N., "Fluvial sediment supply, mud banks, cheniers and the morphodynamics of the coast of South America between the Amazon and Orinoco river mouths", in MARTINI I.P., WANLESS H.R. (eds), *Sedimentary Coastal Zones from High to Low Latitudes: Similarities and Differences*, Geological Society, London, 2014.

[ANT 15] ANTHONY E.J., GARDEL A., DOLIQUE F. *et al.*, "Mud banks, sand flux and beach morphodynamics: Montjoly Lagoon beach, French Guiana", in ROBIN M., MAANAN M. (eds), *Coastal Sediment Fluxes*, Springer, Dordrecht, 2015.

[AUG 04] AUGUSTINUS P.G.E.F., "The influence of the trade winds on the coastal development of the Guianas at various scale levels: a synthesis", *Marine Geology*, vol. 208, pp. 145–151, 2004.

[BAG 04] BAGHDADI N., GRATIOT N., LEFEBVRE J.-P. *et al.*, "Coastline and mudbank monitoring in French Guiana: contributions of radar and optical satellite imagery", *Canadian Journal of Remote Sensing*, vol. 30, pp. 109–122, 2004.

[BER 05] LE BERRE I., HÉNAFF A., DEVOGÈLE T. *et al.*, "SPOT5 : un outil pertinent pour le suivi du trait de côte ?", *Norois*, vol. 196, pp. 22–35, 2005.

[BLA 96] BLASCO F., SAENGER P., JANODET E., "Mangroves as indicator of coastal change", *Catena*, vol. 27, pp. 167–178, 1996.

[BRU 16] BRUNIER G., FLEURY J., ANTHONY E.J. *et al.*, "Close-range airborne Structure-from-Motion Photogrammetry for high-resolution beach morphometric surveys: Examples from an embayed rotating beach", *Geomorphology*, vol. 261, pp. 76–88, 2016.

[CAR 11] CARVALHO G.A., MINNETT P.J., BANZON V.F. *et al.*, "Long-term evaluation of three satellite ocean color algorithms for identifying harmful algal blooms (Karenia brevis) along the west coast of Florida: a matchup assessment", *Remote Sensing of Environment*, vol. 115, pp. 1–18, 2011.

[CHA 92] CHARRON C., LOINTIER M., PROST M.T. *et al.*, "Étude du littoral et des estuaires guyanais utilisant l'imagerie spot et landsat", in PROST M.T. (ed.), *Évolution des littoraux Guyane la Zone Caraïbe méridionale pendant le Quaternaire*, ORSTOM, Paris, 1992.

[DAG 04] DAGG M., BENNER R., LOHRENZ S. *et al.*, "Transformation of dissolved and particulate materials on continental shelves influenced by large rivers: plume processes", *Continental Shelf Research*, vol. 24, pp. 833–58, 2004.

[FRO 88] FROIDEFOND J.-M., PUJOS M., ANDRE X., "Migration of mud banks and changing coastline in French Guiana", *Marine Geology*, vol. 84, pp. 19–30, 1988.

[FRO 02] FROIDEFOND J.M., GARDEL L., GUIRAL D. *et al.*, "Spectral remote sensing reflectances of coastal waters in French Guiana under the Amazon influence", *Remote Sensing of Environment*, vol. 80, pp. 225–232, 2000.

[GAR 05] GARDEL A., GRATIOT N., "A satellite image–based method for estimating rates of mud bank migration, French Guiana, South America", *Journal of Coastal Research*, vol. 21, no. 4, pp. 720–728, 2005.

[GIR 11] GIRI C., OCHIENG E., TIESZEN L.L. *et al.*, "Status and distribution of mangrove forests of the world using earth observation satellite data", *Global Ecology and Biogeography*. vol. 20, pp. 19–47, 2011.

[GRA 05] GRATIOT N., GARDEL A., POLIDORI L., "Remote sensing-based bathymetry of the Amazon-influenced coast of South America", *Procedings of the 9th International Coastal Symposium*, Iceland, 2005.

[GRA 07] GRATIOT N., GARDEL A., ANTHONY E.J., "Trade-wind waves and mud dynamics on the French Guiana coast, South America: input from ERA-40 wave data and field investigations", *Marine Geology*, vol. 236, pp. 15–26, 2007.

[GRA 08] GRATIOT N., ANTHONY E.J., GARDEL A. *et al.*, "Significant contribution of the 18.6 year tidal cycle to regional coastal changes", *Nature Geoscience*, vol. 1, pp. 169–172, 2008.

[GRE 98] GREEN E.P., CLARK C.D., MUMBY P.J. *et al.*, "Remote sensing techniques for mangrove mapping", *International Journal of Remote Sensing*, vol. 19, pp. 935–956, 1998.

[LEF 04] LEFEBVRE J.P., DOLIQUE F., GRATIOT N., "Geomorphic evolution of a coastal mudflat under oceanic influences: an example from the dynamic shoreline of French Guiana", *Marine Geology*, vol. 208, pp. 191–205, 2004.

[LOI 13] LOISEL H., VANTREPOTTE V., JAMET C. *et al.*, "Challenges and new advances in ocean color remote sensing of coastal waters", in ZAMBIANCHI E. (ed.), *Oceanography Research*, InTech, Rijeka. 2013.

[MAS 95] MASON D.C., DAVENPORT I.J., ROBINSON G.J. *et al.*, "Construction of an inter-tidal digital elevation model by the "Water-Line" method", *Geophysical Research Letters*, vol. 23, pp. 3187–3190, 1995.

[MAR 09a] MARTINEZ E., ANTOINE D., D'ORTENZIO F. *et al.*, "Climate-driven basin-scale decadal oscillations of oceanic phytoplankton", *Science*, vol. 326, pp. 1253–1256, 2009.

[MAR 09b] MARTINEZ J.M., GUYOT J.L., FILIZOLA N. *et al.*, "Increase in suspended sediment discharge of the Amazon River assessed by monitoring network and satellite data", *Catena*, vol. 79, pp. 257–264, 2009.

[MON 14] MONTREUIL A.L., LEVOY F., BRETEL P. *et al.*, "Morphological diversity and complex sediment recirculation on the ebb delta of a macrotidal inlet (Normandy, France): a multiple LiDAR dataset approach", *Geomorphology*, vol. 219, pp. 114–125, 2014.

[PRO 93] PROBST M.T., "L'environnement côtier actuel de la Guyane et quelques aspects de son analyse par télédétection", *Caribena*, vol. 3, pp. 125–158, 1993.

[PRO 07] PROST E., Caractérisation du trait de côte sur le littoral guyanais à partir d'images Spot-5: segmentation préliminaire d'objets littoraux, Master's thesis, Toulouse Le Mirail, 2007.

[ROB 02] ROBIN M., "Télédétection et modélisation du trait de côte et de sa cinématique", in BARRON-YELLÈS N., GOELDNER-GIANELLA L., VELUT S. (eds), *Le littoral, regards, pratiques et savoirs*, Editions Rue d'Ulm, Paris, 2002.

[SPA 97] SPALDING M.D., BLASCO E., FIELD C.D., *World Mangrove Atlas*, The International Society for Mangrove Ecosystems, Okinawa, 1997.

[SPA 10] SPALDING M.D., KAINUMA M., COLLINS L., *World Atlas of Mangroves*, Earthscan, London, 2010.

[TRE 05] TREBOSSEN H., DEFFONTAINES B., CLASSEAU N. *et al.*, "Monitoring coastal evolution and associated littoral hazards of French Guiana shoreline with radar images", *Comptes Rendus Geoscience*, vol. 337, pp. 1140–1153, 2005.

[VAN 11] VANTREPOTTE V., LOISEL H., MERIAUX X. *et al.*, "Seasonal and inter-annual (2002-2010) variability of the suspended particulate matter as retrieved from satellite ocean color sensor over the French Guiana coastal waters", *Journal of Coastal Research*, vol. 64, pp. 1750–1754, 2011.

[VAN 13] VANTREPOTTE V., GENSAC E., LOISEL H. *et al.*, "Satellite assessment of the coupling between in water suspended particulate matter and mud banks dynamics over the French Guiana coastal domain", *Journal of South American Earth Sciences*, vol. 44, pp. 25–34, 2013.

[WAL 15] WALCKER R., ANTHONY E.J., CASSOU C. *et al.*, "Fluctuations in the extent of mangroves driven by multi-decadal changes in North Atlantic waves", *Journal of Biogeography*, vol. 42, no. 11, pp. 2209–2219, 2015.

[WES 12] WESTOBY M.J., BRASINGTON J., GLASSER N.F. *et al.*, "'Structure-from-Motion' photogrammetry: a low-cost, effective tool for geoscience applications", *Geomorphology*, vol. 179, pp. 300–314, 2012.

Glossary

2.5D: A representation of a geometric set in the form of a function z = f(x, y). It is very useful for returning the 3D geometry analysis to functional analysis, 2.5D is limited in the geometries that it can represent.

Absorption: Transformation of bright radiative energy into heat.

Actors (stakeholders): Assembly of natural or legal persons who decide to organize into a community to capture and deal with a priority issue.

Aerodynamic temperature: Theoretical temperature of the average source of sensitive heat flux for a complex surface, corresponding to the extrapolation within the canopy of the air temperature profile observed above the canopy.

Aerosols: Particles suspended in the atmosphere capable of interacting with solar radiation through the processes of absorption and backscattering. They can be solid (dust) or liquid (mist) of organic nature (black carbon) or inorganic nature (rock eroded) and more or less large (ranging from a few tenths of a nanometer to a hundred micrometers).

Angle of incidence: Angle between the incident radar beam on a surface and the normal to the surface at the point of incidence.

Anisotropic surface: Surface presenting a texture pattern oriented according to a preferred direction.

Anisotropy: Property of being dependent on direction and having different characteristics depending on the orientation.

Anthropogenic: Created or built byhumans.

Artificialization: Transformation of natural areas such as forestry or agricultural areas into artificial built-up areas.

Astronomical tidal level: Highest theoretical tidal level of the sea depending on the considered location.

Atmospheric correction: Method developed to eliminate the disruptive effects of the atmosphere (absorption and scattering of solar radiation) that contribute to the signal received by a satellite.

Autonomous navigation: Capacity of a vehicle to move by its own means, generally based on the analysis of data acquired by on-board sensors.

Backscatter radar coefficient (denoted σ^0): Designates the RCS by observed surface unit, and is expressed in (m^2/m^2). It is preferred instead of the RCS in the case of large surfaces. Due to its strong dynamics, it is usually expressed in decibels: $\sigma^0_{dB} = 10 * \log_{10}(\sigma^0)$.

Barycenter: Center of gravity of a polygon, also called centroid.

Bathymetric LiDAR: Active remote sensing system dedicated to the analysis of fluvial and coastal environments. It is dedicated to the measurement of water depth by measuring the travel time between two laser pulses at two different wavelengths: infrared to describe the surface of the water and the green for the background.

Beach profile: Topographic section of the beach along an axis perpendicular to the coastline.

Bidirectional Reflectance Distribution Function (BRDF): Ratio of outgoing radiance and incident irradiance.

Biodiversity: In a broad sense, biodiversity, or biological diversity, defines the variety and variability of the living world in all its forms. It is further defined in Article 2 of the Convention on Biological Diversity as "the variability of living organisms from all sources including, among others, terrestrial, marine and other aquatic ecosystems and the ecological complexes of which they are part of; this includes diversity within species, between species, and between ecosystems"[1].

Biomass: In ecology, biomass is the amount of living material in a given habitat. Often, it is estimated in a surface unit or volume rather than absolute mass. The vegetation cover is usually estimated by removing some plants which are weighed fresh and dry (after passing in an oven) in order to obtain dry biomass. In practice, we distinguish aboveground biomass (trunk, branches, leaves) from below-ground biomass (roots).

Brightness temperature: Temperature obtained by converting the radiance measured according to Planck's Law at the level of the sensor, weighted by its transfer function; it can be defined for a measurement in a given direction (called directional brightness temperature).

CALIOP: Cloud-Aerosol LiDAR Pathfinder Satellite Observation is the LiDAR system of the CALIPSO spatial mission.

CALIPSO: Cloud-Aerosol LiDAR Pathfinder Satellite Observation is a French-American spatial mission which was put into orbit on April 28, 2006, by the NASA.

Canopy Height Model (CHM): For a given portion of territory, the mapping (in raster format) of the maximum height of the forest area compared to the ground within each pixel. It is a surface resulting from the calculation of the difference between a Digital Surface Model (DSM) and a Digital Terrain Model (DTM). Excluding forest environments, it is commonly called Digital Elevation Model (DEM). Subtracting the DTM makes it possible to eliminate the effect of the terrain, and any area not composed of objects above-ground has zero height.

Cartographic baseline: Coordinate system allowing location of a point in an absolute manner to the surface of the Earth.

Classification (method of): Technique to determine the affiliation of an object (for example, a pixel of an image) to a group (class) using the object's attributes (for example, its spectral signature). It can be supervised or unsupervised.

---

1 See https: //en.wikipedia.org/wiki/Biodiversity.

Coastline: Border subject to variations of planimetric positions representative of accretion or coastal erosion which is defined in practice by different types of indicators based, either visually on observations and/or field measurements or from the intersection between a tide gauge reference level and the topography.

Confusion matrix: Double entry table used to analyze the quality of a classification; the rows correspond to the class of the evaluated objects and the columns correspond to the assigned class; the boxes in the table indicate the number of objects.

Contract plan: Document through which aState and a region are committed to programming and funding multi-year important projects such as the creation of infrastructure and support of sectors for the future.

Conurbation: Multipolar agglomeration gathering together several towns.

Country: Territory characterized by geographical, cultural, economic or social cohesion, on the scale of a living or employmentarea, often part of the development of a common sustainable development project (country development character).

Crown: Upper part of a tree which is formed of all the branches and leaves. It is the main support of photosynthetic organs. Also referred to tree crown.

Decimation: For a triangulated surface, action that helps reducethe number of triangles by minimizing errors made.

Delaunay tetrahidralization: From a set of points in space (3D), a construction of a set of tetrahedra based on these points so that no sphere circumscribed to a tetrahedron contains another point.

Delaunay triangulation: From a set of points in the plane (2D), the building of a set of triangles based on these points and in a way that no circle circumscribing to a triangle contains another point.

Dempster–Shafer: Theory that allows modeling and combining uncertain information.

Descriptor: Vectors of real numbers that characterizes a data (image, point cloud, and therefore it is called global descriptor) or one of its elements (point, pixel, thus local descriptor).

Differential Global Positioning System (DGPS): Instrument that allows measuring geographic positions in a differential manner using a constellation of satellites and a network of fixed ground-based reference stations.

Diffuse reflection: Corresponds to a reflection of the wave in different directions and with a distribution of energy following these directions. The distribution between specular and diffuse reflection processes is determined by the roughness and the dielectric properties of that surface.

Digital Elevation Model (DEM): 3D representation of a terrain's surface. It represents the Earth's surface and all objects on it.

Digital Height Model (DHM): 3D description of our environment. It describes the ground profile while integrating what is on the ground surface (buildings, vegetation, etc.).

Digital Surface Model (DSM): A topographic surface model describing the buildings, vegetation, roads (above-ground) and/or the characteristics of the terrain (ground). For simplicity, the DSM is the generic term for models that describe the ground and/or above-ground.

Digital Terrain Model (DTM): Representation of the topography of a land area in a matrix form (image), where each cell (pixel) contains the height information generally in relation to a terrestrial reference system (usually expressed in meters above sea level). For simplicity, the DTM is the only ground without vegetation or building.

Direct modeling: Direct modeling aims to predict the return of an electromagnetic wave to the sensor in using a set of physical characteristics representing the observed target (shape, size, dielectric properties).

Dynamics (of an image): Defines the interval in which lies the radiometry of an image. A strong dynamic indicates a significant ratio between the lowest values and the strongest ones. An important dynamic can lead to large errors in sampling (saturation, rounding errors).

Ecosystem service: Benefits derived by mankind from their current or future use of the various functions of ecosystems.

Electromagnetic radiation: An energy wave that is propagated without any hardware support (in vacuum) and is characterized by a period (cycle length or wavelength – in seconds) and a frequency (number of cycles per second – in hertz). The electromagnetic spectrum is the decomposition of electromagnetic radiation according to its wavelength or frequency (Hz). The long wavelengths (low frequencies) are the radio and microwaves while smaller wavelengths (high frequencies) are nuclear or gamma rays.

Entropy (Cloude decomposition): Represents the random nature of the mechanisms of scattering present in the pixels whose average was calculated.

Epipolar: I1 and I2 are two images of the same scene of the optical centers O1 and O2 and P a point of the scene. The projections of lines (O1P) and (O2P) in I2 and I1, respectively, are called epipolar lines. Knowing only the projection of P in an image, the epipolar is the location of possible projections of P within the other image.

Equalization: Action of modifying the overall image radiometry so that the values of identical elements of the imaged scene becomes the same.

External spatial representations: Image, map, figure, symbol, sign that represents a phenomenon, an idea related to the geographical space.

Extrinsic calibration: Action of attaching the reference of a sensor to the reference of the platform on which it is mounted.

Field of view: Solid angle around a sensor in which it acquires data.

Fluorescence: Inelastic scattering process. The incident wavelength is spontaneously re-emitted to a different wavelength.

Food safety: Situation which guarantees all people, at all times, the physical, social and economic opportunity to obtain adequate and healthyfood.

Forest canopy: Highest part of the forest, consisting of the crowns of the dominant and emerging trees.

Full waveform: For every LiDAR pulse, a signal reflected by the Earth's surface and the objects that compose it, received and digitized by the LiDAR system.

GALILEO: European satellite positioning system on a global scale.

Geographic Information System (GIS): A geographic information system or geographical information system (GIS) is a system designed to capture, store, manipulate, analyze, manage, and present all types of spatial or geographical data (source Wikipedia).

Geo-referencing: Operation which consists of properly positioning geographical data (like that from a satellite image) in a geographic coordinate system.

Global Positioning System (GPS): A space-based navigation system that provides location, velocity and time information in all weather conditions, anywhere in the world.

Gradient descent: Method of minimizing a function E from a vector x consisting of moving an initial solution and move a common solution in the direction opposite to the gradient (in which E decreases).

Hydrosols: Materials within the marine medium that are able to influence the optical properties (absorption, diffusion, attenuation) of water. Hydrosols involved in the field of ocean color are typically phytoplankton, materials of mineral origin, colored dissolved organic matter and non-algal detrital materials. All of these materials have microscopic sizes less than typically 200 μm).

Hyperspectral remote sensing: Remote sensing from a sensor allowing observation in many narrow spectral bands (a few nanometers) in the wavelength range from the visible, near infrared, mid-infrared to the infrared thermal.

Ice, Cloud and land Elevation satellite (ICEsat): NASA's spatial mission launched in 2003. The mission embarked a backscatter LiDAR system for the study of surfaces, primarily polar caps (GLAS).

Infrared: Electromagnetic radiation of a wavelength between 0.7 μm and 100 μm.

Inherent Optical Properties: Correspond to the inherent scattering and absorption properties of a medium.

Instantiate: Produce a realization of a parametric model.

Internal spatial representations: That through which an object in mind: a perception, a mental image, etc., whose content refers to a space, a place, a situation, a scene from the world in which the subject lives.

Intrinsic calibration: For a sensor, a method allowing connection of a raw data element acquired by the sensor (point, pixel) to a geometric entity (point, line) in the referential of the sensor.

Inverse modeling: The used model allows to obtain the physical characteristics of the medium observed from data extracted from radar, optical or other remote sensing signals.

Isoline: Line linking points of the same value on a map.

Iterative closest point: Registration method alternating matching between data sets to reset, and estimatingtransformations to apply to the data in order to align them.

Kappa coefficient: A quality estimator. It measures the level of agreement between the classification and the actual data. It varies between 0 and 1. The closer it is to 1, the more the agreement is important.

Lambertian: A material that reflects light in a uniform manner in all directions regardless of the direction of the incident light.

Land cover mapping: Representation of the Earth or portion of land space. Land cover refers to "biophysical coverage of the land surface" and therefore the type of use (or nonuse) of lands by humans. The landscape mosaic is mapped by identifying homogeneous types of environments (for example, artificialized areas, agricultural areas, forests and moors, wetlands, etc.).

Laser (Light Amplification by Stimulated Emission of Radiation): Emits coherent light thanks to an amplified stimulated emission. The time between transmission and reception is converted into a distance by multiplying by the speed of light.

Laser intensity: Amount of radiation of the laser beam.

Laser rangefinder: Instrument that allows measuring distances from the emission of a laser beam to a target that returns the light beam.

Laser scan angle: Angle between the laser beam and the nadir of the airplane.

Leaf area index (LAI): A dimensionless quantity that is defined as the one-sided green leaf area per unit ground surface area (LAI = leaf area/ground area, m^2/m^2); the green LAI is the part of the green vegetation LAI.

LiDAR (Light Detection and Ranging): Active instrument that measures the optical properties of a medium by scanning a laser beam returned to its transmitter.

Litto3D: A precise altimeter digital model of land-sea continuum carried out jointly by SHOM (French Naval Hydrographic and Oceanographic Service) and IGN (French National Institute of Geographic and Forest Information).

Local incidence angle: Angle between the direction of the point of interest (or incidence point) and the perpendicular to the illuminated surface in case of an inclined surface.

Mangroves: Forests which grow in the tidal bays and estuaries of tropical and inter-tropical regions.

Megaripples: Large, sand waves or ripple-like features having wavelengths greater than 1 meter or a ripple height greater than 10 centimeters; Megaripples are formed in a subaqueous environment, and they are also known as subaqueous dunes.

Metrological: Related to the measurement. A metrological method seeks to obtain a geometric measurement (in m) of its objects of interest.

Microwave: High spectral wavelength band (wavelength in the range of 1 mm – 1 m).

Mobile mapping: Acquisition of image/laser data by sensors onboard a terrestrial vehicle in order to extract geo-located information.

Monoscopic image: A single optical image.

Mosaic: For overlapping images, the action of only capturing a portion of each image so as to generate a single image whose support is the union of supports of the images used.

Multiple scattering (optical): A photon is released outside of the medium after its interaction with several particles.

Multispectral remote sensing: Remote sensing from a sensor allowing for the observation of several spectral bands (of the order of several tens of nanometers) of wavelengths ranging from the visible, near infrared, mid-infrared to the thermal infrared.

Multi-view images: Several images of the same scene acquired with different viewing angles with an overlapping area (e.g. stereoscopy, tri-stereoscopy).

Non-rigid transformation: Bijective continuous transformation of the space not having (necessarily) the property of being rigid.

Normalized Difference Vegetation Index (NDVI): This vegetation index is calculated from the reflectances in the Near Infrared (NIR) and in the red (R): NDVI = (NIR–R)/(NIR+R). It measures the intensity of the photosynthetic activity of the surface. It is efficient to distinguish the types of covers provided that they are not too dense as tropical rainforests.

Object-oriented classification: Image classification method aiming to segment an image into distinct polygonal objects. The objects are classified using the radiometric properties of the image as well as the geometric properties of the objects such as their size or form.

Observatory: Device in response to the information needs of a group of diverse actors which share the same concern. It is materialized through a set of tools serving, on one hand, for the collection, storage and management of data, and on the other hand, for the production, the analysis and the retrieval of information and new knowledge.

Occultation: Masking of an element in a scene by an element closer to the sensor.

Ocean color: Research topic of studying the interaction of sunlight with hydrosols present in a marine medium in order to achieve the detection of these hydrosols from a satellite or airborne platforms.

Ocean color inversion algorithm: Method that allows for retrieving the optical properties (absorption and scattering) of hydrosols as well as concentrations of suspended solids (phytoplankton, minerals) from a reflectance measurement acquired by a satellite sensor or from a research vessel.

Ocean primary production: Total mass of organic matter produced by photosynthesis in the ocean in grams of carbon (gC) $m^{-2}day^{-1}$.

Optical center: As a first approximation, a camera performs a conical or spherical projection of the surrounding environment in its image space. The optical center is the center of this projection (top of the cone or the center of the sphere).

Optical depth: The optical depth of a layer (atmosphere, vegetation) describes the degree of transparency of the medium. It characterizes the fraction of electromagnetic radiation (or light) scattered or absorbed by the components of the traversed layer.

Optical properties: Physical parameters used to quantify the process of absorption and scattering of light by materials in the atmosphere–ocean system (molecules, aerosols, hydrosols). It mainly encompasses absorption coefficients, scattering and attenuation (unit: m^{-1}).

Optical reflectance: Proportion of incident light (solar illumination) reflected by the Earth's surface, expressed as a percentage.

Orthoimage: Image corrected geometrically in a projection system given so that all of its pixels perfectly coincide with a geographic map. Deformations due to terrain, the roundness of the Earth and the conditions of viewing were corrected: each pixel seems to have been acquired by a sensor performing measurements perpendicular to the ground at any point in the observed area.

Panchromatic: A panchromatic image is a comprehensive record of the radiation reflected by the Earth's surface in the range of wavelengths of the visible spectrum without color separation. For example, a SPOT 5 panchromatic image is acquired between 0.48 and 0.71 μm. This data has higher spatialresolution than the multispectral image acquired by the same satellite.

Panoramic: Data whose field of view covers an angular sector of 360° around the sensor in the horizontal plane.

Participation: Set of measures to mobilize actors in collective actions.

Photorealistic: Representation of reality produced by data processing and whose rendering is similar to that of an image.

Photosynthesis: Photosynthesis is a process used by plants and other organisms to convert light energy, normally from the Sun, into chemical energy that can be later released to fuel the organisms' activities (energy transformation).

Photosynthetically Active Radiation (PAR): Designates the spectral range (wave band) of solar radiation from 400 to 700 nm that photosynthetic organisms are able to use in the process of photosynthesis.

Phytoplankton: Vegetal unicellular microorganism living in the aquatic environment. These microorganisms are the first link in the marine food chain. They also play a role in the absorption of atmospheric carbon dioxide by the oceans through the process of photosynthesis and primary production.

Pixel: Individual element constituting an image (deriving of the English phrase "picture element"). Depending on the image type and the encoding used, the pixel "contains" one or more values corresponding to the digital pixel accounts in different image channels.

Platform: A mechanical device to control the position of one or more sensors, for example: a satellite, plane, car, cart, tripod, etc.

Polarimetric decomposition: Tool that helps to physically interpret the behavior of a target measured by into simpler components.

Polarization: The orientation of the electric field vector in the plane orthogonal to the wave propagation direction.

Polyhedral: Consists of planar polygons.

Pose: Generally refers to the time interval during which a camera acquires an image. By extension, it may designate the transformation linking the referential of a sensor to a referential to which it relates during this interval. By extension, for a system of synchronized cameras, it designates all the images acquired at the same trigger time.

Precision agriculture: Designates a set of agricultural land management practices oriented towards optimizing returns on inputs while preserving resources. These practices are based on technological innovations (precision GPS positioning, proxy sensors - or remote sensing of soil and crops properties. Tools of soil work or the spreading of input variable rate, etc.) which makes it possible to take into account the intra-parcel heterogeneity of yield potential (due to the heterogeneity of soils, topography, etc.)

Primitive: Geometric object defined by a mathematical equation with a fixed number of parameters (line, segment, circle, sphere, etc.).

Principal Components Analysis (PCA): A statistical procedure that uses an orthogonal transformation to convert a set of observations of possibly correlated variables into a set of values of linearly uncorrelated variables called principal components. The number of principal components is less than or equal to the number of original variables (source Wikipedia).

Processing chain: Set of operations oriented to chain raw data to draw a more structured data, more exploitable by applications. A processing chain is generally represented by an oriented graph where each node represents a treatment, and the edges represent incoming and outgoing data treatment.

Property law: Relative to the rules governing access to land and related resources.

Radar bands X, C, L, P: Frequency bands used in the microwaves, from 8 to 12.5 GHz for the X-band, 4 to 8 GHz for the C-band, from 1 to 2 GHz for the L-band and from 0.3 to 1 GHz for the P-band.

Radiative surface temperature: Emission temperature of the surface considered as a gray body; it can be defined for a measurement in a given direction (referred to as directional radiative temperature).

Radiative transfer: Theory based on the conservation of energy flux through an elementary volume. It allows to describe the behavior of an electromagnetic wave passing through a scattering medium composed of dielectric particles.

Radiometric resolution: It describes the ability of a sensor to distinguish small differences in the intensity of the received energy. The maximum number of levels of intensity depends on the number of bits used. For example, a RapidEye image is encoded in 16 bits or 65,536 numeric values available to transcribe the recorded intensity; a SPOT image is encoded in 8 bits or 256 digital values.

Radiometry: Field that studies the measurement of electromagnetic radiations. For an image, it refers to this measurement of radiation for each of its pixel.

Raman effect: Physical phenomenon, corresponding to an energy exchange between the light beam and the medium, changing the frequency of light that circulates.

Rasterization: Process consisting of performing a raster image.

Registration: Determination of a transformation between data sets in order to align them, for example, to superimpose the corresponding elements in these datasets.

Reference ellipsoid: In geodesy, a reference ellipsoid is a mathematically defined surface that approximates the geoid, the truer figure of the Earth, or other planetary body. Because of their relative simplicity, reference ellipsoids are used as a preferred surface on which geodetic network computations are performed and point coordinates such as latitude, longitude, and elevation are defined (source Wikipedia).

Reflectance: Proportion of incident light (solar illumination) reflected by the Earth's surface, expressed as a percentage.

Remote sensing: The acquisition of information about an object or a phenomenon without making physical contact with the object.

Rigid transformation: Space transformation preserving distances. A rigid transformation consists of a translation and a rotation.

Sampling: In signal processing, sampling is the reduction of a continuous signal to a discrete signal.

Scale: The ratio of the distance on a map to the corresponding actual distance. Often expressed as a fraction, a large scale is a ratio which is approximated to the integer 1.

Scattering (optical): Angular redistribution of solar radiation after its interaction with a component present in the atmosphere-to-surface system (i.e. air or water molecules, aerosols, hydrosols, etc.).

Schorres (salt marshes): Is a coastal ecosystem in the upper coastal intertidal zone between land and open salt water or brackish water that is regularly flooded by the tides.

Segmentation: Process oriented to separating a given set in homogeneous subsets according to given criteria, in order to identify them.

Semantics: Denotes the fact of being categorized, to be of a defined type.

Sensor (satellite): Device on board of the satellite that collects and measures the incident radiation coming from the observed surface.

Shimmer (or speckle): Shimmering is produced by random interference, constructive and destructive, originating from multiple scattering that occurs in every cell of resolution of a radar image. This effect manifests itself as a salt and pepper texture on images.

SIFT: Very popular and effective method of extraction of points of interest, calculation of descriptors on these points of interest and matching between points of interest described.

Simple scattering (optical): A photon is disseminated outside of the medium traversed as a result of its interaction with a single particle.

Spatial planning: Set of actions conducted by public actors on a given territory in order to transform it.

Spatial resolution: The spatial resolution of a geographical document is the actual size of the smallest element shown in that document. On a satellite image, the spatial resolution corresponds to the size of a side of a pixel.

Spectral radiance of a surface dA: The energy in the elementary spectral range $[\lambda + d\lambda]$ confined in the elementary solid angle $d\Omega$ around the direction of propagation (θ, φ) relative to the normal to the surface dA, expressed in $W.m^{-2}.sr^{-1}$(watts per square meter per steradian).

Spectral resolution: Describes the finesse of a spectral band used by a sensor. If the spectral resolution is finer, the windows of the different channels of the sensor are narrower.

Spectroradiometer: Instrument for measuring the intensity of the FLUX of electromagnetic radiation depending on the wavelength. It usually operates in ultraviolet, visible and infrared wavelengths.

Specular reflection: The mirror-like reflection of light (or of other kinds of wave) from a surface, in which light from a single incoming direction (a ray) is reflected into a single outgoing direction (source Wikipedia).

Spherical geometry: Method of representing geometric entities from a central point. In particular, the points in spherical geometry will be represented by two angles and their distance to the center.

Station: Fixed platform in translation on which the sensor or sensors are driven in rotation. Device often used to make panoramics from non panoramic view of field sensors.

Stereoscopic base: Denotes an assembly of two cameras pointing in the same direction, ideally in such a way that the epipolar lines correspond to the two image lines. This type of installation simplifies the problem of correlation allowing determination of the depth of the scene for each pixel and draws heavily on the depth perception by the human visual system.

Sunglint (or glitter): Reflection of direct solar radiation on the surface of the sea in the presence of waves.

Supervised classification: Automatic learning technique that produces a classification map from a ground truth.

Supervised learning: Automatic learning method which uses reference data, also called assisted learning.

Surface-from-Motion: Structure from motion (SfM) is a range imaging technique; it refers to the process of estimating three-dimensional structures from two-dimensional image sequences which may be coupled with local motion signals.

Suspended Solids (SS): Particles of organic or mineral origin found in the water that cause more or less interference in terms of passage of light.

Synthetic Aperture Radar (SAR): Side-looking radar instrument designed to provide radar images of fine spatial resolutions (from 1 to 30 m for satellite data). This spatial resolution is obtained from a coherent summation (taking the phase into account) of raw echoes. Improved spatial resolution is obtained to the detriment of the radiometric resolution.

Tacheometer: A device that measures the horizontal and vertical angles between two targets and the distance of these targets. The data acquired from these measures allow, through the characterization of a geodesic triangle, establishing a plan, a map, or a terrain model.

Target: Designates any element of an imaged scene susceptible to sending or returning electromagnetic radiation to a sensor.

Temporal resolution: Frequency at which a remote sensing image acquisition system can acquire an image of the same point on the Earth's surface.

Terrestrial laser scanner: Equipment operating on the basis of a laser allowing us to measure and digitize objects or large complex spaces in three dimensions.

Territorial engineering: Set of concepts, tools and devices made available from regional stakeholders to support the design, implementation and evaluation of their territorial projects.

Territorial policies: Policies supporting, in participatory approaches, the actors of a territory in order to adopt a strategic development project.

Texture (of an image): A certain spatial arrangement of the image signal levels.

Texture index (image): Local statistic calculated in the neighborhood of a pixel of an image which takes into account spatial organization.

Tidal, flood tidal or ebb deltas: Sedimentary accumulations of several km^2 in the mouth of a coastal river, sea side (external) for the ebb delta and land side (internal) relative to the coastline, for the flood delta.

Topographic LiDAR: Active remote sensing system dedicated to the analysis of land surfaces. It mainly operates in the near infrared. The distances are obtained by measuring the roundtrip time between the sensor and the target.

Ultraviolet: Electromagnetic radiation of a wavelength between 10 and 380 nm, shorter than that of visible light, but longer than that of X-rays.

Unsupervised learning: Automatic learning method, sometimes referred to as "clustering", which divides a heterogeneous group of data in subsets of similar data.

Urban dynamics: Change resulting from a play of forces. For a city, dynamics are read and analyzed in the changes of location of activities, people and infrastructure.

Urban heat island: Urban heat islands (UHI)are localized elevations of temperature recorded in urban areas compared to adjacent rural or forested areas or compared to regional average temperatures.

Urban matrix: The city considered as a particular environment, nurturing, protective, but also encompassing or binding; by analogy with "nature that provides the raw material, the seed and soil, the feeder matrix and the elements of the plants that man studies, models, paints and sculpts at will"[2]

Urban morphology: Study of plans and built forms of the city.

Urban scenes: Part of territory in which anthropogenic structures are dominant.

---

2 From Huysmans J.-K., *À rebours*, Charpentier, 1884.

Urban sprawl: Urban sprawl occurs when the rate of change of occupation of lands exceeds the growth rate of the population, which means that, in a given area, the progression of urbanized land exceeds the growth in population.

Urbanization: Development, expansion of the urban population whose processes and formsare studied.

Visible: Part of the electromagnetic spectrum visible to the human eye, between 380 and 780 nm.

Voxel: A voxel is a unit of graphic information that defines a point in three-dimensional space.

Waveform: Temporal correlation between the received signal and the transmitted signal. In altimetry, this corresponds to a synthetic echo radar formed by averaging elementary echoes (called bursts).

Wavelength: Each electromagnetic wave is defined by its wavelength, which represents the spatial periodicity of the oscillations (maximum distance between two oscillations, for example). The wavelength is inversely proportional to the frequency and is expressed in meters.

List of Authors

Edward J. ANTHONY
CEREGE
Aix-MarseilleUniversity
Aix en Provence
France

Nicolas BAGHDADI
TETIS
IRSTEA
Montpellier
France

Jean Stéphane BAILLY
LISAH
AgroParisTech
Montpellier
France

Xavier BRIOTTET
ONERA
Toulouse
France

Malik CHAMI
LOV
University of Pierre et Marie Curie
Villefranche sur Mer
France

Nesrine CHEHATA
Géoressources & Environnement
Bordeaux-INP
Bordeaux Montaigne University
Pessac
France

Jean-Baptiste FÉRET
TETIS
IRSTEA
Montpellier
France

Laurent FROIDEVAL
M2C
University of Caen-Basse Normandie
France

Franck GARESTIER
M2C
University of Caen-Basse Normandie
France

Jean-Philippe GASTELLU-ETCHEGORRY
CESBIO
University of Paul Sabatier
Toulouse
France

Nicolas GRATIOT
Centre Asiatique de Recherche sur l'Eau (CARE)
IRD
Ho-Chi-Minh City
Vietnam

Tristan HARMEL
LOV
University of Pierre et Marie Curie
Villefranche sur Mer
France

Nicolas LAURET
CESBIO
Paul Sabatier University
Toulouse
France

Arnaud LE BRIS
MATIS
IGN
Saint-Mandé
France

Franck LEVOY
M2C
University of Caen-Basse Normandie
France

Clément MALLET
MATIS
IGN
Saint-Mandé
France

Pierre MAUREL
TETIS
IRSTEA
Montpellier
France

Olivier MONFORT
M2C/CREC
University of Caen-Basse Normandie
France

Martin MONTES-HUGO
ISMER
University of Québec at Rimouski
Canada

Rosa OLTRA-CARRIO
ONERA
Toulouse
France

Yves PASTOL
SHOM
Brest
France

Emilie POULLAIN
M2C
University of Caen-Basse Normandie
France

Christophe PROISY
AMAP
IRD
Montpellier
France

Jean Philippe TONNEAU
TETIS
CIRAD
Montpellier
France

Bruno VALLET
MATIS
IGN
Saint-Mandé
France

Christiane WEBER
TETIS
CNRS
Montpellier
France

Romain WALCKER
EcoLab
University of Toulouse
France

Mehrez ZRIBI
CESBIO
CNRS
Toulouse
France

Index

A, B

C, D, E

S, T

U, V, W

Scientific Committee

Jacques POPULUS, IFREMER, France

Christophe PROISY, IRD, France

Marc ROBIN, University of Nantes, France

Alain POLIDORI, Engineering School for Land Surveying and Topography (ESGT), France

Virginie LAFON, I-SEA, France

Nicolas BAGHDADI, Irstea, France

Mehrez ZRIBI, CNRS, France

FSC
www.fsc.org
MIX
Papier | Fördert gute Waldnutzung
FSC® C083411

Zeitfracht Medien GmbH
Ferdinand-Jühlke-Straße 7
99095 Erfurt, Deutschland
produktsicherheit@kolibri360.de